THE HEAVENS ON EARTH

SCIENCE AND

CULTURAL THEORY

A Series Edited by

Barbara Herrnstein Smith

and E. Roy Weintraub

THE HEAVENS ON EARTH

Observatories and

Astronomy in Nineteenth-

Century Science and Culture

Edited by DAVID AUBIN,

CHARLOTTE BIGG, and

H. OTTO SIBUM

DUKE UNIVERSITY PRESS

Durham and London

2010

© 2010 Duke University Press
All rights reserved.
Printed in the United States
of America on acid-free paper ∞
Designed by Amy Ruth Buchanan
Typeset in Janson by Tseng
Information Systems, Inc.
Library of Congress Cataloging-
in-Publication Data appear on the
last printed page of this book.

CONTENTS

LIST OF ILLUSTRATIONS

ACKNOWLEDGMENTS

This book is the considerably revised outcome of a workshop held and made possible by the Experimental History of Science Independent Research Group of the Max Planck Institute for the History of Science in Berlin. We are very grateful to the participants of this workshop, especially Simon Schaffer, who have contributed in numerous ways, both intellectual and material, and been unfailing in their support for this project. We would also like to thank the scholars who read and commented upon early versions of this manuscript, in particular Robert Smith, M. Norton Wise, Dominique Pestre, Marie-Noëlle Bourguet, Kapil Raj, and Suman Seth.

The book benefited from funding from the French CNRS, the French Ministry of Research, and the Agence Nationale pour la Recherche for the two follow-up projects to the workshop, entitled *Savoirs et techniques de l'observatoire, fin XVIIIe–début XXe siècle* and *De Humboldt à Gaïa: histoire des sciences du système-terre*. We thank participants to these projects, especially Jérôme Lamy, Fabian Locher, Stéphane Le Gars, and Arnaud Saint-Martin.

We were greatly assisted by Bettina Schütz in organizing this workshop. For their help in putting this book together we would also like to thank Nina Ruge, Gemma Trival, Madeleine Dietrich, and Ulla-Britt Jansson, as well as Jacob Orrje and Jens Eriksson. Sara Diaz, Melissa Ellis Martin, and Fred Kameny assisted with editing.

Note on the Bibliography

We have paid special attention to the bibliography in the hope of making this book a useful resource for future studies in the history of the observatory sciences. All titles cited more than once feature in the general bibliography at the end of the book, together with some important contributions to the field not cited in the text. All works cited in the bibliography appear only in short form in the chapters' notes.

Introduction: Observatory Techniques in Nineteenth-Century Science and Society

DAVID AUBIN, CHARLOTTE BIGG,

AND H. OTTO SIBUM

> Observatories—Temples of the most sublime of the sciences, . . .
> mysterious sanctuaries where, in the silent night and away from the
> busy hum of men, philosophers are in intimate communication with
> the innumerable worlds which people the Universe.
> —AMÉDÉE GUILLEMIN (1864).[1]

> The fact is that when an astronomer goes into his observatory for his
> night's work he finds it usually convenient to leave all the ecstatic and
> most of the poetic portions of his constitution outside.
> —ROBERT BALL (1892).[2]

> The object of the Observatory is the accurate and systematic observation
> of the heavenly bodies, for the advancement of Astronomical Science;
> co-operation in Geodetic and Nautical Surveys; meteorological and mag-
> netic investigations, and the improvement of Tables useful in Navigation.
> —WILLIAM. C. BOND (1894).[3]

What is an observatory? The term conjures up images of a neoclassical
monument surrounded by delightful gardens, a makeshift camp on a desolate beach, a wooden shack on a university campus, or a refuge on an icy
mountaintop; a place where astronomers scrutinize faraway stars through
gigantic and delicate telescopes, calculating and recording their motions
on endless series of folio volumes. There, celestial mechanics and nautical
almanacs are elaborated, reference clocks and maps are kept, and magnetic
surveys and colonial expeditions are planned and coordinated. There, more
or less distinguished audiences gather to hear lectures on the history of the
universe and the progress of science.

The observatory is all of this and much more. This book is a first attempt at surveying the observatory's multiple roles in nineteenth-century scientific, economic, and cultural life. Without pretending to be exhaustive, we want to highlight the range of activities carried out in the observatory in the hope to stimulate further investigation. The observatory, we argue here, was essential in ensuring the growing social and cultural significance of the mathematical, physical, and cosmological sciences in the nineteenth century. It was simultaneously indispensable in constructing elements of the modern western state and society—among others, European colonial expansion and the emergence of a public enthusiastic about scientific and technological developments.

Most nineteenth-century observatories were concerned with both the heavens and the earth. Astronomy of course played a prominent part in observatories: it was their main purpose and, for all involved, the model of what science should be. But astronomy belonged to a larger group of sciences that we refer to as "observatory sciences." Just as historians have recently attempted to characterize the "laboratory sciences" or the "field sciences," we want to focus on another emerging family of nineteenth-century sciences that besides astronomy includes cartography, geodesy, meteorology, and to an extent physics and statistics. While universities and academies tended to split science along disciplinary lines, a number of pursuits coexisted at the observatory, and we make a first attempt to investigate them as a coherent whole.

In the course of the nineteenth century, according to one estimate, the number of astronomical observatories worldwide rose from fewer than three dozen to more than two hundred—and this excludes the observatories devoted to meteorology, geomagnetism, geodesy, navigation, or statistics.[4] From Göttingen to Königsberg, from Brussels to St. Petersburg, from Rio de Janeiro to Petchaburi (Thailand), the endowment of expensive observatories became an inescapable requirement for any modern state intent on preserving its political independence and securing its integration into the world-system (figure 1).[5] Observatories were dear to modern states not only for the different services they rendered—most obviously in domains connected with the control and administration of space, cartography, geodesy, and navigation—but also for ideological purposes. Simultaneously, observatories were a central focus for expressing public curiosity about science and enthusiasm for it.

The picture of the observatory that we paint here therefore moves away from conventional representations of the solitary astronomer at his eye-

1. Over the course of the nineteenth century the number of observatories in the world skyrocketed. This print gives a conventional representation of some of the observatories built in the United States during the national observatory movement, a popular drive supported by universities, philanthropists, and to a lesser extent the federal government to endow the country with up-to-date scientific institutions for navigation and astronomy. From top left to bottom right: Observatories of Cambridge, Dudley, Sheffield scientific school, Hamilton College, and the University of Michigan. *La Nature* 4, no. 1 (1876): 249.

piece. We present the observatory as a specific space of scientific practice and as an essential element in a number of industrial, technological, and military undertakings, in many ways a pillar of the state (figure 2). The frequent public and official celebrations of the observatory and of the observer accordingly emphasized both the abstract and remote character of the science they produced *and* its very material utility. In all its manifestations, the observatory united heavenly with earthly concerns.

This diverse set of practices ought to be considered together because they share two fundamental unifying characteristics: they are observatory sciences in that both occupy the *space* of the observatory and, more fundamentally, because they are based on a set of common *techniques*. In the following, we explain what we mean by observatory techniques and show some of the ways they were employed.

Observatory Techniques

The exuberant juxtaposition of activities at the observatory might at first seem bewildering. A nineteenth-century guidebook documents this coexistence of worlds at the Paris Observatory: "The Observatory . . . is a most curious piece of architecture, having in it neither wood nor iron; it is not a large building, but has fine appearance; . . . it is vaulted throughout, and a geometrical staircase, having a vacuity of 170 feet deep, merits particular notice. There is a circular universal chart upon the pavement of one of the apartments. By means of mechanical arrangements . . . every night, the weather permitting, astronomical observations are taken. M. Arago, the most celebrated astronomer of France, lectures here, where there is every facility, and every instrument to be found requisite to the promotion of the science of astronomy; there are two pluvia-meters, for ascertaining the quantity of rain that falls in Paris during a year. There is a general map of France, containing 182 sheets, a marble statue of Cassini (the author of the *work*)."[6] Etymology suggests that an observatory is a place dedicated to observation; and indeed, astronomy was traditionally considered the most accomplished observation-based science. In particular, celestial mechanics long provided a model for systematizing scientific observations by using mathematics. For the astronomer Pierre-Simon Laplace, "the only way to know nature is to question it through observation and computation."[7]

But while they insisted on the importance of observation in the pursuit of scientific truths, scientists and philosophers widely differed on what they meant by it. The *encyclopédistes* defined observation as "the attention of the

2. After his nomination as director of the Paris Observatory in 1854, Urbain-Jean-Joseph Le Verrier was inspired by Airy's precedent at Greenwich to transform the main pavilion, built in the late seventeenth century, into a technical space entirely devoted to science. This print was published when the new observatory was dedicated. *Le monde illustré*, 1 February 1862, n.p.

mind [*l'esprit*] turned toward the objects offered by Nature." As opposed to experimentation, which could only teach investigators about artificial circumstances, observation was construed as the nonintrusive, albeit methodical, process of perceiving and recording that was the true basis of science. With the rise of the laboratory sciences, observation took on new connotations and often came to be portrayed as a somewhat weaker and overhauled counterpoint to the interventionist methods of the new experimental sciences. Though experimentation now often replaced observation as the idealized methodological foundation of science, laboratory scientists continued to regard the practice of observation as central to experimental work. The value placed by laboratory scientists on the "mechanical objectivity" guaranteed by self-recording apparatuses appears in this light as an appropriation of earlier discourses on scientific observation within a new, experimentalist framework; and underlines the continuing ambivalence of the experimental sciences toward observation. Conversely the observatory sciences' avowed adherence to the observation ideal cannot hide their eminently interventionist ethos; one need only mention land surveying or navigation. Far from being a stable, self-evident, universal category, "observation" is a situated set of practices, methods, and ideals whose complex history remains to be written.[8]

The observatory plays a central role in the history of observation as a practice, method, and value. But to understand this history we need to look beyond the individual astronomer. Observing is a socially and culturally determined experience; the astronomer "looks with his own eyes, but sees with the eyes of the collective."[9] And conversely, what goes on in the observatory contributes to change in the wider experience of observation. We therefore want to broaden the focus to include the material and conceptual environment of scientists working in the observatory, the cultural and economic constraints to which they are subjected, and the significance of their work in this wider context.

To tackle the complex technical space that was the observatory, we find it useful to focus on what we call "observatory techniques." We take the whole set of *physical*, *methodological*, and *social techniques* rooted in the observatory as our focus of inquiry. Techniques, according to Marcel Mauss, are "a set of generally and primarily manual, organized and traditional, motions and acts concurring to reach a known goal."[10] What we call observatory techniques included the set of practices required to perform successfully at the telescope eyepiece: the calibration, manipulation, and coordination of precision instruments for making observations and taking measurements.

They embraced methods of data acquisition, reduction, tabulation, and conservation, along with complex mathematical analyses (error analysis and celestial mechanics). They also included various techniques for producing maps, drawings, and photographs, but also of material, numerical, and textual—indeed poetic—representations of the heavens and the earth that ultimately shaped the way the world, society, and science itself could be construed. Finally, these techniques incorporated the social management of personnel within the observatory as well as international collaborations.

The techniques we examine here were developed inside and outside observatories—by instrument makers in their workshops, navy officers on ships, civil engineers in the field, and physicists in their cabinets. But in the observatory they were uniquely assembled. Thereby these techniques helped define a *space* of knowledge: the observatory. The distinction between place and space was introduced by Michel de Certeau in *The Practice of Everyday Life*.[11] Place is understood geometrically, in the sense that two things cannot be in the same place at the same time. Space, on the other hand, is a "practiced place." It is "actuated by the ensemble of movements deployed within it," determined by historical subjects, by the users of place. Thus the streets geometrically defined by urban planners are transformed into space by walkers. The physical, geometrical, and geographical way in which Certeau has understood places owes much to the work of observatories in the eighteenth and nineteenth centuries. Technologies closely associated with observatories had a major impact in reconfiguring understandings of space and time whose measurement was changed by theodolites and high-precision clocks, while telegraphy and photography helped transform the way they were culturally experienced.[12] But more fundamentally, Certeau helps us to consider how observatory scientists created a knowledge space that transcended the boundaries of the observatory.

The observatory techniques as we conceive them reveal the perpetually reengineered cohesion of the observatory sciences. They formed a consistent foundation to a unified science of the heavens and the earth practiced by observatory scientists in the first part of the nineteenth century and later publicized in widely popular works such as François Arago's *Popular Astronomy* and Alexander von Humboldt's *Cosmos*. Humboldt's presentation of his scientific travels is revealing. He insisted on precision and examined the entire terrestrial globe with intense scrutiny. He envisioned the earth as a celestial body to be measured with theodolites and weighed with pendulums, a gigantic magnet whose field could be accurately determined by the sophisticated instruments designed by his friend Carl Friedrich Gauss. Humboldt saw the earth as a thermodynamic system, which could be studied using the tools of geography, chemistry, and statistics, and as a meteorological system whose precise empirical study could be furthered by using the geomathematical isotherms he had invented. As much a stimulus to as an outcome of the scientific practices deployed by the observatory scientists of his day, Humboldt located the unity of science in the necessary connection between the various forms and phenomena of nature, and he eschewed hierarchies among the sciences.

The unity of science was an important concern for nineteenth-century scientists. But one should always remember that the unity of science has been construed in a variety of ways. Some argued, as Auguste Comte did in his *Traité philosophique d'astronomie populaire* (1844), that a single way of knowing united the sciences. Others, like the director of the Roman College Observatory, Angelo Secchi (as Massimo Mazzotti recalls here), later preferred to found the unity of science on a unique abstract principle such as the conservation of energy, as would later become the norm. As John Tresch shows in this book, Humboldt's unity rested more on the complex interaction between active instruments and trained users than on abstract principles. Even if by 1900 most of the observatory sciences had evolved beyond recognition and in some cases had broken away from the astronomical observatory, Humboldt's vision continued to inspire discourses on the unity of science, not least in the popular science narratives of the last decades of the century (Ole Molvig, Charlotte Bigg).[13]

A Science of Precision

Early on astronomers had insisted on the precision of their measurements and computations. The transformation of industrial and scientific cultures in the nineteenth century has been traced in part to the rise of "the values of precision."[14] Laboratories in universities and industries have traditionally been seen as central locations in this process. The observatory may have been as important, if only because a number of these values already pervaded observatory culture by the end of the eighteenth century. Many fundamental experiments—Léon Foucault's speed-of-light measurements or Gauss's electromagnetic experiments, to name just two—were carried out in observatories. "The most precise observations," Gauss assured his sponsors in 1833, "can be expected only of those mathematicians

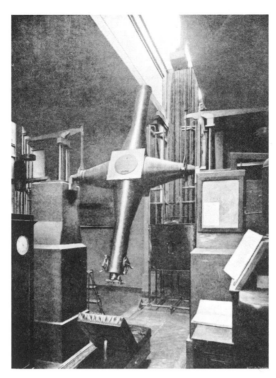

3. Gambey's transit refracting telescope at the Paris Observatory was built around 1819. On this picture showing its setup in the late 1870s, the various components of telescopic observation are clearly displayed: the astronomical clock on the left, the reclining chair, and the registry books on the right. Joh. A. Repsold, *Zur Geschichte der astronomischen Messwerkzeuge von Purbach bis Reichenbach 1450 bis 1830* (Leipzig: Wilhelm Engelmann, 1908), fig. 171 (after p. 126).

who are familiar with the finest means of observation, namely the practical astronomer."[15] At the heart of the observatory's material culture lay a family of scientific instruments; most, though by no means all, were optical. The telescopes, polariscopes, spectroscopes, magnetometers, clocks, and thermometers populating the observatory expressed the concern of their users—to achieve the highest possible level of precision in the (mostly quantitative) measurement of natural phenomena. Astronomy was the first precision science.[16]

Take the telescope (figure 3). The construction of this high-precision optical instrument required skilled labor on the part of artisans able not only to etch fine, evenly distributed lines on a brass circle but also to polish glass blanks into aberration-free lenses. These skills were developed in close collaboration with astronomers: standards for making circles were kept at the observatory, and the optical laws for producing achromatic lenses were derived from experiments often conducted by observatory scientists.[17] In observatories, telescopes were often set up on stone pillars to isolate them from vibrations; they rotated along an axis precisely oriented with respect to the meridian. Telescopic observation relied on precise routines, such as the eye-and-ear method: the observer listened to the tick of a sidereal clock (another high-precision instrument), noting on paper slips the exact time when a star passed above a metallic thread in the telescope's ocular. Mastering this procedure required a long apprenticeship. And what could not be disciplined by training was quantified. The systematic differences in observers' measurements of the same object were identified and investigated both mathematically and experimentally by astronomers early in the nineteenth century. This knowledge was condensed into the personal equation, which gave each observer's deviation from the average (correct) measurement.[18] Finally, as Simon Schaffer explains here, the consignment of observatory measurements and their subsequent treatment was a highly ritualized operation. Raw data had to be reduced by taking into account such factors as the time of the night, the longitude and latitude of the observatory, the position of the earth, and even meteorological conditions which affected the refraction index of the atmosphere. Numbers were produced in vast quantities in the observatory and subjected to extensive manipulations before being carefully preserved for posterity in well-kept archives and long rows of beautifully bound volumes. The computing room and the library were essential components of all large observatories (figure 2).

"Every instrument," Friedrich Bessel wrote, "is made twice: once in the workshop of the maker out of brass and steel, and a second time by the astronomer on paper."[19] In the observatory, instruments were submitted to intense scrutiny. The observatory was a workshop where a wide range of technological devices—optical instruments, electrical apparatuses for telegraphy, clocks—were developed, tested, calibrated, and put to extensive use. This was done in association with instrument makers whose workshops were usually close to the observatory, frequently their main customer.[20] Conversely, instrument makers' workshops in the nineteenth century were sometimes used as observatories, where observations were made on instruments rather than stars. *Instrumentenkunde*, the art of studying the properties of instruments to design better ones, became characteristic of much nineteenth-century experimental physics. At times physicists have seemed to pay more attention to their instruments than to the natural phenomena they purportedly studied.[21] Concern with instrumentation was certainly shared by observatory scientists from the early 1800s and remained a structuring element of the interaction between physics and astronomy well into the last decades of the century, as Richard Staley shows in his discussion of Michelson's ether-drift experiments.

The observatory helped to further the culture of precision that transformed scientific practices in the nineteenth century. Humboldt's crusade to survey the Earth's magnetic field is an example of how observatory techniques were adopted for electromagnetic research. In 1828 Humboldt built a small magnetic observatory in Berlin and initiated a program of coordinated observation at various locations at prearranged times. This required a precise knowledge of time and of the geographical locations, each of which was determined by astronomical means. Gauss, as director of the Göttingen Observatory, took a major part in this survey. Mathematical equations had previously been used to account for electromagnetic phenomena, but Gauss was the first to quantify them.[22] Characteristically for an astronomer, he gave thorough descriptions of the instruments he had taken from the observatory panoply and adapted to geomagnetic surveys. His addition of a telescope to Gambey's dip magnetometer allowed the scientist to observe the needle at a distance and avoid its disturbance though air currents and bodily heat (figure 4). With his bifilar magnetometer, Gauss claimed that "the horizontal part of the earth's magnetic field can now be observed as precisely as the stars in the sky."[23]

Even the physical laboratory borrowed from the observatory. Before large physical laboratories were established in the 1860s and 1870s, it was common to speak of "physical observatories."[24] Recognizing the need for a specially designed environment for his experiments, Gauss had an iron-free building set up on the grounds of the astronomical observatory. This was one of the first modern physics laboratories. In Germany, his collaborator Wilhelm Weber wrote to Edward Sabine, "until now there existed only collections of physical instruments without permanent facilities for their use; there were no physical *laboratories* or *observatories*."[25] Throughout the nineteenth century the astronomical observatory continued to be an important setting for conceiving and performing experiments in optics, magnetism, and physiology.

4. Observatory techniques in the physics laboratory: Gauss measuring the magnetic field of the Earth in Göttingen. Antoine César Becquerel, *Traité d'électricité et de magnétisme: atlas pour la seconde partie du 5ᵉ volume, le 6ᵉ et le 7ᵉ volume* (Paris: Firmin Didot Frères, 1840), vol. 6, part 2, plate 5, fig. 6.

Managing Numbers: Statistics

In observatories a wide array of mathematical skills were fostered and developed. The observatory sciences were heir to the eighteenth-century "mixed mathematics," sciences such as astronomy, music, optics, and mechanics in which mathematics figured prominently.[26] In the hands of observatory scientists from Leonhard Euler, Alexis Clairault, Jean Le Rond d'Alembert, later Laplace and Joseph Lagrange, and to Louis-Augustin Cauchy and Henri Poincaré, the celebrated three-body problem stimulated the development of mathematical analysis. The division of computing labor was first carried out in close collaboration with the Bureau of Longitudes, much as a team led by Gaspard Prony had computed logarithmic tables during the French Revolution. It has also been argued that Gauss's work on non-Euclidean geometry was related to his geodetic work.[27] We will illustrate the complex relationship between mathematics and observatory techniques by taking the example of statistics.

In the early nineteenth century statistics was defined as "all that can contribute to give an exact idea of an empire, enlighten the competent administrator, and train the true statesman."[28] By 1900 statistical science had been quantified and tied to probability theory, and this was in part the work of astronomers. As Simon Schaffer emphasizes, the astronomer's unique ability to manage numbers had wide-ranging implications. In his discussion of early-nineteenth-century Sweden, Sven Widmalm shows how geodetic and statistical surveys were linked to coordinated efforts for mobilizing the nation as a whole for military purposes. Drawing attention to techniques for managing numbers throws new light on the participation of observatory scientists in efforts to mathematize the social sciences.

"Quantification," according to Theodore Porter, "is a social technology."[29] The techniques used by statisticians to gather, tabulate, and manipulate data with the goal of making them universally comparable bear close resemblance to those designed by observatory scientists. Authors of almanacs had long recognized this when they published ephemeredes, meteorological, and statistical data side by side.[30] The evolution of statistics might indeed be compared to the development of longitude determinations for navigation, as discussed by Guy Boistel. In both cases the conflicts between expert quantifiers in the observatory and actors on the field who relied on traditional techniques led to a continual conceptual and social adjustment.

Observatory scientists were among the first scientists to face the "avalanche of printed numbers" studied by Ian Hacking for the social sciences.[31] Astronomers pioneered probabilistic and statistical tools for analyzing and assessing the significance of their data. Laplace compiled comet statistics and attempted to evaluate the number of observations necessary to ascertain the influence of the moon on the Parisian weather.[32] Astronomers such as Laplace, Gauss, Francis Baily, John Herschel, and Adolphe Quetelet developed principles of probability theory to deal with the most intricate astronomical questions. It is therefore hardly a surprise to find that observatory scientists often paid close attention to statistics and contributed significantly to their development in the nineteenth century. As an example one might mention the *Annuaire du Bureau des Longitudes* for 1869, in which statistical demography occupies a large place, alongside astronomical, geographical, meteorological, and physical data and memoirs (figures 5–6).

Quetelet, the director of the Brussels Observatory, developed his social theory of the "average man" on the basis of a formal analogy with astronomy: "In my opinion, statistics must therefore tread the same path as the sciences of observation."[33] Among historians of statistics, the analogy has been debated at length and caused some uneasiness, owing to the perceived

INTERVALLE semi-diurne.	CORRECT. POUR LES LEVERS ET COUCHERS DE LA LUNE.						
	LATITUDE : 42°			43°			
	30'	40'	50'	0'	10'	20'	30'
3h30m	36m1	35m3	34m5	33m7	32m9	32m0	31m2
40	33,4	32,6	31,9	31,1	30,3	29,6	28,8
50	30,8	30,1	29,3	28,6	27,9	27,2	26,5
4. 0	28,2	27,6	26,9	26,3	25,6	24,9	24,3
10	25,8	25,2	24,6	24,0	23,4	22,8	22,2
20	23,4	22,9	22,3	21,8	21,2	20,7	20,1
30	21,1	20,6	20,1	19,6	19,1	18,6	18,1
40	18,9	18,4	18,0	17,5	17,1	16,6	16,2
50	16,7	16,3	15,9	15,5	15,1	14,7	14,3
5. 0	14,5	14,2	13,8	13,5	13,2	12,8	12,5
10	12,4	12,1	11,8	11,6	11,2	10,9	10,7
20	10,4	10,1	9,9	9,6	9,4	9,1	8,9
30	8,3	8,1	7,9	7,7	7,5	7,3	7,1
40	6,3	6,1	6,0	5,8	5,7	5,5	5,4
50	4,3	4,2	4,1	4,0	3,9	3,8	3,7
6. 0	2,3	2,2	2,2	2,1	2,1	2,0	2,0
10	0,3	0,3	0,3	−0,3	0,3	0,2	0,3
	+	+	+		+	+	+
20	1,7	1,7	1,6	1,6	1,5	1,5	1,4
30	3,7	3,6	3,5	3,4	3,3	3,2	3,2
40	5,7	5,6	5,4	5,3	5,1	5,0	4,9
50	7,7	7,5	7,3	7,2	7,0	6,8	6,6
7. 0	9,7	9,5	9,3	9,1	8,8	8,6	8,3
10	11,8	11,5	11,3	11,0	10,7	10,4	10,1
20	13,9	13,6	13,3	12,9	12,6	12,2	11,9
30	16,0	15,7	15,3	14,9	14,5	14,1	13,7
40	18,2	17,8	17,4	16,9	16,5	16,1	15,6
50	20,4	20,0	19,5	19,0	18,5	18,0	17,5
8. 0	22,7	22,2	21,7	21,1	20,6	20,0	19,5
10	25,1	24,5	23,9	23,3	22,7	22,1	21,5
20	27,5	26,9	26,2	25,6	24,9	24,3	23,6
30	30,0	29,3	28,6	27,9	27,2	26,5	25,8
40	32,6	31,8	31,1	30,3	29,6	28,8	28,1
50	35,3	34,5	33,7	32,9	32,1	31,2	30,4

Correction + : ajoutez au lever, retranchez du coucher.
Correction − : retranchez du lever, ajoutez au coucher.

TABLE I.
Loi de la mortalité en France suivant la Table de Deparcieux, complétée dans les premières années.

AGES.	VIVANTS à chaque âge.	SOMME des vivants.	DURÉE DE LA VIE	
			Moyenne.	Probable.
			Ans. Mois.	Ans. Mois.
0	1286	51467	39 8	42 0
1	1071	50181	46 4	53 2
2	1006	49110	48 4	54 11
3	970	48104	49 1	55 4
4	947	47134	49 4	55 2
5	930	46187	49 2	54 10
6	917	45257	48 10	54 4
7	906	44340	48 5	53 9
8	896	43434	48 0	53 2
9	887	42538	47 5	52 6
10	879	41651	46 11	51 10
11	872	40772	46 3	51 1
12	866	39990	45 7	50 3
13	860	39034	44 11	49 6
14	854	38174	44 2	48 9
15	848	37320	43 6	47 11
16	842	36472	42 10	47 2
17	835	35630	42 2	46 5
18	828	34795	41 6	45 8
19	821	33967	40 10	44 11
20	814	33146	40 3	44 2
21	806	32332	39 7	43 5
22	798	31526	39 0	42 9
23	790	30728	38 5	42 0
24	782	29938	37 9	41 3
25	774	29156	37 2	40 6
26	766	28382	36 7	39 10
27	758	27616	35 11	39 1
28	750	26858	35 4	38 4
29	742	26108	34 8	37 7
30	734	25366	34 1	36 10

5–6. Observatory techniques for managing numbers: a table for correcting the times of the rise and setting of the moon given for Paris and a table giving mortality in France as a function of age. Tables such as this combining observation, extrapolation, and computations were designed as bases for further computation. *Annuaire du Bureau des Longitudes*, 1869, 48 and 265.

epistemological rift between Laplacian determinism and the notion of statistical causation in mathematical statistics.[34] But Quetelet might be described as an observatory scientist intent on adapting the techniques he had learned in the observatory to the process of understanding and controlling social phenomena.

Developed largely for observatory purposes, Gauss's and Laplace's theory of errors is justly famous for having set the foundations for Quetelet's mathematical statistics. Error theory has been compared to the technologies of mechanical objectivity: "like photography, [it] was a strategy for eliminating interference by subjects."[35] Stéphane Callens has noticed that in elaborating his theory of errors, Gauss seemed more concerned with the order inherent in the mathematical analysis of celestial motions than with natural order.[36]

Again, it is no surprise to find that in France statistical institutions were often called observatories. In *Hard Times* Charles Dickens had already imagined such a space, "a stern room, with a deadly statistical clock in it, which measured every second with a beat like a rap upon a coffin-lid." "As if an astronomical observatory should be made without any windows, and the astronomer within should arrange the starry universe solely by pen, ink, and paper, so Mr. Gradgrind, in his Observatory (and there are many like it), had no need to cast an eye upon the teeming myriads of human beings around him, but could settle all their destinies on a slate, and wipe out all their tears with one dirty little bit of sponge."[37] Martina Schiavon's and Sven Widmalm's contributions detail how geodetic surveys led to the militarization of observatory techniques. Writing to Sabine, Weber explained: "One finally increasingly recognizes the importance that the education of *exact observers of nature* has for science and for practical life. So far only astronomy has offered an opportunity . . . for the education of exact observers."[38] By 1900 the number of "exact observers" of society and nature had skyrocketed; the significance of observatory techniques in this process remains to be established. This does not imply that the circulation of observatory techniques between different communities was always straightforward. Guy Boistel's close study of nautical astronomy shows that the transfer of techniques from the astronomer to the seafarer was always a contested affair in which each community tried to extend its acknowledged domain of expertise at the expense of the other.

Observing with "Science's Eye": Networks

Statistical enterprises relied on extensive data-gathering networks that were of central importance in the history of nineteenth-century observatories. Twice in the space of one year in 1853, Quetelet welcomed delegates from western nations to the first international congresses on navigation and on statistics in Brussels. This experience inspired him to dream up "the *vastest* observation system ever conceived by the human mind: to cover the whole globe, in all its accessible parts, with a vast network of observers . . . so that science's eye remains ceaselessly open, so to speak, to all that happens on the surface of our planet."[39] Networks have been a favorite metaphor for the way science works outside of controlled environments and in relation to political power.[40] One origin of this metaphor, and indeed the model that inspired late-nineteenth-century and twentieth-century technoscientific and imperial networks, was the observatory. This book sheds light on our contemporary understanding of science by examining another typical observatory technique: the construction and maintenance of extensive networks in which observatories were key nodes.

Observatory scientists were at the forefront of scientific networking. In 1800 an international group of astronomers led by Franz Xaver von Zach had set up the Vereinigte Astronomische Gesellschaft to look for what they thought was a missing planet between Mars and Jupiter. They established an early international vehicle for communication by publishing regular observations in Zach's *Monatliche Correspondenz* (1800–14) and Heinrich Christian Schumacher's *Astronomische Nachrichten* (from 1821).[41] In the case of geomagnetic campaigns, observatory instruments and techniques were designed at the same time as the networks in which they functioned.[42] In his chapter on Humboldt's metaphysics, John Tresch argues that the wide range of instruments and their users formed a "republic." They existed only as parts of coextensive networks that had to be organized both socially and technologically.

A case in point is meteorology (figure 7).[43] Observatories traditionally measured, centralized, and published data about the weather. When systematic surveys were first envisioned under Napoleon I, they were designed on the basis of exemplary observatory techniques—instrumental technologies and tabulating techniques. In meteorogical surveys, Jean-Baptiste Lamarck specified, "careful, detailed, simultaneous and comparable observations" should be tabulated in standard ways, and all the tables thereby produced should be reunited in a central site where they would be examined and compared.[44]

Observatory networks were specially crafted to integrate particular techniques. Observatories collected local data in standardized ways and centralized them at a few selected, carefully regulated centers. The meteorological networks established by the Paris Observatory director Urbain-Jean Le Verrier in France, the Admiral FritzRoy in Britain, and the Smithsonian Institution in the United States were grafted to telegraphic lines in just that way, while the director of the U.S. Naval Observatory, Matthew Fountaine Maury, compiled naval logbooks by the thousands. To take their place in the network, observers had to be disciplined or the observations had to be mechanized, as Massimo Mazzotti shows with regard to Secchi's meteograph. Extracting numbers from precision instrumentation, tabu-

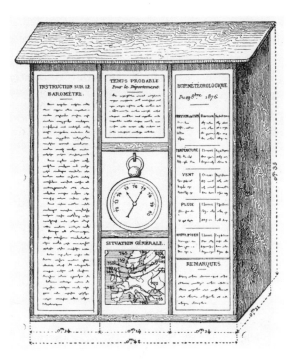

7. A box for posting meteorological warnings cabled from the Paris Observatory in 1876. The barometer, map, and data are displayed together with instructions for reading the barometer and a "probable" forecast. "Les avertissements météorologiques agricoles," *La Nature* 4, no. 2 (1876): 397.

lating them, and making forecasts and theories on that basis followed the model provided by astronomy, not metaphorically but by mobilizing the very techniques that defined observatories. No wonder meteorological stations came to be called observatories!

While meteorological networks replaced traditional qualitative assessments of the weather with quantified data wired telegraphically to, and compiled in, a central node, observatory scientists used similar resources for the precise measurement of the earth. Geodesists who relied on what topographical information they could gather from peasants in the fields or *notables* in the villages they visited to select their observation sites transformed local knowledge about the landscape into standardized astronomical coordinates, as Sven Widmalm and Martina Schiavon show. Observatories also established time distribution networks, supplying public clocks with standard time set to the Greenwich meridian (Ole Molvig). Thanks to photography, even astronomical data could be dispatched in a worldwide network of observers and analyzers of plates.[45]

Observatories were key sites for the technological networks of nineteenth-century industrial nations. Standards of measurement such as the yard and meter were defined by observatory scientists and safeguarded in observatories. As astronomers had long realized, universal standards are an absolute requirement for communication within a distributed network of observers. Gauss's introduction of absolute units in electromagnetism was said to have "extraordinary significance for physics."[46] The observatory was an acknowledged site of technical expertise where a broad range of technologies, clocks, electrical devices, thermometers, and computing machines were calibrated and subjected to stringent testing. In the United States they were even deemed essential to business, in part because they provided time—"An accurate knowledge of time is important to all business men, but especially to banking."[47]

Not all networks are equivalent, however. In his discussion of the Russian Central Observatory in Pulkovo, Simon Werrett shows that the center can sometimes stand in for the whole network. In this case the network's pragmatic function of extending control over Russian territory was overshadowed by the tsar's desire to represent this network theatrically. Similarly, both Theresa Levitt and John Tresch reveal that different conceptions of the autonomy of individual participants (whether observers or instruments) in collective networks of observation could be a potential source of conflict. Indeed, all the contributions to this book in one way or another emphasize the utterly political nature of the observatory scientists' networking activities.

Observatory Techniques on the World Stage

"Outside the scientific services," George Orwell noted in *Burmese Days*, "there is no particular need for a British official in India to do his job competently."[48] Historians are increasingly becoming aware of the importance of scientific institutions for imperialistic enterprises. It is hardly worth insisting that observatory techniques—and not just the techniques of medicine and natural history that are often studied by historians of colonial science—also proved indispensable in the constitution of large overseas empires. Navigation and cartography, as Guy Boistel recalls here, motivated the establishment of the Paris and Greenwich observatories: "it is in such institutions that science became part of the infrastructure of the mod-

8. François Perrier's station in M'Sabiha, French Algeria. The mastery of observatory and laboratory techniques in the field was crucial for making the geodetic junction between Algeria and Spain (and beyond, France) for establishing accurate maps of the colony. F. Perrier, "Jonction géodésique et astronomique de l'Algérie avec l'Espagne," *La Nature* 8, no. 1 (1880): 97–99, 152–54, at 97.

ern state."[49] In the eighteenth century the longitude problem structured many of the scientific and technological questions raised in the academy, the observatory, and the workshop, and at the highest level of government. In the following century the grafting of permanent or makeshift observatories in conquered territory went hand in hand with extensive efforts in cartography, geodesy, hydrography, and meteorology in service to the discovery, conquest, and settlement of new colonies. A precise study of observatory techniques in empires is bound to produce a more comprehensive picture of the resources needed for imperialistic pursuits (figure 8). In his study of the Paramatta Observatory in Australia, Simon Schaffer shows the overlap between observatory techniques of number crunching and the techniques of colonial administration.

The colonial project transplanted the metropolitan situation into foreign lands. Local complexities in turn had a crucial impact on the science itself—on the knowledge it accumulated and the way it was organized socially, as well as on its political significance. By expanding observatory techniques and networks and taking them into new territories (geographic and social), imperialistic projects sometimes overstretched both techniques and networks. These challenges provided opportunities for overhauling techniques and networks and served to enhance the global prestige of observatory techniques.

The participation of observatory scientists in imperial projects also had important consequences for the metropolis. In the nineteenth century, as Alex Pang has shown in his study of eclipse expeditions, astronomers came to rely on imperial structures to extend the range of their observations.[50] As David Aubin and Simon Schaffer suggest, such enterprises in turn could have a profound impact on the astronomical science carried out in British and French central observatories.

Recent postcolonial studies have shown that traditional dichotomies that form the basis for nineteenth-century colonial rhetoric—Occidental and Oriental, civilized and primitive, scientific and superstitious—were constantly "crossed and hybridized." Claims about progress, civilizing missions, and rationality were blatantly negated by the practices of exploitation in the colonies. The historian Gyan Prakash has explained that the "contortions of the discourse were endemic to colonialism not because of the colonizer's bad faith but due to the functioning of colonial power as a form of transaction and translation between incommensurable cultures and positions."[51] The boundary between Orient and Occident is blurred further when one considers the political uses made of astronomical events, such as the eclipse in Thailand in 1868 discussed by David Aubin. His analysis shows that the pragmatic and ideological uses of the eclipse were very similar in the European and Southeast Asian contexts.

Representations: Instruments, Images, and Imagination

A final example of observatory technique is representation.[52] The observatory was a representation factory in three main overlapping senses. Observatory scientists were very much concerned with developing imaging technologies. Part of their job was to process raw data and construct elaborate—and socially potent—representations of the cosmos in various forms (maps, tables, equations, narratives, and pictures). A true "laboratory of visuality," the observatory was a favorite site for reconfiguring the visual experience in the nineteenth century charted by cultural historians.[53] Recent studies have shown that seeing is a culturally determined act, shaped by deliberate strategies and by more diffuse but nevertheless powerful forces. Particular architectures (e.g. the panopticon or the urban perspective) discipline the sight as well as individual and communal behavior. New visual technologies influenced nineteenth-century aesthetics, giving rise to discourses surrounding the objectivizing techniques of mechanically rendered images, especially photography.

From the mid-nineteenth century onward, optical technologies became ubiquitous in modern western societies—in the observatory, the laboratory, and the scientific field, in the theater, in the museum, and at the exhibition, in public institutions, and at the factory. Optical instruments have contributed to transforming the nature and practice of representation, and thereby have created new ways of seeing the world and society. Optical devices, such as stereoscopes, spectroscopes, photographic cameras and all sorts of deforming lenses and motion apparatus, became as widespread as opera glasses. They were to be found on the streets, in fun fairs, and at exhibitions and were purchased by an increasingly wide range of customers (figure 50, page 311; figure 53, page 316).

Optical instruments helped to bring skills, techniques, and technologies to a wider public. The significance of the observatory in the elaboration, circulation, and adoption of the material techniques that partly conditioned the nineteenth-century reconfiguration of vision has been barely discussed in the historical literature. All optical devices were characterized by shared hardware (screws, divided circles, prisms, lenses, etc.) and shared methods of producing and interpreting evidence, all of which were important concerns of the observatory. These devices have mostly been considered in isolation from each other, obscuring their common origin.[54] Popular optical devices were often produced by successor workshops of observatory suppliers. The Soleil-Duboscq-Pellin dynasty, a major supplier of the Paris Observatory throughout the century, also produced projection apparatuses for popular use.[55]

As the preeminent locus of expertise and innovation in optics, the observatory closely interacted with instrument makers in the development of optical technologies. Makers sought to put their instruments to extensive use in the observatory before city planners and social reformers would turn them into actual or emblematic tools for disciplining space and society. Astronomers and makers were concerned with the nature of light and vision, and of error and illusion in perception, as well as the theory and practice of optical arrangements. As Richard Staley and Martina Schiavon show, the observatory, as a repository of expertise, was a source of advice to surveyors, armies, navies, and scientists from other fields, but also to artists, popular lecturers, and showmen who all relied on optical observatory techniques.

New technologies transformed science as well. Astronomers and historians have discussed the impact that new instrument technology had on the emergence of astrophysics in the late nineteenth century by examining the role of the photographic camera and particularly the spectroscope (Richard Staley and Massimo Mazzotti).[56] By transforming the means of observation available to astronomers, these instruments led them to new questions, such as the chemical constitution of celestial bodies, which could not previously be answered satisfactorily. Astronomers were quick to realize that such investigations had a wide popular appeal in particular because of their connection to the question of the plurality of inhabited worlds.[57]

Early spectroscopists decisively contributed to transforming into displays of political and scientific power the optical instruments that they used in the observatory, in the laboratory, and on overseas scientific expeditions. A site for popularizing science, the observatory pioneered many optical techniques which later made their way to fun fairs, exhibitions, and theater halls. Ole Molvig and Charlotte Bigg discuss some of the innovative mechanisms developed by astronomers and popularizers to convey their narratives, which were often closely patterned on scientific devices. The emergence of the concept of the "public" in the second half of the nineteenth century is related to the optical techniques of projection, of photography as a widespread amateur pursuit, and of both combined to create a large press industry. From the mid-nineteenth century this set of techniques helped to launch a broad-based popular science that produced representations of astronomical knowledge, along with visions of society and of science's place within it.

Representation therefore should also be taken in the extended sense, for the observatory produced not only representation techniques but also powerful representations of science (see figure 9).[58] Central observatories, with imposing architecture and ancient traditions, occupied a privileged place in the life of the city.[59] They had always served representative purposes for their patrons, whether as symbols of a monarch's enlightened power (Simon Werrett), a nation's industrial and scientific excellence (Ole Molvig), or republican progressive values (Theresa Levitt). They were expected to be the site of soirées and visits by dignitaries, state representatives, and the wider public: treating visitors to evening lectures, tours, or peeks through telescopes were important duties for professional astronomers. Public participation could take different forms, from an impromptu, late-night visit to Pulkovo by the tsar, who turned up with guests after a dinner party, to Arago's deliberately democratic lectures to the Parisian population on astronomy.

As several contributions to this volume suggest, cultural and political representation was central to observatory practices, that is, the question of who is entitled to represent science and simultaneously to offer a vision of the polity to be expressed by popular scientific representations. The observatory was a theater in which the state, science, and empire were displayed. There cosmological narratives were crafted and delivered, thematizing the place of man in the universe, the conquest of new territories and popula-

9. The public display of the heavens: visitors to Mr. Perini's "Planetary" in London could see the sun illuminated by gas and orbited by the planets suspended by thin wires. *La Nature* 8, no. 2 (1880): 17.

tions, and scientific and technological progress. Theresa Levitt and John Tresch suggest that cosmological narratives first voiced in fashionable public lectures put forward a vision of how science and the state could work together in the conquest and administration of societies, territories, and nature. Charlotte Bigg shows that as the different observatory sciences grew apart in the late nineteenth century, the unity of science was reaffirmed in grand narratives wedding positional astronomy with meteorology, physics, astrophysics, and geology. Cosmic stories flowing out of the observatory acquired a level of authority that put them on par with competing theological and evolutionary worldviews.

Tensions could arise from the observatory's dual role: emblematic and pragmatic. Simon Werrett shows that Pulkovo functioned as a pristine showcase for the imagined order of the Russian observatory sciences and the Russian Empire, while the actual networks withered away. Similar tensions between scientific and representational work existed at the Greenwich and Paris Observatories. Yet both were equally important aspects of observatory life. As Ole Molvig argues in this volume, with its cosmic theater, popular observatory, and museum, the Urania was an essential piece in the political economy of Berlin science. His story helps us to understand better why public and private funds were made available for the Berlin Observatory and the laboratories of the Physikalisch-Technische Reichsanstalt, and also what work was carried out there.[60]

In their public exertions, observatory scientists could meet with opposition, even hostility. Their narratives were criticized as being unacceptable for political or religious reasons; they had to meet the challenges raised by the natural sciences and by debates triggered by geology, evolutionary theory, or paleoanthropology. The astronomers' power to broadcast their cosmologies was also resented by competitors on the popular-science market. A growing number of popular lecturers expressed worldviews alternative to those put forward by established scientists, a sign of the utterly political nature of the observatory's public activities. In various outlets, popular science often pictured scientific power as a benevolent force at the very moment when many in the lower classes were left aside by technological change. Cosmological narratives provided forceful arguments in an ongoing negotiation between different social groups about the nature of science, the world, and society.

The cosmos was contested territory. Who could legitimately speak *about*, or *for*, the cosmos was a matter of debate. Both David Aubin's and Simon Schaffer's contributions to this volume track the conflicts between various clans intent on asserting their rights to speak for the heavens and—ultimately—for science and modernity. Both further show that numerical, literary, pictorial, and political representations of science cannot be dissociated from knowledge production and scientific practice. Just as we have argued that imperialistic enterprises cannot be confined to the periphery of the history of nineteenth-century science, several chapters suggest that it would be a mistake to view popularization as sitting on the margins of the history of science. In contrast to the view that scientific results are stabilized first in the professional observatory and only then shown to the public, John Tresch, Theresa Levitt, and Charlotte Bigg insist that inter-actions with the public must be counted among the elements that structure scientific pursuits.

In Conclusion

How then to account for the strong tension between the ubiquity of ob-servatory techniques in modern society and the prevalent image of astron-omy as an ethereal pursuit? By the end of the nineteenth century state-of-the-art astronomical observatories were no longer urban monuments but structures on remote mountaintops. Astronomy seemed to have been symbolically cut off from earthly concerns, to the point where its promi-nent social position seemed inconsistent with its object of study, even to major scientists such as Poincaré: "Governments and parliaments must find that astronomy is one of the sciences which costs most dear: the least in-strument cost hundreds of thousands of dollars, the least observatory costs millions; each eclipse carries with it supplementary appropriations. And all that for stars which are so far away, which are complete strangers to our electoral contests, and in all probability will never take any part in them."[61] This tension can be accounted for by the increased specialization of the observatory users during this period. A need for specialization was felt by an increasingly diversified pool of users adopting observatory techniques. Geodesists and meteorologists in particular deserted the astronomical ob-servatory to found disciplines of their own. In the early twentieth century the observatory more narrowly focused on positional astronomy and as-trophysics.[62]

Nevertheless, the observatory and its personnel continued fulfilling a social function into the twentieth century. In times of emergency as in times of peace, observatory techniques—the manipulation of optical in-struments, precision measurement, number management, social organi-zation—continued being vital.[63] During the First World War observatory scientists were called upon to perform a great variety of services ranging from active duty on the front to technical and scientific war work.

The observatory was no longer the main site where the reliability of these techniques was put to the test—it was now only one among a wide range of institutions. Bureaus of standards safeguarded metrological units; meteorological stations were established independently of astronomical ob-servatories; large laboratories for physics and psychology investigated the laws of nature and of sensory perception on which observatory practices depended; national statistical bureaus and the international institutes of statistics compiled data and turned their science into a thoroughly mathematical branch; popular astronomy was organized in learned societies where professional astronomers often played no more than figurative roles; separate institutions were set up for popularizing science; photography and soon cinema were by and large turned into forms of art or reporting, losing touch with their scientific origins.

Still, as the techniques fostered in the observatory continued to spread, there were potent reasons to maintain observatories as safe havens for astronomy. Untainted by lowly collusion with industry, war, and politics, the astronomical observatory, more than any other scientific institution, was seen to embody the ideals of disinterested science. Laplace had used astronomy to define the omniscient intelligence that human rationality, striving toward the truth, must approach: "The human mind offers, in the perfection which it has been able to give to astronomy, a feeble idea of this intelligence."[64] This was the view that Arago sought to popularize. His friend Honoré de Balzac portrayed astronomers as paragons of precision and lawful order.[65] In the last third of the century professional popularizers, poets, and novelists turned astronomy into an endless source of romantic bewilderment. But one wonders whether most professional astronomers, spellbound by their numbers, mathematics, and instruments, were not out of touch with the world. From the misanthropic Palmyrin Rosette, handcuffed by his companions in an effort to force him to return to the Earth in Jules Verne's *Off a Comet!* (*Hector Servadac*), to Professor Phostle, the crazed scientist who shocked Tintin by predicting the end of the world in Hergé's *Shooting Star* (*L'étoile mystérieuse*), the astronomer had become the lunatic described by Anatole France (figure 10): "Engulfed in the celestial spaces he knew not what occurred upon the surface of the earth."[66]

Steven Shapin has argued that seclusion is the seed of error.[67] In Thomas Hardy's *Two on a Tower*, the nightly solitude of the astronomer—especially if troubled by the presence of a married woman—looked suspicious. Set up on mountains, astronomical observatories escaped direct public scrutiny, but the knowledge that they produced retained an aura of certainty, perhaps because their personnel seemed to have so little to do with mundane concerns. The very idea of pure science depended on preserving this untainted site of knowledge production.

Forced to acknowledge that navigation alone provided enough justifi-

10. In Jules Verne's *Off a Comet!*, the astronomer Palmyrin Rosette is a misanthropic character who will not leave his observatory unless forced to do so. Jules Verne, *Hector Servadac: voyages et aventures à travers le monde solaire* (Paris: J. Hetzel, 1874), 280.

cation for the support of observatories, Poincaré nonetheless insisted that astronomy was valuable in its own right, because it taught philosophical lessons about the proper place of human beings in the universe and provided a methodological guide to the conduct of science: "it is useful because it is grand; that is what we should say."[68]

Notes

1. Amédée Guillemin, *The Heavens: An Illustrated Handbook of Popular Astronomy*, 2nd edn, ed. J. Norman Lockyer (London: Richard Bentley, 1866), 485.
2. Robert Ball, *In Starry Realms* (London: Isbister, 1892), 99. We are grateful to Simon Schaffer for suggesting this reference.
3. W. C. Bond to Josiah Quincy, 11 October 1849, quoted in Carlene E. Stephens,

"Astronomy as Public Utility: The Bond Years at the Harvard College Observatory," *Journal for the History of Astronomy* 21 (1990): 21–35, 29.

4. See the data provided by Dieter B. Herrmann in "An Exponential Law for the Establishment of Observatories in the Nineteenth Century," *Journal for the History of Astronomy* 4 (1973): 57–58; Herrmann, *The History of Astronomy from Herschel to Hertzsprung*, 179. On the multiplication of observatories see Bigourdan, *Histoire de l'astronomie d'observation et des observatoires en France*; Hahn, "Les observatoires en France au XVIIIᵉ siècle"; Howse, "The Greenwich List of Observatories"; Steven J. Dick, "National Observatories: An Overview," *Journal for the History of Astronomy* 22 (1991): 1–4; and Dick, "Pulkovo Observatory and the National Observatory Movement."

5. On the American observatory movement see Craig Howard White, "Natural Law and National Science: The 'Star of Empire' in Manifest Destiny and the American Observatory Movement," *Prospects: An Annual of American Cultural Studies* 20 (1995): 119–60; and Marlana Portolano, "John Quincy Adams's Rhetorical Crusade for Astronomy," *Isis* 91 (2000): 480–503.

6. Francis Hervé, *How to Enjoy Paris in 1842, Intended to Serve as a Companion and Monitor Indicating All That Is Useful and Interesting in the French Metropolis: Containing Historical, Political, Commercial, Artistical, Theatrical and Statistical Information, as Also a Description of the Manners and Customs of the Parisians of the Present Day, with Instructions for the Stranger in Respect to Economy, and Advice as to His General Proceeding with the French* (Paris: Amyot, 1842), 185–86.

7. Laplace, *Exposition du système du monde*, 6th edn (Paris: Bachelier, 1835), 207. In a recent biography Laplace is shown to have been in many ways the last savant of the Enlightenment. Hahn, *Pierre Simon Laplace*.

8. See however Marie-Noëlle Bourguet, "Landscape with Numbers: Natural History, Travel and Instruments in the Late Eighteenth and Early Nineteenth Century Centuries," *Instruments, Travel, and Science*, ed. Bourguet, Licoppe, and Sibum, 96–125. In his study of observation logbooks in eighteenth-century France, Nicolas Lesté-Lasserre has recently shown this noninterventionist ethos to be a very partial view of the observational practices of the period. See Lesté-Lasserre, "Le journal d'observations astronomiques au XVIIIᵉ siècle." On mechanical objectivity see Daston and Galison, "The Image of Objectivity."

9. Fleck, "Schauen, sehen, wissen," 154.

10. Mauss, "Les techniques et la technologie," 252. For a historiographical discussion of "techniques" in the history of science and technology see Dominique Pestre and Yves Cohen, "Présentation," *Annales HSS*, nos. 4–5 (July–October 1998), 721–44; and Sibum, "Narrating by Numbers."

11. Certeau, *The Practice of Everyday Life*, 91–130. For the historiography of space in science studies see e.g. Hannaway, "Laboratory Design and the Aim of Science"; Shapin, "The House of Experiment in Seventeenth-Century England"; Ophir and Shapin, "The Place of Knowledge"; David N. Livingstone, "The Spaces of Knowledge: Contributions towards a Historical Geography of Science," *Environment and Planning D: Society and Space* 13 (1995): 5–34; Crosbie Smith and Jon

Agar, eds., *Making Space for Science: Territorial Themes in the Shaping of Knowledge* (Houndmills: Macmillan, 1998); and Peter Galison and Emily Thompson, eds., *The Architecture of Science* (Cambridge: MIT Press, 1999).

12. On cultural changes in notions of time and space see Kern, *The Culture of Time and Space*, and the more technically informed Bartky, *Selling True Time*.

13. On this issue see also Peter Galison and David J. Stump, eds., *The Disunity of Science: Boundaries, Contexts, and Power* (Stanford: Stanford University Press, 1996).

14. Wise, *The Values of Precision*.

15. Gauss to Göttingen University curator, 29 January 1833, quoted in Jungnickel and McCormmach, *Intellectual Mastery of Nature*, 1:64.

16. Historians of astronomy can be counted among those who have pioneered the exploration of nontheoretical issues in the history of science, in particular paying considerable attention to instruments of high precision and their makers. See for example Henry C. King, *The History of the Telescope* (Mineola, N.Y.: Dover 2003 [1955]); and Chapman, *Dividing the Circle*.

17. See e.g. Jackson, *Spectrum of Belief*.

18. On the "personal equation," which has given rise to extensive literature, see Schaffer, "Astronomers Mark Time"; Canales, "Exit the Frog," and her contribution to the History of Venus Transits Conference in Paris, 4 June 2004 (forthcoming in *Cahiers François Viète*); Schmidgen, "Time and Noise." On the social organization of an observatory see Smith, "A National Observatory Transformed."

19. Friedrich Wilhelm Bessel, *Populäre Vorlesungen über wissenschaftliche Gegenstände*, ed. H. C. Schumacher (Hamburg: Perthes-Besser und Mauke, 1848), 432. We are grateful to Simon Schaffer for this reference.

20. On instrument makers see Turner, *Nineteenth-Century Scientific Instruments*; and Bennett, *The Divided Circle*. See also Paolo Brenni's series "19th-Century French Scientific Instrument Makers" in the *Bulletin of the Scientific Instrument Society*, in particular "I: H.-P. Gambey," 38 (1993): 11-3; "III: Lerebours et Secretan," 40 (1993): 3-6; "XI: The Brunners and Paul Gautier," 49 (1996): 3-8; "XII: Louis Clément François Breguet and Antoine Louis Breguet," 50 (1996): 19-24; and "XIII: Soleil, Duboscq, and Their Successors," 51 (1996): 7-16.

21. Dörries, "Balances, Spectroscopes, and the Reflexive Nature of Experiment"; Blondel, "Electrical Instruments in Nineteenth Century France, between Makers and Users"; Blondel, "Les physiciens français et l'électricité industrielle à la fin du XIXᵉ siècle"; Jackson, *Spectrum of Belief*; Bigg, "Behind the Lines"; and Terry Shinn and Bernward Joerges, eds., *Instrumentation between Science, State, and Industry* (Dordrecht: Klüwer Academic, 2001).

22. Darrigol, *Electrodynamics from Ampère to Einstein*.

23. Gauss to Olbers, 2 September 1837, in *Briefwechsel zwischen C. F, Gauss und H. W. M. Olbers* (Hildesheim: Georg Olms, 1976), 2:649. See also David Aubin, "Astronomical Precision in the Laboratory: The Role of Observatory Techniques in the History of the Physical Sciences," *Grundsätze über die Anlage neuer Sternwarten*, by Borheck, 31–36; Fabien Locher, "The Observatory, the Land-Based

Ship and the Crusades: Earth Sciences in the European Context, 1830–50," *British Journal for the History of Science* 40, no. 4 (2007): 491–504.

24. Aubin, "Orchestrating Observatory, Laboratory, and Field."

25. Weber to Sabine, 20 February 1845, quoted in Jungnickel and McCormmach, *Intellectual Mastery of Nature*, 1:77. Our emphasis.

26. Gary I. Brown, "The Evolution of the Term Mixed Mathematics," *Journal of the History of Ideas* 52 (1991): 81–102.

27. Ernst Breitenberger, "Gauss's Geodesy and the Axioms of Parallels," *Archive for History of Exact Sciences* 31 (1984): 273–89. Ivor Grattan-Guinness, "Work for the Hairdressers: The Production of de Prony's Logarithmic and Trigonometric Tables," *Annals of the History of Computing* 12, no. 3 (1990): 177–85.

28. D. F. Donnant, *Théorie élémentaire de la statistique* (Paris: Valade, 1805), xii.

29. Porter, *Trust in Numbers*, 49.

30. Perkins, *Visions of the Future*. See Marie-Noëlle Bourguet, "Décrire, compter, calculer: The Debate over Statistics during the Napoleonic Period," *The Probabilistic Revolution*, ed. Lorenz Krüger, Lorraine J. Daston, and Michael Heildeberger (Cambridge: MIT Press, 1989), 1:305–16.

31. Ian Hacking, "Biopower and the Avalanche of Printed Numbers," *Humanities in Society* 5 (1982): 279–95; Hacking, *The Taming of Chance*.

32. Stephen M. Stigler, "Napoleonic Statistics: The Work of Laplace," *Biometrika* 62 (1975): 503–17; O. B. Sheynin, "On the History of Statistical Methods in Astronomy," *Archives for History of Exact Sciences* 29 (1984): 151–99.

33. Adolphe Quetelet, *Recherches statistiques sur le Royaume des Pays-Bas* (Brussels: Hayez, 1829), iv.

34. Stephen M. Stiegler, "Adolphe Quetelet: Statistician, Scientists, Builder of Intellectual Institutions," *Actualité et universalité de la pensée scientifique d'Adolphe Quetelet: actes du colloque organisé à l'occasion du bicentenaire de sa naissance, Palais des Académies, 24–25 octobre 1996* (Brussels: Académie Royale des Sciences, 1997), 51. See also Porter, *The Rise of Statistical Thinking*, 47ff.

35. Porter, *Trust in Numbers*, 201.

36. Stéphane Callens, *Les maîtres de l'erreur: mesure et probabilité au XIXᵉ siècle* (Paris: Presses Universitaires de France, 1997), 60. On the history of statistics and its relation to astronomy see also Alain Desrosières, *The Politics of Large Numbers: A History of Statistical Reasoning*, trans. Camille Naish (Cambridge: Harvard University Press, 1998); and Michel Armatte, "Histoire du modèle linéaire: formes et usages en statistique et économétrie jusqu'en 1945" (doctoral thesis, École des Hautes Études en Sciences Sociales, Paris, 1995). On error theory see also Martina Schiavon's contribution to this volume.

37. Dickens, *Hard Times*, chapter 15. We thank Simon Schaffer for this reference. See also his contribution to this volume for further discussion of the same episode.

38. Weber to Sabine, 20 February 1845, quoted in Jungnickel and McCormmach, *Intellectual Mastery of Nature*, 1:77.

39. Adolphe Quetelet, *Sciences mathématiques et physiques au commencement du XIXᵉ siècle* (Brussels: C. Muquardt, 1867), 23.

40. Latour, *Science in Action.*
41. On the history of collective organization in astronomy see, for Germany, Jürgen Hamel, "H. C. Schumacher: Zentrum der internationalen Kommunikation in der Astronomie und Mittler zwischen Dänemark und Deutschland," and Gudrun Wolfschmidt, "Internationalität von der VAG (1800) bis zur Astronomischen Gesellschaft," *Acta Historia Astronomiae* 14, 89–120, 182–203; and Dieter B. Herrmann, "Das Astronomentreffen im Jahre 1798 auf dem Seeberg bei Gotha," *Archive for the History of Exact Sciences* 6 (1969–70): 326–44. For the United States see Lankford with Slavings, *American Astronomy*, and DeVorkin, *Henry Norris Russell.*
42. Cawood, "Terrestrial Magnetism and the Development of International Collaboration in the Early Nineteenth Century"; and Cawood, "The Magnetic Crusade."
43. On the history of meteorology see Fleming, *Meteorology in America*; Anderson, *Predicting the Weather*; and Locher, "Le nombre et le temps."
44. Jean-Baptiste Lamarck, *Annuaire météorologique pour l'an XI de l'ère de la République française* 4 (1804): 153.
45. Bigg, "Photography and the Labour History of Astronomy."
46. Ernst Dorn, "Anmerkungen" to C. F. Gauss's *Die Intensität der erdmagnetischen Kraft auf absolutes Maass zurückgeführt* (Leipzig: Engelmann, 1894), 50, quoted in Jungnickel and McCormmach, *Intellectual Mastery of Nature*, 1:71. On standards in observatories see e.g. Frängsmyr, Heilbron, and Rider, eds., *The Quantifying Spirit in the 18th Century*; Schaffer, "Metrology, Metrication and Victorian Values"; and Alder, *The Measure of All Things.*
47. Elias Loomis, "Astronomical Observatories in the United States," *Harper's New Monthly Magazine* 13 (1856), 25–52, at 51–52.
48. George Orwell, *Burmese Days* (San Diego: Harcourt Brace, 1962 [1934]), chapter 5, 68.
49. Michael S. Mahoney, "Charting the Globe and Tracking the Heavens: Navigation and the Sciences in the Early Modern Era," *The Heirs of Archimedes: Science and the Art of War through the Age of the Enlightenment*, ed. Brett D. Steele and Tamara Dorland (Cambridge: MIT Press, 2005), 221–30, at 228. On longitude see Howse, *Greenwich Time and the Discovery of the Longitude*; Andrewes, *The Quest for Longitude*; Boistel, "L'Astronomie nautique au XVIIIe siècle en France"; and Jullien, ed., *Le calcul des longitudes.*
50. Pang, *Empire and the Sun.*
51. Both quotes from Gyan Prakash, ed., *After Colonialism: Imperial Histories and Postcolonial Displacements* (Princeton: Princeton University Press, 1995), 3. Specifically on observatory techniques see Raj, *Relocating Modern Science*; and Pratt, *Imperial Eyes.*
52. On representation in observatory sciences see Levitt, "The Shadow of Enlightenment."
53. Benjamin, "The Work of Art in the Age of Mechanical Reproduction"; Foucault, *Discipline and Punish*; Crary, *Techniques of the Observer*; Daston and Galison, "The

Image of Objectivity"; Schivelbush, *The Railway Journey*; and Schivelbush, *Disenchanted Night*. On the role of the observatory sciences see Schaffer, "On Astronomical Drawing"; and Canales, "Photogenic Venus."

54. See Bigg, "Behind the Lines."

55. Mauss: "un mécanicien de précision opère des visées, lit des verniers, qui, autrefois, étaient le privilège des astronomes." "Les techniques et la technologie," 254.

56. Hentschel, *Mapping the Spectrum*; Bigg and Staubermann, eds., "Spectroscopic Histories"; Bigg, "Behind the Lines."

57. Crowe, *The Extraterrestrial Life Debate*.

58. The history of popular science in the nineteenth century has mostly focused on natural history at the expense of the observatory sciences, despite the prominence of cosmological themes and the enthusiasm for optical instruments in such forums as museums, international exhibitions, and *conversazione*. Among recent publications see Secord, *Victorian Sensation*; and Daum, *Wissenschaftspopularisierung im 19. Jahrhundert*.

59. Laurie, "The Board of Visitors of the Royal Observatory"; and Aubin, "The Fading Star of the Paris Observatory in the Nineteenth Century."

60. Cahan, *An Institute for an Empire*.

61. Poincaré, *The Value of Science*, 84.

62. See Dominique Pestre, "The Moral and Political Economy of French Scientists in the First Half of the Twentieth Century," *History and Technology* 13 (1997): 241–48.

63. See Schiavon, "Des savants-officiers entre science, armée, état et industrie de précision."

64. P.-S. Laplace, *A Philosophical Essay on Probabilities*, trans. Frederick W. Truscott and Frederick L. Emory (New York: Dover, 1951), 4.

65. See *Le Père Goriot*, *Eugénie Grandet*, and *Mémoires de deux jeunes mariées*, chapter 25.

66. Anatole France, *L'île des pingouins* (Paris, 1908).

67. Steven Shapin, "'The Mind Is Its Own Place': Science and Solitude in Seventeenth-Century England," *Science in Context* 4 (1991): 191–218.

68. Poincaré, *The Value of Science*, 115.

Chapter 1

The Astronomical Capital of the World:

Pulkovo Observatory in the Russia of Tsar Nicholas I

SIMON WERRETT

> A high appreciation of astronomy seems inherent in the Romanov race.
> —WILHELM STRUVE (1862).[1]

After its foundation in 1839, Pulkovo observatory near St. Petersburg quickly became a much-admired model for the organization of observatory sciences. Astronomers deemed acquaintance with Pulkovo and its first director, the Baltic German astronomer Friedrich Wilhelm Georg Struve, as paramount to successful observing. George Biddell Airy, British astronomer royal, visiting Pulkovo in 1847 wrote: "No astronomer can feel himself perfectly acquainted with modern observing astronomy . . . who has not well studied the observatory at Pulkowa." The American astronomer Benjamin Gould (the younger) said of Struve that his "biography might almost be said to be the history of stellar astronomy during the period of his labors."[2] By the 1860s Struve and Pulkovo had gained a glowing reputation across Europe and America for contributions to the pressing astronomical concerns of the nineteenth century, including the accurate positional surveying of the heavens, searches for double stars and optical pairs for determining stellar parallax, and practical geodesy in pursuit of precision cartography. Astronomers everywhere emulated Pulkovo, the "astronomical capital of the world" (figure 11).[3]

What was distinctive about Pulkovo as an observatory in the mid-nineteenth century? What made Pulkovo the astronomical capital of the world? Examining Pulkovo's local Russian context may provide an answer. Alexander von Humboldt described Pulkovo as an "astronomical city whose form of government may be difficult to conceive."[4] His remark is suggestive of Pulkovo's unique place within the Russian polity, ill fitting the various

11. View of the Pulkovo Observatory fifty years after it was built. *Himmel und Erde* 1, no. 11 (1888).

models that historians have proposed for observatories elsewhere. Pulkovo was not a "Hevelian" observatory, with civic funding and an urban location. With princely patronage and a focus on stellar astronomy, Pulkovo was closer to Tycho's observatory Uraniborg.[5] Jean-Baptiste Biot called it a "Uranian colony," a group of Baltic Germans working in Russia.[6] But Pulkovo also engaged in major state-funded geodesic surveys and was intended to serve a network of imperial observatories. Unlike the British or Dutch empires, the Russian empire grew out of military rather than commercial demands, since in Russia industrialization was never the priority that it was elsewhere. Consequently, Pulkovo did not "keep the books" in the same way as economically minded British astronomers, and it ill fits the model of Greenwich as a center for imperial astronomy focused on the accounting of time to aid overseas colonial exploitation.[7] As a contemporary noted: "Pulkowa is like the palace of an astronomical autocrat, who has but to *will*, and men and money appear at his call to take the heavens by storm. Greenwich resembles the counting-houses of some of our opulent city merchants, showing more brick than marble, but whose cellars are stored with the accumulated wealth of generations."[8] If Pulkovo did have a precedent, it was perhaps the Paris Observatory, founded by Louis XIV as an act of both symbolic prestige and practical, cartographical administration. This had been the model for Russia's first observatory, in the St. Petersburg

Academy of Sciences founded by Peter the Great in 1725, and directed by a French astronomer, Joseph Delisle. But by the early nineteenth century the analogy no longer held. The Paris Observatory's nature had changed significantly. As Theresa Levitt shows, Arago tried to make his institution the very opposite of a centralized, absolutist observatory. Pulkovo, meanwhile, expressed the concerns of a continuing autocratic regime in Russia.[9]

In contrast to western European states, Russia adopted new modes of life comparatively recently, beginning in the late seventeenth century and accelerating under Peter the Great. This profoundly shaped the country in the eighteenth and nineteenth centuries, making it, according to the Russian semiotician Yuri Lotman, a fundamentally theatrical culture, as Russians imitated foreign manners, traditions, and institutions, from dress and appearance up to government ministries and military organization.[10] Understanding this theatricality is useful in assessing the nature of Pulkovo observatory and Russian imperial astronomy.

Theatricality made Russia seem curiously illusionistic. Mikhail Epstein has said of Russia that "models of reality replace reality itself, which then becomes irrecoverable."[11] It was commonplace in the nineteenth century to characterize Russia as a land where the illusion of order often stood in for real order. Foreigners condemned this theatricality as deception or fakery. After his visit of the country in 1839, the year of Pulkovo's founding, the Marquis de Custine wrote: "Russians have only names for everything, but nothing in reality. Russia is a country of façades. Read the labels—they have 'society,' 'civilization,' 'literature,' 'art,' [and] 'science'—but as a matter of fact, they don't even have doctors. . . . Russia is an Empire of catalogues: if one runs through the titles, everything seems beautiful. But . . . open the book and you discover that there is nothing in it. . . . How many cities and roads exist only as projects. Well, the entire nation, in essence, is nothing but a placard stuck over Europe." Custine proposed there were really "two nations" in Russia, "one of these nations is the Russia as she is in reality, the other is the Russia as they would like her to seem to Europe."[12]

Custine was being critical, but he misinterpreted the theatrical nature of Russian life. Russians were not trying to deceive, and so demanded no veracity—theatricality was inherent to the culture. Nevertheless, Custine's comments are suggestive as a means for approaching imperial Russian astronomy under Nicholas I. In what follows, two contrasting scenes of Russian imperial astronomy are revealed. The first refers to the Russian empire, a vast, ungovernable territory which Nicholas spent much of his time attempting to discipline through an elaborate system of administration and surveillance. To make the empire more manageable, Nicholas launched surveying and astronomy projects, organized by the military. Pulkovo was intended to coordinate this imperial astronomy, but it was beset with difficulties. In the provinces of empire, Russian astronomy floundered.

To the British, as Schaffer observes of Paramatta observatory, such oblivion would have been loathsome. But for Nicholas what mattered most was the staging of good order, and here may be found the other scene in Russia's astronomical theater—the "Central Observatory" at Pulkovo. Thus Nicholas's second reason for founding Pulkovo was to make a demonstration of imperial patronage of the sciences, to impress audiences in Western Europe and America, and in this he succeeded well. Luxurious patronage enabled Struve to build the best observatory in the world and to create a showcase of the finest observatory techniques then available. This was Russia's second reality, part of the nation "as they would like her to seem to Europe." Later we explore the theatrical side of Pulkovo's activities. Pulkovo is situated in its original context, built on the tsar's estate, where diverse showcase technologies of surveillance were put on display, and to which foreigners were frequently invited to witness the spectacle of Pulkovo's techniques. These visits shaped the observatory's techniques and affected choices over observational conduct. Russian theatricality, it is proposed, was an essential element in making Pulkovo the astronomical capital of the world.

The Empire: Attempts to Build an Imperial Astronomy

Russia's vast expanse has often been characterized as a "burden," and the history of its governance as a continuous struggle to impose regularity on an essentially amorphous and contradictory territory.[13] Unlike the nations of western Europe, with neatly defined borders, metropolitan centers, and relative ease of communications, Russia's continuously expanding empire was defined more by intensities of population and culture than natural or clear political boundaries. Territories were a series of shifting peripheries peopled by Christians, Chinese, Siberians, and Muslims. Tsar Peter I first imported the terms "imperia" and "imperator" into Russia in the early eighteenth century. Between 1800 and 1860 the empire swallowed Poland, Georgia, Baku, Daghestan, Finland, regions of the Caucasus, and Turkish Bessarabia. Only the Crimean War (1853–56) briefly halted Russian expansionism. The boundary of Russia was constantly redefined. Traveling costs and difficulties only added to the vagueness of territory. To command such an extensive empire, Russia's tradition of centralized, autocratic government has been deemed necessary. Administration, rather than history, culture, or tradition, became the chief means to comprehend—and control—Russia's vast territory.

These problems led Russian governments of the nineteenth century to place particular importance on the army as a tool of administration and police. From the time of Peter I, Russia was idealized as a "well-ordered police state," similar to Prussia or Sweden, best administered by the army, which epitomized good order.[14] Little had changed by the reign of Nicholas I, who still conceived of militarism as the best way to manage his empire: "Here [in the army] there is order, there is strict unconditional legality, no impertinent claims to know all the answers, no contradiction, all things flow logically one from the other; no one commands before he has learned to obey, . . . everything is subordinated to one definite goal, everything has its purpose."[15] Nicholas became obsessive about creating an orderly state and set about a massive program of centralization and police, intending to turn every branch of government and culture into a machine under constant central surveillance. This policy was enshrined in the reactionary slogan of Nicholas's minister for education Sergei Simeinovich Uvarov, also the minister responsible for founding Pulkovo: "Orthodoxy, Autocracy, Nationality." To this "militarism" and "police" could be added. Most government ministers were military officers. Even the Orthodox Church was headed by a colonel. The model of Russia's secret police right up to the KGB, the Third Department of His Majesty's Own Chancellery, was founded by Nicholas. Inspecting everywhere, the tsar himself constantly traveled across Russia, and became so notorious for turning up unexpectedly that citizens devised alarms to warn of his approach. Nicholas wrote to his wife, Alexandra: "One sees everything, and they never know when or where I am going to arrive. They expect me everywhere, and if everything is not well, they at least try and make it so." Even Pulkovo experienced these inspections.[16]

Since it was of course impossible for the tsar to supervise everything personally—and obituary writers thought that Nicholas died trying to do so—he showed much interest in developing technologies of surveillance. Members of the Imperial Academy of Sciences at St. Petersburg eagerly helped. Moritz Jacobi designed electromotors to power Russian naval vessels and with Pavel Schilling devised an electric telegraph in the 1830s. Demands for militarized order also led to improved communications, with the construction of Russia's first solid road in 1834 and its first railway in 1837, intended to accelerate troop movements.[17] Astronomy could also improve imperial surveillance. Like the army, astronomers valued order and precision, and their discipline evidently appealed to the tsars, as the epigraph by Struve to this chapter suggests.

Pulkovo's foundation was the culmination of an effort dating back to the reign of Alexander I. Napoleon I's invasion had highlighted the need for accurate maps of the empire and prompted the formation of a Military-Topographical Depot and Hydrographic Bureau to survey Russia's lands and seas and train expert geodesists and hydrographers.[18] At Dorpat University, young Wilhelm Struve trained naval officers (the "General Staff") in astronomy and surveying while orchestrating a survey of Livland in 1815–16. He planned to combine this with further measurements of some 25° 20' of an arc of the meridian between northern Finland and the mouth of the Danube on the Black Sea, a significant geodetic and astronomical achievement which would also be crucial for mapping new imperial territories.

Although Dorpat was equipped with a fine observatory, this together with the topographic offices was deemed insufficient for Nicholas's purposes. As the tsar acknowledged in a decree of 24 February 1830, "the interior of the Empire and especially the Asiatic regions has a vast expanse, having a complete lack of astronomical determinations, as a result of which both the private surveys and the works of the General Staff cannot be carried out with the desired advantage and cannot procure for us fundamental geographical knowledge of many parts of the empire."[19] Nicholas ordered the General Staff, the Military-Topographic Depot, and the Academy of Sciences to step up imperial surveys. A series of geodesic expeditions backed by the military followed, determining points on the Baltic coasts (1833), between Lake Baikal and China (1830), between the Ural Mountains and Irkutsk (1832), and between the Black and Caspian seas (1836).[20]

Founded to contribute to this practical astronomy and geodesy, Pulkovo would increase "the efficiency of the ordinary governing of the land," as Wilhelm Struve put it. "For . . . it is at Pulkovo that the officers of the General Staff and Topographical Corps are educated in higher geodesy, and there also that questions of surveying are settled in mapping the country alike for taxation, and railways, canals and agricultural improvements."[21] In 1830 a commission of prominent military officers, academicians, and ministers including Admiral Aleksei Greig, founder of the naval observatory at Nikolaev, and Count Uvarov, Nicholas's chief propagandist, were called together to devise plans for Pulkovo, and nine years later the observatory was opened.[22] Struve, with an established personal reputation in surveying, as well as in the training of military surveyors, was put in charge, to be assisted by his son Otto.

Conferred a "ministerial" role similar to that of the Academy of Sciences, Pulkovo would coordinate imperial astronomy and act as the "central" observatory (*Glavnaia observatoriia*; *observatoire central*) in a network of hitherto unconnected regional observatories scattered across the empire.[23] Pulkovo's remit included coordination of observatories in Moscow (founded in 1805), Kharkov (1808), Nikolaev (1820), Warsaw (1825), Kazan (1833), and Kiev (1845). These sites had often been created in the wake of colonial conquest. After Finland fell to Russia in 1808, for example, Russian observatories were established in Helsingfors and Åbo. Russian astronomers valued this centralized administration. When they gathered from across the empire to attend the opening of Pulkovo in 1839, astronomers were asked "henceforth to join their efforts with those of the Central Observatory to make astronomers prosper in Russia."[24] The same sentiment permeated Pulkovo's statutes, according to which the Central Observatory was to "take care that the activities in other observatories are in accordance with the current condition of astronomy, that their actions, as far as possible, are coordinated with one another, and that the observations undertaken in them are of as much benefit as possible for science."[25]

Such were the intentions to create an imperial astronomy. Yet in practice Pulkovo's objectives were scarcely met. Like the British at Paramatta, Russians found managing imperial astronomy a difficult business. At first things went well in the metropolis, with astronomy flourishing in Moscow and St. Petersburg after Pulkovo's foundation. Struve trained astronomers in cooperation with Professor A. N. Savich of St. Petersburg University. In Moscow a new observatory was built in 1846–47 under the direction of A. N. Drashusov according to plans inspired by Pulkovo and approved by Struve. The Moscow astronomers B. Ya. Schweitzer, F. A. Bredekhin, and M. F. Khandrikov reiterated Pulkovo's program of stellar astronomy and parallax measurements and introduced new astrophysical techniques to Russia. These were considered successful in the European context.[26]

Successful too were Struve and the Military-Topographical Depot's efforts to measure twenty-five degrees of the meridian arc passing through Dorpat. The military provided essential resources. Colonel Carl Friedrich Tenner made a survey of Lithuania which was combined with Struve's Livland measurements to give the first 8° 2′ of the meridian. Then, between 1816 and the early 1850s, Tenner and other officers made most of the measurements needed to calculate the meridian arc, with occasional supervision from Struve. Other imperial surveys of the Caucasus and chronomet-ric expeditions were also managed from the Central Observatory. In 1843 and 1844 Struve conducted expeditions from Pulkovo to Altona and from Altona to Greenwich, establishing the longitudinal connection between Pulkovo and Greenwich.[27]

Much effort was required to make these expeditions work. Standardized measurement proved difficult to make amid diverse imperial cultures: "The whole system, transferred here from Russian provinces, did not conform to the level of civic awareness of the local inhabitants, who have completely different concepts, beliefs, customs, and life habits."[28] Local inhabitants under Russian rule resisted imperial surveillance. After the Polish uprising of 1831, it became imperial policy to confiscate locally drawn maps in order to force indigenous astronomers and geodesists to adopt Russian maps.[29]

The empire's landscape was another problem for geodesists, as for the administration in general. Since environmental conditions varied greatly across the Russian empire, extensive labor was needed to secure steady measures. A single surveying party might travel between the barren desert hills of Bessarabia through Baltic forests into the frozen landscape of Fin-land and Sweden. As the English found in their Indian surveys, such con-ditions demanded ad hoc innovation. Thus Colonel Tenner used Kurgan barrow-mound graves in the steppes as geodesic viewing stations, "where visual rays are elevated a little above the surrounding terrain."[30]

Instruments suited to observatory conditions had to be discarded in the field. Having used Baumann and Troughton circles in his early surveys, Struve grew dissatisfied with repetition instruments after discovering a discrepancy when the same point was measured twice with the telescope reversed. Repetition was "admirable . . . in theory, [but] one cannot always make measures of the highest exactitude, because it supposes conditions which, strictly speaking, do not exist." Instead he used William Mudge's reiteration method, employed by Everest in his Indian survey, whose results then needed to be combined with adjusted results from the earlier repeti-tions (figure 32, page 202).[31]

Instruments needed to be adjusted to field conditions. To reduce the swaying in the wind of forty-meter-high viewing stations, smaller instru-ments were used. Ertel's and Reichenbach's universal instrument failed in the field "because of its excessive sensitivity." Consequently, only well-trained personnel were able to use it, and this did not include most of the soldiers working on surveys in the field: "Though admirable mathemati-cally in its totality and in its parts . . . it demands a rational observer who has studied it scrupulously."[32] These troubles led to the establishment of extensive training programs for provincial surveyors.[33] Finally, even if instruments were utilizable, it was not always possible to buy in large quantities from the best German makers. Imperial patronage from Nicholas was ideal for ordering unique state-of-the-art telescopes from foreign masters, but cheap instruments were harder to obtain. Distance from skilled German makers forced Struve to create an instrument-making workshop at Pulkovo to produce theodolites and other instruments for military students. This had already been a problem in the Dorpat era. For his Baltic surveys Tenner used a repeating theodolite built in the workshops of the Russian General Staff, "less exact than the other [German] instruments."[34]

Similar problems beset Russia's smaller imperial observatories. Lavish funds for the Central Observatory were not matched in the provinces. Despite producing much valuable work in stellar astronomy and geodesy, many observatories languished in the nineteenth century. Distance, accidents, poor funding, and ironically the attraction of working at Pulkovo led provincial observatories into ruin. Under its new director Johann Mädler, Dorpat Observatory fell into decline as soon as Struve left. Directed by E. N. Fuss and a string of Pulkovo astronomers, Vilnius observatory had early successes with the second photoheliograph ever made in 1866 and used extensively by P. M. Smyslov to photograph the sun. Ten years later, however, the observatory burned down. The Academy of Sciences refused to pay for a replacement, and eight hundred of Smyslov's photographs were carried off to Pulkovo. In Warsaw the observatory remained so small that its directors often focused on popularizing astronomy rather than doing it. One of them, Jan Baranowski, published the first Polish-language edition of Copernicus's *De Revolutionibus* in 1854 and translated Humboldt's *Kosmos* into Polish. In Kiev the observatory fell into such disorder during the 1850s that it needed to be rebuilt by A. P. Shidlovskii in 1856–62. Then the first director of the new observatory, Bredekhin, left for Moscow only two months after the start of his tenure in Kiev, so strong was the attraction of working in the imperial center. Despite Struve's enthusiasm, Kharkhov did not have a permanent observatory until the 1880s. Even when an observatory was finally built, it was paid for not by the government but by the independently wealthy astronomer G. V. Levitskii, and work followed his own agenda.[35]

Perhaps the worst case was the Naval Observatory of Nikolaev in the recently annexed region of Tauride, north of the Black Sea, which was headed, following Struve's recommendation, by K. F. Knorre. Set up by Admiral Grieg in 1821, the Naval Observatory established coordinates of points on the Black Sea coast and trained Russian officers in astronomical navigation. But reproducing observatory techniques at a distance proved extremely difficult. Eight years were required to build an observatory so far from the imperial center, and the finished building leaked and had a shaky roof. When Grieg left the Black Sea region, Nikolaev was quickly ruined and essential labor siphoned off for naval work. Knorre complained: "The astronomer is left without any help, so he has to be a secretary and a mechanic himself. He has lots of other problems."[36]

The Observatory: Dramatizing Astronomy

Russia's imperial astronomy was a mixture of successes and failures. The "first reality" of the empire was too large and unwieldy for a well-coordinated imperial astronomy to develop. Geodesy produced results but had to endure many difficulties, and provincial observatories often failed. But this was only one of Russia's "nations." Astronomy in Custine's "Russia as they would like her to seem to Europe" was much more successful. So the focus now turns to Pulkovo Observatory itself.

The second concern of Nicholas I in founding Pulkovo Observatory was to bring prestige to Russia through patronage of the sciences. Struve was promoted as a leader in the "abstract" study of stellar astronomy, as well as for his skills in geodesy. Diffused across Europe and America, achievements in stellar astronomy would bring honor to Russia. Additionally, the observatory itself would be a showcase of talent and observational excellence. Shown off continuously to foreign dignitaries and professional astronomers, Pulkovo would impress upon them Nicholas's generosity and Russia's power over the heavens. With this as his goal, the troubles of Russia's other reality—the failings of astronomy in the empire at large—were, at least for Nicholas and foreign observers, of less consequence than they might otherwise have been. Pulkovo helped make Russian science and patronage world-famous: "Who has not heard of Pulkowa?," Gould asked in 1849.[37]

How did Pulkovo become so famous? Let us examine the links between the production of first-rate stellar astronomy at Pulkovo and its role as a theater for the tsar. Custine's negativity misunderstood Russian theatricality. In Russia state-building followed a recognizable pattern in which elaborate models and theaters of order stood in for a broader system, which might in fact never be built. In what remained a court culture, spectacular shows before the emperor's gaze were what mattered, as Nicholas himself observed on his inspection tours. Courtiers made sure that the emperor witnessed ideals of good order and successful work, but when he was gone these ideals could be abandoned. Such was the fate of the famous villages of the 1780s, "stage sets" set up to impress Catherine the Great on a tour of the Crimea but abandoned thereafter by their creator Prince Grigorii Potemkin.[38] A similar logic pervaded the sciences. The St. Petersburg Academy had been founded to impress Europe with Russia's civility, initiated with lavish funds yet with no charter or long-term plans. The academy partly rose and fell on the whims of the tsars. What looked like enlightened state-building there-fore was often little more than theatrical gesture, though remnants of that gesture partially built an enlightened state: "Catherine II ordered for the observatory of St. Petersburgh the finest and most expensive instruments which London artists . . . could produce; but having done so, and [having] filled the journals with announcements of her liberality and love of science, they were allowed to rust in their cases."[39] Even so, *some* astronomy did advance as a result of this kind of gesture.

A similar pattern is evident in Russia's imperial astronomy and the Cen-tral Observatory. Its official name notwithstanding, Pulkovo functioned mostly, and most successfully, as a spectacular emblem of Russian prestige in the sciences, standing in before foreign audiences for an imperial astron-omy which was far less prestigious than the central observatory. As usual, foreigners were employed in these theaters and abundant patronage given them to ensure success. Nicholas's whims were crucial—when the emperor died, patronage quickly diminished and the observatory was left struggling. Nevertheless, Nicholas's astronomical theater left its mark, securing for Russian astronomy a fine reputation. In the remainder of this chapter Pul-kovo observatory is read as a theater of astronomy, and the drama of its techniques elucidated.[40]

It is common to note that Pulkovo rested on extensive patronage from Nicholas I, and this point should not be underestimated. When Struve met Nicholas to discuss the possibility of a new central observatory, Nicholas gestured the new observatory into existence, for the "honor of the coun-try," and had it "furnished with everything . . . that was most perfect."[41] This level of patronage was exceptional and the key to making Pulkovo an exceptionally well designed observatory. Besides giving up land, Nicholas spent 600,000 silver rubles on his new fancy, worth about $480,000 in 1845, or about $11 million today.[42] In comparison, funds spent setting up contemporary observatories in the United States were much less, usually in the tens of thousands of dollars. None received anything close to Struve's funds. Bessel in Königsberg Observatory was unable to afford expensive instrumentation for many years after becoming director in 1810. In contrast, Struve was given carte blanche to buy anything he needed.[43] Struve chose to take advantage of imperial patronage in building an entirely new specialized observatory, directed in this case to stellar astronomy and geodesy. Airy noted the contrast to Greenwich, where astronomers saddled with government demands for navigational astronomy had to make continuous solar and lunar observations. At Pulkovo, Struve's freedom to create the task-specific observatory he wished for "powerfully aided" his work.[44]

The clearest indication that Nicholas's observatory was intended as a showcase was its location—on Pulkovo hill, about twelve miles south of St. Petersburg, within the tsar's estate of Tsarskoe Selo, where the emperor had his summer residence and received dignitaries and foreign ambassadors. This special location has been overlooked in most discussions of Pulkovo, yet the observatory was not built on the tsar's estate by coincidence. Nicholas chose to put it there. The private estate was a fundamental element in Russia's theatrical noble culture—a place where the Russian nobility could escape from formal duties, play the foreigner, and imagine utopian schemes.[45] Estates provided an opportunity to experiment with social, economic, and intellectual practices, often borrowed from the West, in a place where nobles mastered all their surroundings. There they built gardens, factories, and theaters; they dressed serfs in exotic costumes; and they staged innovative forms of life. Tsars did the same, only on a grander scale. St. Petersburg itself was such an experiment—a new form of city and social order built by Peter the Great on empty marshland to command respect for Russia in Europe. Tsarskoe Selo continued this trend as Nicholas populated it with model western institutions and technologies that dramatized Russian prestige even if their large-scale implementation was erratic and constrained. For instance, Nicholas made Tsarskoe Selo the terminal for Russia's first permanent road and railway. Both were model technologies, though the subsequent establishment of road and railway systems was haphazard. Pulkovo was founded shortly after, at the southern end of the tsar's new *chaussée*.

From the beginning the Central Observatory was presented to a variety of audiences in carefully staged performances. Its inauguration included a distribution of medals by the tsar and a blessing by Russian Orthodox priests, fitting the event into official policy of "Orthodoxy, Nationality, and Autocracy." Local announcements proclaimed that Pulkovo would "annihilate Russia's intellectual dependence on foreigners."[46] Yet this was stage-managed Slavophilia, and Pulkovo remained a "little German colony," as one visitor termed it.[47] Links to other European observatories, crucial to securing its reputation abroad, remained tight. Print helped this reputation, one of the earliest works produced by the observatory being a lavish book about the observatory itself. In 1845 some 17,500 rubles paid for the *Description de l'Observatoire Astronomique Central de Poulkova*, a detailed and richly illustrated description of Pulkovo's architecture and instruments written by Struve and distributed to most European and American libraries and observatories. The *Description* was an ostentatious display of imperial munificence for the sciences. In it Struve gushed over imperial patronage: "The history of science cannot furnish a single example of protection as liberal as that which has been accorded to it by the Emperor Nicholas." Surely such claims influenced readers' perceptions of the observatory. Arago called Pulkovo "un véritable monument astronomique." The *Description* was also scientifically important, helping to spread Pulkovo's methods across the astronomical world. It became, according to Gould, "the best source from which students can derive an accurate knowledge of the refinements of modern astronomical instruments and observations of the modern school."[48]

Virtual tours of the observatory could not stand in for the real show. Pulkovo's publicity was above all achieved through the reception of visitors. A steady stream of prominent foreign astronomers paid it a visit through the nineteenth century, including the director of Altona Observatory, Heinrich Christian Schumacher (1840), the Greenwich director Airy (1847), the American surveyor Benjamin A. Gould (1848), the Harvard astronomer George Phillips Bond (1851), the Scottish astronomer Royal Piazzi Smyth (1859), the Martian specialist Giovanni Schiaparelli (1860), and the Dartmouth astronomer Charles A. Young (1887).[49] Astronomers were given "princely hospitality" and most published accounts that eulogized Pulkovo as an essential and desirable place for an astronomer to visit.[50] Smyth's account prompted a visit by the surveyor Cleveland Abbe in 1864, who in turn enticed the American astronomer Simon Newcomb to go a few years later.[51] Such visits highlight the importance of Pulkovo as a *space* in Michel de Certeau's sense—not just the place where its astronomers achieved their theoretical results but a site that needed to be experienced.[52]

Visitors routinely remarked upon the fine construction of the Central Observatory buildings. As Mari Williams has argued, Pulkovo's architecture was fundamental to its astronomical successes.[53] The main building, designed by Alexander Briulov, consisted of a central dome housing a fifteen-inch equatorial refractor by Fraunhofer and two east-west side wings, or Meridian rooms, containing transit instruments and circles. Topped with "imposing" turrets, the wings accommodated a heliometer and comet-seekers. A southern wing housed an extensive library and a prime-vertical transit instrument. Briulov's severe and neoclassical building was supposed to indicate the observatory's scientific purpose. Airy admired its simplicity. Gould noted: "The chaste and elegant architecture . . . shows in every part and every form the Temple of Science."[54]

Architecture and fittings performed key astronomical tasks — protecting and maintaining the positions and temperatures of sensitive instruments. Visitors noted how trapdoors in the roof could be opened and closed mechanically for observations, thus protecting instruments when not in use. The central refractor was supported in a dome set in masonry which could be turned, opened, and closed by a pulley system. Other instruments were sufficiently separated to eliminate temperature changes caused by the walls, which consisted of thin wooden partitions. Floors and ceilings were made from polished oak, easily cleaned and dust-free, and instruments were encased in mahogany cabinets moved on rails to protect them. "What luxury . . . those instruments seemed to be enjoying!," exclaimed Smyth.[55] Not all the architecture was viewed as ideal, however. The basement housed extensive brick supports originally designed by Struve for Dorpat. But Fearon Fallows, constructing the Cape Observatory in 1827, preferred stone to the brick pillars designed by Struve: "I should sooner cut my right hand off than do anything of the kind."[56]

Such comments highlight the degree to which Dorpat was an *innovative* observatory and Pulkovo a *showcase* for established techniques. In Dorpat instruments were tried out for the first time, while in Pulkovo well-tried-out techniques were presented as in a show, as if those responsible had an audience in mind. The instruments in Dorpat were "new and peculiar," notably Fraunhofer's great nine-inch refractor. This telescope was equipped with an original clockwork mechanism for maintaining stars in the field of view, with a new high-quality achromatic lens and, as Fallows suggests, a controversial support. It was also reversible about the polar axis, making the whole sky accessible and leading Struve to devise one of the first revolving "cupolas" to house the instrument. These were innovations. Pulkovo, on the other hand, was rarely criticized for its innovation, no more than it was lauded — visitors rather marveled that what was already being done in other observatories was being done better in Russia. Thus Pulkovo's telescope was a larger, fifteen-inch Fraunhofer refractor, "mounted much like the Dorpat telescope, but with some farther improvements," specifically a construction of lightweight pinewood rather than metal. Struve became famous for such refinements: "The formulae by which the errors of a transit instrument may be corrected or allowed for are to be found in Delambre, Bessel, &c. but they appear to have been more used and more strictly attended to by M. Struve than by other astronomers." The Harvard astronomer Bond wrote: "Tenths of a second of arc take the position here that seconds have hitherto done elsewhere."[57]

Improvements, rather than innovations, were also more consensual, and Pulkovo's visitors responded well to what they witnessed. Nineteenth-century astronomers approached their instruments as connoisseurs paying attention to the finest details. Piazzi Smyth dismissed everything at Pulkovo as "merely preparatory matter of comparatively low order," compared to the instruments. Airy thought they were "without exception, first class." The Prime Vertical instrument by Repsold was described by Gould as "perhaps the greatest triumph of mechanical art in existence." It was clear to Gould that makers were crucial players in Russia's surveillance projects: "Without them, Struve could never have verified his declaration, *Alles Sichtbare ist messbar*, 'whatever can be seen may be measured.'" Struve's own attention to instrumental precision was publicized at Pulkovo, reminiscent of Nicholas's personal panopticism. As Airy recalled, "I believe not only the general plans but even the details to the smallest screws at least for the meridional instruments were submitted to Mr. Struve's approval before the instruments were made."[58]

Also on display at Pulkovo was a social order underlying the observatory's instrumental precision. Mirroring Nicholas's imperial state, this order asserted rank and hierarchy by gestures of power and surplus talent. Reports thus celebrated Struve's feats of tacit skill and the failures of instrument makers, noting how Struve had assembled the great refractor at Dorpat and its "great number of small pieces" without any help from its maker, who had "forgotten to send the directions for doing it." At Pulkovo, Gould observed how the astronomer's hand was critical in observations that instruments alone could not have achieved. The observatory, he went on, derived its advantage from the absence of any "instrument in which the accuracy of the observer is entirely dependent upon the degree of skill exercised by the mechanician." Airy recognized this when asked to witness a performance by Wilhelm Struve with the Prime Vertical instrument: "I had the pleasure twice of witnessing complete observations made by him; and I trust that he will not be offended by the testimony . . . to the caution, the delicacy, the steadily waiting till the proper time, the promptitude at the proper time, which distinguish Mr. Struve's observation. . . . In other hands I should have no such confidence in [the vertical circle's] accuracy."[59]

These practices evoke the "free play" of instruments envisioned by Humboldtians at this time and expertly described by John Tresch elsewhere in this volume. But where Humboldt saw egalitarianism, at Pulkovo the flexible mechanisms reflected the rigidity of social hierarchy. This perhaps explains Humboldt's view of Pulkovo as an observatory whose form of government was difficult to conceive. Indeed in Nicholas's empire, similar gestures signified the autocrat's power. The Russian people were often conceived as a dutiful machine effortlessly guided by the skilled hand of an all-powerful tsar. In 1838 Mikhail Pogodin explained his doctrine of the state thus: "All [Russia's] forces, physical and moral, compose one enormous machine, arranged in a simple, convenient manner, and directed by the hand of a single man, the hand of the Russian Tsar who can at any moment with a single motion start it, give it any direction and attain any speed he wishes."[60] This of course was not true, but in Russia's second reality, the small-scale theater of the observatory, the tsar's astronomers ably demonstrated its meaning: "Pulkowa is like the palace of an astronomical autocrat, who has but to *will*, and men and money appear at his call to take the heavens by storm." Wilhelm and Otto Struve repeatedly made similar shows, as Piazzi Smith recalled: "Our learned friend [Otto Struve] had kindly taken a great deal of trouble in showing how well the arrangement [of mechanisms for opening Pulkovo's dome] must act for carrying their vast burden, by applying one finger only to the final handle . . . and lo! it moved."[61]

Gestures of this sort secured precise observation and displayed the hierarchy inside the observatory. Observing astronomers stood above instruments, superior to mere machines in their tacit skills, manifested by handy gestures. Equating more lowly personnel in the observatory with machines reinforced the image. The majority of workers in Pulkovo were soldiers, probably members of the Internal Defense regiments of retired soldiers garrisoned all over Russia to guard banks and state institutions.[62] Some 180,000 of these policed the Russian empire, and at Pulkovo they carried out the physically laborious and repetitive tasks of the observatory. These, combined with many observatory-trained army geodesists, conferred a military character to the institution. Pulkovo, visitors noted, displayed its own form of discipline—thus Schumacher found it hard to describe "in words the spirit of order . . . that Staatsrath v. Struve has introduced into this great whole." He "knew how to keep it" too. Struve was identified variously by Piazzi Smyth as Pulkovo's "Lord and Master" and by Wilhelm von Humboldt as "the tyrant of Pulkovo," with Otto Struve his "grand prince."[63]

The dynamics of quasi-military discipline made human labor part of the observatory, exhibited as a single mechanism easily manipulated by the hand of its director. Smyth brought this out in comments on Pulkovo's refractor. He conflated the stability of the observatory with the docility of its workforce, and understood high precision as a consequence of both. Smyth noted how soldiers prepared every arrangement for making observations except the observation itself. Describing what he thought was peculiar to Russian astronomical practice, Smyth wrote that the sergeant assisting Otto Struve on a night observing double stars stood

> like an impassive pillar . . . but the instant that the last motion had been given to point the huge optic-tube in its true direction,—the faithful servitor, without . . . needing any further word of explanation or instruction . . . firstly, chose from a number of seats, one that exactly suited the height of the eye-piece above the floor, and placed it appropriately; secondly, he picked out from a collection of hook-joint handles, the one which was proper to the distance of said eye-piece from the right ascension circle, connected it therewith and brought its end to the hand of the observer, already seated in the chair; thirdly, the sergeant turned the windlass of the dome movement, until he saw the object-glass looked fairly out of the opening; and fourthly, he brought a small table with pens, ink, paper, and a lamp, close to the observer's elbow; and then returned once more into that passive columnar attitude.[64]

Smyth effortlessly equated the soldier's "support" with the observatory's. Both appeared as instruments set in motion by the director's will. At this point Struve began his observations, with "nothing left him to employ his own higher powers on, than, after deciding what should be observed, to observe it." Only slight adjustments of the micrometer screw were required, and these were achieved by "a dexterous final touch from thumb and finger." Thus the will of the "autocrat" set in motion the observatory machine— at least in the performance staged for Smyth. Everything seemed to run smoothly. Soldiers were disciplined into acting like "pillars" and the observatory employed a permanent inspector (*smotritel'*), "an officer who looks after the commissariat, the police, and generally the due performance of all the *muscular duties* of the place."[65] Thus Pulkovo made a small-scale theater of just the model of order which was supposed to characterize Nicholas's empire.

A final technique of great importance to Pulkovo's functioning and its theater was the archive. The Central Observatory contributed to making Russia an "empire of catalogues," since this was its essential product. Observations were meticulously recorded in pencil in notebooks of different sizes according to the instrument being used. Originals were archived in a wooden case in the director's office or in the "computing-room" marked with the words "Works of the Central Observatory," while copies were used for making reductions and distributed to all members of the observatory.[66] Struve's approach to reduction perhaps reflected the grandiose gestures of accumulation which typified the observatory—he liked to accumulate measurements for a whole project before reducing the data, something which Airy complained about in typical economic terms as an "arrears." Airy preferred to make reductions regularly as observations were continuing. Struve, assisted by more army recruits—in this case young officers from the General Staff—spent nearly nine years reducing the meridian arc data for publication, but only after the surveying had been completed. In the process he needed to handle some seventeen thousand documents.[67]

Results were then published, always geared to foreign audiences, predominantly in German journals like the *Astronomische Nachrichten*, or in books written in French, German, or Latin. Most publications were catalogues of stars, mostly double stars, which Struve and his associates meticulously recorded over their careers. Three catalogues appeared between 1827 and 1852.[68] Struve's *Arc de Méridien* presented two more, published in 1857–62, giving all the positions and distances of points on the meridian arc passing through Dorpat.[69] These thousands of observations then provided the resource for more theoretical speculations. In 1837, after extensive surveys of double stars, Struve measured the stellar parallax of α *Lyrae*, just after Bessel determined the parallax of 61 *Cygni* and Henderson the parallax of α *Centauri*.[70] At Pulkovo, Struve pursued this work and used star catalogues to determine if a statistical dependence existed between the brightness of stars and their distances. His resulting conclusions on stellar distribution, including the assertion that the sun was not at the center of the field of the Milky Way but above its main plane, were summarized in his *Études d'astronomie stellaire* of 1847.

Another grandiose archive was Pulkovo's library, the largest astronomical collection of the mid-nineteenth century.[71] As with instruments, Struve employed "agents" to scour bookshops across Europe for appropriate works, which were inventoried and catalogued by a librarian. Collections included Heinrich Olbers's and Johannes Kepler's papers, the latter purchased in one of Catherine the Great's displays of enlightened patronage. Smyth called the library a "collection of curiosities," though one "invaluable in tracing the history of ideas, and the actual progress of the human mind in casting off . . . scholastic dogmas or false metaphysical subtleties . . . learning to investigate nature directly and for rational ends."[72] Pulkovo was a spectacle, but a spectacle productive of first-rate science.

The contrast between Knorre's ailing observatory in Nikolaev and the luxurious observatory at Pulkovo highlights how Russia in Nicholas's reign was, as Custine imagined, composed of "two nations." Pulkovo, whose excesses greatly contributed to the fortunes of Struve and stellar astronomy, was on the front stage, a spectacular demonstration of Russian scientific patronage on display for foreign observers. Backstage was a dispersed network of erratic astronomical development and provincial observatories that often failed. This contrast was partly the product of Russia's theatrical imperial culture, which in the mid-nineteenth century helped to make Pulkovo the astronomical capital of the world. In the reign of Nicholas I, surveillance technologies replaced personal supervision in governance of the Russian empire. But the operation of these technologies often reflected, and was part of, a long tradition of noble theater in which idealized and orderly local arrangements stood in for empire-wide systems that in practice could be disorderly and hard to manage. Astronomy in Nicholas's Russia reflected this distinctive culture. Imperial surveys succeeded, but only with difficulty. Provincial astronomy floundered. But Pulkovo was a showcase to local and foreign audiences, and enjoyed lavish personal patronage from the tsar. Nicholas's "astronomical palace" made a show of imperial order, even while simultaneously helping to construct that order at large. Pulkovo's theater helped to shape stellar observatory techniques and became a model for observatories around the world.

Nevertheless, the degree to which Pulkovo was a plaything of the tsar should not be underestimated. Its fate was tied to Nicholas's enthusiasm, and after his death in 1855 funds began to shrink. As in other court cultures, grandiose public gestures often mattered more for patrons than continuous support did, and when clients fell from grace or a patron died the consequences could be catastrophic. Otto Struve, who took over the running of the observatory in the mid-1860s, understood this well. Smyth reported his saying in 1862: "The records of . . . science . . . do show that it has had magnificent patrons from time to time . . . A truly great patron appears as seldom in a hundred years as a really great philosopher. . . . How shall we contrive to cross these broad gulfs which exist between the appearance of one Augustan Maecenas and his next similar representative?"[73] Otto told George Airy that he was envious of Greenwich Observatory's funding.[74] Russian government financers used clever arguments to reduce funds. The 1860s were the decade when Tolstoy, Turgenev, and Dostoevsky had their early successes, and the government wags argued, "Science and Literature . . . are twin sisters. . . . If literary men support themselves, and see how famously they do . . . why should science be calling out for State assistance?"[75] In 1865 Otto tried to argue the observatory's case by publishing a summary in Russian of Pulkovo's achievements.[76] So the reign of Pulkovo as the astronomical capital of the world was not to be a long one, though the observatory did have a measure of success into the twentieth century.[77] Its real fortunes were bounded by the life and enthusiasms of Nicholas I and the emperor's dual concerns for surveillance and spectacle.

Notes

1. Friedrich Georg Struve quoted by Smyth, *Three Cities in Russia*, 2:179.
2. Airy, "Schreiben an den Herausgeber," *Astronomische Nachrichten*, 1848, 353, and Gould, "The Observatory at Pulkowa," 146. On Wilhelm Struve see also Otto Struve, *Wilhelm Struve: Zur Erinnerung an den Vater den Geschwistern*, and Novokshanova (Sokolovskaia), *Vasilii Iakovlevich Struve*. Note that in this chapter transliteration is according to the Library of Congress method, which renders "Pulkovo."
3. Simon Newcomb attributed this to Gould in Newcomb, *Reminiscences of an Astronomer* (Cambridge, Mass.: Houghton, Mifflin, 1903), 309.
4. Quoted in Batten, *Resolute and Undertaking Characters*, 67.
5. On "Hevelian" and "Tychonic" observatories see Aubin, "The Fading Star of the Paris Observatory in the Nineteenth Century"; on Uraniborg see Hannaway, "Laboratory Design and the Aim of Science," and Shackleford, "Tycho Brahe, Laboratory Design, and the Aim of Science."
6. Gould, "The Observatory at Pulkowa," 159.
7. "Bookkeeping" in British imperial observatories, and at Paramatta especially, is the topic of Simon Schaffer's article in this volume; for more on imperial observatory networks see also David Aubin's and Massimo Mazzotti's contributions to this collection.
8. James Forbes with George Airy, "National Observatories: Greenwich," *Edinburgh Review* 184 (1850): 299–357, at 316.

9. See Levitt's contribution to this volume, as well as Guy Boistel's concerning earlier navigational concerns at the Paris Observatory.

10. Ju. M. Lotman and B. A. Uspenskij, *The Semiotics of Russian Culture*, ed. Ann Shukman (Ann Arbor: Dept. of Slavic Languages and Literatures, University of Michigan, 1984).

11. Mikhail Epstein, *After the Future: The Paradoxes of Postmodernism and Contemporary Russian Culture*, trans. Anesa Miller-Pogacar (Amherst: University of Massachusetts Press, 1995), 188–210.

12. Adolphe de Custine, *Nikolaevskaia Rossiia* [excerpts from *La Russie en 1839* (Paris: Amyot, 1843)], trans. I. A. Gessena and L. Domgera (Moscow: Terra, 1990), 94, 155–56.

13. Sergei Medvedev, "A General Theory of Russian Space: A Gay Science and a Rigorous Science," and other essays in *Beyond the Limits: The Concept of Space in Russian History and Culture*, ed. Jeremy Smith (Helsinki: Studia Historica, 1999), 15–48. For a geodesic negotiation of ambiguous Russian boundaries see Raj, *Relocating Modern Science*. For recent interpretations of Russian imperialism see Michael Khodarkovsky, *Russia's Steppe Frontier: The Making of a Colonial Empire, 1500–1800* (Bloomington: Indiana University Press, 2002); Jane Burbank and David L. Ransel, eds., *Imperial Russia: New Histories for the Empire* (Bloomington: Indiana University Press, 1998); Daniel R. Brower and Edward J. Lazzerini, eds., *Russia's Orient: Imperial Borderlands and Peoples, 1700–1917* (Bloomington: Indiana University Press, 1997); and Mark Bassin, *Imperial Visions: Nationalist Imagination and Geographical Expansion in the Russian Far East, 1840–1865* (Cambridge: Cambridge University Press, 1999).

14. Marc Raeff, *The Well-Ordered Police State: Social and Institutional Change through Law in the Germanies and Russia, 1600–1800* (New Haven: Yale University Press, 1983); and Werrett, "An Odd Sort of Exhibition."

15. Quoted in Riasanovsky, *Nicholas I and Official Nationality in Russia*, 1; on the Army under Nicholas see Curtiss, *The Russian Army under Nicholas I*.

16. See Batten, *Resolute and Undertaking Characters*, 78. Nicholas quoted in Riasanovsky, *Nicholas I and Official Nationality in Russia*, 195.

17. Richard M. Haywood, "The 'Ruler Legend': Tsar Nicolas I and the Route of the St. Petersburg-Moscow Railway, 1842-1843," *Slavic Review* 37 (1978): 640–50, and Haywood, *The Beginnings of Railway Development in Russia in the Reign of Nicholas I*. On Jacobi's work see Sibum, "Experimentalists in the Republic of Letters," and M. F. Khartanovich, *Uchenoe soslovie Rossii: Imperatorskaia Akademiia nauk vtoroi chetveri XIX v.* (St. Petersburg: Nauka, 1999), 76–82.

18. Vorontsov-Vel'iaminov, *Ocherki istorii astronomii v Rossii*, 241–42; Z. K. Novokshanova, *Kartograficheskie i geodezicheskie raboty v Rossii v XIX–nachale XX v.* (Moscow, 1967); Z. K. Novokshanova, *Fedor Fedorovich Shubert: voennyi geodezist* (Moscow: Izd-vo geodezicheskoilit-ry, 1958).

19. *Sbornik postanovlenii po Ministerstva Narodnogo Prosveshcheniia*, vol. 2, part 2 (St. Petersburg, 1875), no. 129, p. 348.

20. Struve, *Description de l'observatoire astronomique central de Poulkova*, 21.

21. Struve's reported speech in Smyth, *Three Cities in Russia*, 2:184–85; and Novokshanova, *Vasilii Iakovlevich Struve*, 97–98, 170–91.

22. Novokshanova, *Vasilii Iakovlevich Struve*, 48–79; and Cynthia H. Whittaker, *The Origins of Modern Russian Education: An Intellectual Biography of Count Sergei Uvarov, 1786–1855* (DeKalb: Northern Illinois University Press, 1984).

23. On the academy see Werrett, "An Odd Sort of Exhibition," 56–58.

24. Struve, *Description de l'observatoire astronomique central de Poulkova*, 47.

25. Quoted in Vorontsov-Vel'iaminov, *Ocherki istorii astronomii v Rossii*, 142; see also Novokshanova, *Vasilii Iakovlevich Struve*, 80–83.

26. Vorontsov-Vel'iaminov, *Ocherki istorii astronomii v Rossii*, 145–49, 172–91.

27. Earliest surveys are discussed in Batten, *Resolute and Undertaking Characters*, 36, and Struve, *Arc du méridien de 25°20′ entre le Danube et la Mer Glaciale*. On Tenner see Z. K. Novokshanova, *Karl Ivanovich Tenner- voennyi geodezist* (Moscow, 1957). The connection between Greenwich and Pulkovo was published as Struve, *Expédition chronométrique exécutée par ordre de Sa Majesté l'Empereur Nicolas 1er*.

28. The citation is from a government report published in 1840 on attempts to introduce Russian legal codes to the Caucasus, cited in E. Willis Brooks, "Nicholas I as Reformer: Russian Attempts to Conquer the Caucasus, 1825–1855," *Nation and Ideology: Essays in Honor of Wayne S. Vucinich*, ed. Ivo Banac, John G. Ackerman, and Roman Szporluk (Boulder: East European Monographs, 1981), 227–63, at 243.

29. Stanislaw Eile, *Literature and Nationalism in Partitioned Poland, 1795–1918* (New York: Palgrave Macmillan, 2000).

30. Struve, *Arc du méridien de 25°20′ entre le Danube et la Mer Glaciale*, 2:3; see also Raj, *Relocating Modern Science*.

31. Struve, *Arc du méridien de 25°20′ entre le Danube et la Mer Glaciale*, 1:21. The reiteration method was used in Volynie and Bessarabia1; see 21–33. This method is explained in Martina Schiavon's contribution to this volume.

32. Both quotes in Struve, *Arc du méridien de 25°20′ entre le Danube et la Mer Glaciale*, 1:16, 18; Ertel's universal instrument, a form of altitude and azimuth instrument, was first used in surveying by Struve. See F. G. W. Struve, "Über das Universalinstrument von Reichenbach und Ertel als Horizontalwinkelmesser," *Astronomische Nachrichten* 2 (1824): 431–40, 451–54, 457–64.

33. For Finland see Alexei V. Postnikov, "Contact and Conflict: Russian Mapping of Finland and the Development of Russian Cartography in the Eighteenth and Early Nineteenth Centuries," *Fennia* 171 (1993): 63–98, esp. 85–88; and Novokshanova, *Vasilii Iakovlevich Struve*, 208–11.

34. Struve, *Arc du méridien de 25°20′ entre le Danube et la Mer Glaciale*, 1:19.

35. Vorontsov-Vel'iaminov, *Ocherki istorii astronomii v Rossii*, 143–66.

36. Quoted in G. M. Petrov, "Storinki istorii," *Nikolaevskaia astronomicheskaia observatoriia: Zvezdnyi put' dlinoiu v 175 let* (Atol: Nikolaev, 1998), 20–77, at 33. See also Vorontsov-Vel'iaminov, *Ocherki istorii astronomii v Rossii*, 226–27.

37. Gould, "The Observatory at Pulkowa," 143.

38. A. M. Panchenko, "Potemkinskie derevni kak kulturnii mif," *XVIII Veka* 14 (1983):

93–104; Simon Sebag-Montefiore, *Prince of Princes: The Life of Potemkin* (New York: Macmillan, 2000), 380–83.

39. Forbes with Airy, "National Observatories."

40. Theatrical aspects of observatory techniques are further explored in Charlotte Bigg's and Ole Molvig's contributions to this volume.

41. Struve, *Description de l'observatoire astronomique central de Poulkova*, 29; and M. Mardar'ev, "Imperator Nikolai I i akademik Parrot," *Russkaia starina* 29 (1898): 385–96.

42. By CPI method; Struve, *Description de l'observatoire astronomique central de Poulkova*, 53; the railway from Tsarskoe Selo to St. Petersburg cost one and a half million rubles, or about twice as much as Pulkovo; see Haywood, *The Beginnings of Railway Development in Russia in the Reign of Nicholas I*, 131.

43. Struve, *Wilhelm Struve*, 49.

44. Airy, "Schreiben an den Herausgeber," 356.

45. Priscilla Roosevelt, *Life on the Russian Country Estate: A Social and Cultural History* (New Haven: Yale University Press, 1995); and Roosevelt, "Emerald Thrones and Living Statues: Theater and Theatricality on the Russian Estate," *Russian Review* 50 (1991): 1–23.

46. D. M. Perevoshchikov, "Glavnaia Pulkovskaia observatoriia," *Otechestvennye zapiski* 5, no. 9 (1839): 21–27. Russian audiences learned about Pulkovo from several publications, including a newspaper article by Struve, "Über die wissenschaftliche Bedeutung der Kaiserl. Hauptsternwarte auf Pulkowa," *St. Petersburg Zeitung*, 13 September 1839, also published in Russian in *Sankt-Peterburgskie vedomosti* no. 209, and later A. N. Savich, "Opisanie Glavnoi v Rossii, Nikolaevskoi observatorii, na Pulkovskoi gore bliz S.Peterburga," *Morskoi sbornik* 27, no. 8 (1855), part II.

47. Eugène Melchior de Vogüé, "Social Life in Russia," *Harper's New Monthly Magazine*, May 1889, 833–55, at 851.

48. Struve quoted by Gould, "The Observatory at Pulkowa," 162; Arago, *Astronomie populaire*, 4:784; Gould, "The Observatory at Pulkowa," 144. For Struve's printing budget and his views on publication see Struve, *Description de l'observatoire astronomique central de Poulkova*, 53, and Novokshanova, *Vasilii Iakovlevich Struve*, 90–92. Struve also published *Fondation de l'observatoire central de Russie* (St. Petersburg, 1855).

49. H. C. Schumacher, "Nachrichten über die Instrumente der kaiserlichen Hauptsternwarte Pulkowa," *Astronomische Nachrichten* 18, no. 411 (1840): 33–44; Smyth, *Three Cities in Russia*, 1:70–175, 2:160–95, 343–52; Batten, *Resolute and Undertaking Characters*, 89–112; Eufrosina Dvoichenko-Markov, "The Pulkovo Observatory and Some American Astronomers of the Nineteenth Century," *Isis* 43 (1952): 243–46; and Charles A. Young, "An Astronomer's Summer Trip," *Scribner's Magazine* 4, no. 1 (July 1888), 82–101.

50. Gould, "The Observatory at Pulkowa," 143.

51. Batten, *Resolute and Undertaking Characters*, 105, 107; Nathan Reingold, "Cleveland Abbe at Pulkowa: Theory and Practice in the Nineteenth Century Physical Sciences," *Archives internationales d'histoire des sciences* 17 (1964): 133–47; Reingold,

"A Good Place to Study Astronomy," *Library of Congress Quarterly Journal of Current Acquisitions* 20 (1963): 211 17.

52. See Certeau, *The Practice of Everyday Life*, and the discussion of this concept in the introduction to this volume by David Aubin, Charlotte Bigg, and H. Otto Sibum.

53. Mari E. Williams, "Astronomical Observatories as Practical Space: The Case of Pulkowa," *The Development of the Laboratory: Essays on the Place of Experiment in Industrial Civilization*, ed. Frank A. J. L. James (London: Macmillan, 1989), 118–36.

54. Gould, "The Observatory at Pulkowa," 160. For a treatise on observatory architecture written at the turn of the century see Borheck, *Grundsätze über die Anlage neuer Sternwarten*. On British observatory architecture see Morton-Gledhill, "The Architecture of Astronomy in the British Isles."

55. Smyth, *Three Cities in Russia*, 1:98.

56. Fallows to John Barrow, 20 July 1827, quoted in Brian Warner, *Royal Observatory, Cape of Good Hope, 1820–1831: The Founding of a Colonial Observatory* (Dordrecht: Kluwer Academic, 1995), 121–25.

57. Gould, "The Observatory at Pulkowa," 152; F. G. W. Struve, "On the Astronomical Observations Made at Dorpat," *Edinburgh Journal of Science* 1 (1824): 26–34, at 27; and *Memorials of William Cranch Bond, Director of the Harvard College Observatory, 1840–1859, and of His Son George Phillips Bond, 1859–1865*, ed. Edaward S. Holden (San Francisco: C.A. Murdock), 100.

58. Smyth, *Three Cities in Russia*, 1:100; Gould, "The Observatory at Pulkowa," 150; Airy, "Schreiben an den Herausgeber," 357–58.

59. Batten, *Resolute and Undertaking Characters*, 46–47, 77; Wilhelm Struve, "An Account of the Arrival and Erection of Fraunhofer's Large Refracting Telescope at the Observatory of the Imperial University at Dorpat, Read March 11, 1825," *Memoirs of the Royal Astronomical Society* 2 (1826): 93–100, at 93; Gould, "The Observatory at Pulkowa," 144; Airy, "Schreiben an den Herausgeber," 358–59.

60. Quoted in Riasanovsky, *Nicholas I and Official Nationality in Russia*, 118.

61. Anon., Book Review, *Living Age*, 440; Smyth, *Three Cities in Russia*, 1:133.

62. See Curtiss, *The Russian Army under Nicholas I*, and Rostislav Fadieev, *Vooruzhennye sily Rossii* (Moscow, 1868), 28.

63. Schumacher, "Nachrichten über die Instrumente der kaiserlichen Hauptsternwarte Pulkowa," 33; Smyth, *Three Cities in Russia*, 1:198, 129; Humboldt quoted in Batten, *Resolute and Undertaking Characters*, 103.

64. Smyth, *Three Cities in Russia*, 1:143–44; Smyth noted much the same of the retired soldier who attended Savich's small observatory in the Academy of Sciences. The "simple-minded veteran . . . Just turned eighty years of age . . . was always ready exactly when wanted." *Three Cities in Russia*, 2:280–81.

65. Quotes above in Smyth, *Three Cities in Russia*, 1:144, 136.

66. Struve, *Description de l'observatoire astronomique central de Poulkova*, 275–76, 280–82; and Smyth, *Three Cities in Russia*, 1:96.

67. Vitali Kaptüg, "General Review of the History of the Manuscripts Related to

the Struve-Tenner Arc Measurements," *Struve Arc 150: Reports of the International Scientific Conference*, by Association of Estonian Surveyors et al. (Tallinn-Tartu, 2002), 68–71.

68. F. G. W. Struve, *Catalogus Novus Stellarum Duplicium et Multiplicium* (Dorpat, 1827); *Stellarum Duplicium et Multiplicium Mensurae Micrometricae* (St. Petersburg, 1837); and *Stellarum fixarum imprimis duplicium et multiplicium positiones mediae pro epocha 1830.0* (St. Petersburg, 1852).

69. F. G. W. Struve, *Arc du méridien de 25°20′ entre le Danube et la Mer Glaciale*.

70. C. A. F. Peter, "Recherche sur la parallaxe des étoiles fixes," *Recueil de mémoires présentés a l'Académie des Sciences par les astronomes de Poulkova* (St. Petersburg, 1853), 1:1–180; see also Batten, *Resolute and Undertaking Characters*, 113–29.

71. On the library see Novokshanova (Sokolovskaia), *Vasilii Iakovlevich Struve*, 94–96; Struve, *Description de l'observatoire astronomique central de Poulkova*, Appendix: "Catalogus librorum in bibliotheca speculae pulcoviensis contentorum."

72. Smyth, *Three Cities in Russia*, 1: 97.

73. Smyth, *Three Cities in Russia*, 2:182–3 (O. Struve's reported speech); see Mario Biagioli, *Galileo, Courtier: The Practice of Science in the Culture of Absolutism* (Chicago: University of Chicago Press, 1993).

74. Meadows, *Greenwich Observatory*, 4.

75. Smyth, *Three Cities in Russia*, 2:182.

76. Otto Struve, *Obzor deiatel'nosti Nikolaevskoi glavnoi observatorii v prodolzhenie pervykh 25 let ee sushchestvovaniia* (St. Petersburg, 1865).

77. Kevin Krisciunas, "The End of Pulkovo Observatory's Reign as 'Astronomical Capital of the World,'" *Quarterly Journal of the Royal Astronomical Society* 15 (1984): 301–5.

Chapter 2

The Jesuit on the Roof: Observatory Sciences, Metaphysics, and Nation-Building

MASSIMO MAZZOTTI

> "Oh Secchi! Look at the Aurora Borealis. It's the Paris Commune
> that's arriving!"
> —A voice from the street, as Father Secchi observes the intensely
> red phenomenon from the roof of the church of Saint Ignatius
> (6 February, 1872).[1]

This chapter explores the emergence of new scientific practices at the pontifical observatory of the Collegio Romano in the mid-nineteenth century and their subsequent appropriation by official Italian science. I shall argue that the presence of pioneering astrophysical and meteorological research in pontifical Rome, far from being accidental, can be related to issues of centralization and control in the papal territories as well as to Rome's cultural battle against atheism and materialism. I then look at the translation of this scientific tradition into a national research program after Rome fell into Italian hands in 1870. I shall argue that this case of technoscientific transfer is best understood as the result of the reinterpretation of what was originally a Jesuit research program in the context of the sociopolitical discourse of Italian nation-building. Observatory sciences as practiced in pontifical Rome could indeed be used to sustain processes of sociotechnical unification while consolidating the hegemony of the increasingly conservative Italian élites.

The Machines of the Pope

In the 1850s and 1860s the observatory of the Jesuit-run Collegio Romano was the site for elaborating and standardizing pioneering techniques in the emerging discipline of physical astronomy or, in later parlance, astrophysics. It may come as a surprise that the pontifical government, notorious for its reactionary cultural policy, would patronize advanced scientific research. Pope Pius IX had been condemning everything modern, including freedom of speech, democracy, liberalism, socialism, and the separation of state and church. His Syllabus from 1864 was a typical expression of the visceral antimodernism that was current in Rome in the years preceding its military occupation and annexation to the kingdom of Italy (1870). One should however not conclude that the Catholic Church opposed modern science and technology *tout court*. Quite the contrary: during the final years of its temporal power the pontifical government promoted technoscientific innovation in an attempt to increase centralization and control. The achievements of this brief period of reform were celebrated in a series of elegant plates, a clear message for the Italian government, whose plans to take over the eternal city were often legitimated by referring to the obscurantism of the church. The pontifical government had an obvious interest in mobilizing every available cultural resource in its deadly struggle with the Italian nation. The case of scientific archaeology is the most striking, but railways, telegraphy, and other technological systems and scientific disciplines were also supported during Pius IX's pontificate (1846–78). This pontiff, it should be noted, had taken science courses at a Scolopian college before entering the church and had graduated with a competent dissertation on the construction of telescopes.[2]

Among the sciences, astronomy played a prominent role in the cultural policy of the late pontifical state. In the aftermath of the revolutionary years 1848–49 Vatican astronomers turned decisively toward the practice of physical astronomy, thus breaking with the earlier tradition of positional astronomy. The new emphasis on physical astronomy marked a conceptual shift as well as a shift in scientific practices and technologies. The Jesuit Angelo Secchi headed the leading pontifical observatory (at the Collegio Romano) through this physical revolution (figure 12). An experimental physicist by training, Secchi had been forced to flee Rome in 1848 and conclude his studies at the Jesuit college of Stonyhurst, Lancashire. He began his scientific career teaching experimental physics at the Jesuit college of Georgetown in Washington, where he published his first papers on

12. Father Angelo Secchi. Riccardo Finzi, *P. Angelo Secchi, astronomo (1818–1879)*. Courtesy of the photographic library of the Biblioteca Panizzi, Reggio Emilia.

the measurement of electrical resistance and the application of electricity to telegraphy. His interest in astronomy and meteorology emerged in this period, during his collaboration with Father James Curley, director of the Jesuit observatory of Georgetown, and Captain Matthew Fontaine Maury, a meteorologist and oceanographer who directed the U.S. Naval Observatory. By the end of 1849 order reigned again in continental Europe and the Jesuits were allowed to return to their home institutions. On his return Secchi was offered a professorship at the Collegio Romano and the directorship of its prestigious observatory. He was to remain in charge from 1850 to 1878, and through this entire period he was able to count on the trust and protection of Pius IX.[3]

Thanks to private donations and direct papal support Secchi was able to acquire instruments such as a Merz equatorial with an aperture of 24.5 cm and a focal distance of 430 cm, the biggest telescope in Italy at the time of purchase in 1853 (figure 13). Secchi built a completely new observatory in less than a year, a true emblem of the renewed astronomical sciences in Rome. The pope also patronized the acquisition of new instruments for the university observatory of the Campidoglio. In 1865 he gave its directorship to Lorenzo Respighi, a devout professor from Bologna who had done some pioneering spectroscopic research and had refused to take the oath of subjection to the Italian king.[4]

For his new observatory Secchi chose a highly symbolic location: the roof of the church of Saint Ignatius at the Collegio Romano (figure 14). Instruments were placed above the huge pillars that had been designed to support a never-to-be built dome with a diameter of eighteen meters. Instead the church was crowned by three opening domes in metal and wood, packed with telescopes and electrical machinery. In addition to the cupola for the Merz equatorial the observatory had a cupola for the Ertel meridian circle and one for the Cauchoix refractor (with an aperture of 16.3 cm), the latter to be used exclusively for solar observation. The observatory included a meteorological section, and in 1858 the first magnetic observatory on the Italian peninsula became operational in adjacent rooms, for the study of terrestrial magnetism in relation to solar activity. Secchi was indeed the director of the pontifical meteorological service, and his observatory functioned as the center of a telegraphic network of meteorological stations spread across pontifical lands. Secchi's plan for coordinating meteorologi-

13. The Merz equatorial at the Collegio Romano observatory. Angelo Secchi, *L'Astronomia in Roma nel pontificato di Pio IX* (Rome: Tip. della Pace, 1877), plate 5.

14. The domes of the Collegio Romano observatory. Angelo Secchi, *L'Astronomia in Roma nel pontificato di Pio IX* (Rome: Tip. della Pace, 1877), plate 3.

cal observations had been orchestrated with the Ministry of Trade and the Pontifical Navy. The output of the observatory and its network of stations was published regularly in the *Bullettino Metereologico dell'Osservatorio del Collegio Romano*, in which astronomical, meteorological, and magnetic observations appeared side by side, framed in a unitary, physical perspective. The design of Secchi's observatory mirrored nicely the new order of physical astronomy, with magnetic, electric, and meteorological sections surrounding and integrating efficiently the great dome of the Merz equatorial, itself filled with spectroscopic instruments for observing the sun and stars.[5]

Secchi's shift from positional astronomy to physical astronomy and meteorology took place at a time of radical social changes for the pontifical state, and indeed for Italy. The revolution of 1848–49 had brought the establishment of the Roman Republic, crushed only after French military intervention. After these events the restored pontifical government enforced new forms of religious discipline in ecclesiastic institutions and among secular priests, established a centralized system of religious training, and strengthened the episcopal hierarchy worldwide, particularly in Protestant countries. The proclamation of the papal infallibility dogma in 1870 was part of this centralizing strategy. This move enabled the curia to impose strict discipline on the ranks of the church worldwide, turning it into a much more compact, uniform, and controllable institution. In this way the curia made it possible for the church to survive and adapt to a rapidly changing social and cultural environment, characterized by strong separatist and nationalistic forces.[6]

As for the pontifical territory, traditional local autonomies were reduced in an attempt to strengthen central control. Meanwhile the Roman people learned to regulate their time by the midday firing of the gun at Castel Sant'Angelo, the notorious papal prison. Time was given by the Swiss guards from an optical signal taken from Secchi's observatory. The pontiff in person had come up with this innovation, which was followed by the synchronization of all national clocks with Roman meridian time. Secchi's control over Roman time went so far as to include the supervision of solar clocks. Standardized national time was necessary both for the functioning of new pontifical infrastructures, like the railway system, and for implementing Secchi's new observatory practice, which required synchronized data collection. Similarly, the network of telegraphic lines and Morse machines that covered the pontifical territory during the 1850s served multiple purposes. Secchi supervised their installation while experimenting with cables and electricity to connect stations and to wire machines for the automatic registration of data. The first pontifical telegraphic line connected the Vatican palaces to the port of Anzio and the papal residence of Castel Gandolfo. The dual use of this line as a key communication system and a scientific instrument for the investigation of terrestrial magnetism reveals the heterogeneous nature of Secchi's scientific practice and the degree of his control on pontifical technological infrastructures.[7]

Emphasis on communication and control was accompanied by the production of a large mass of statistical data on the conditions of the pontifical state, inaugurated by the general census of 1853. The substantial territorial losses of 1859–60 only accelerated processes of this kind. With the papal border set at thirty kilometers from Rome, survival depended on having the eternal city be well integrated with the infrastructures and economies of the neighboring Italian regions. The papal government planned further strategic investments in railways, telegraphic lines, and suspended bridges, and favored the creation of new commercial banks, capitalist companies, and mechanized manufacturing. Appropriately enough, the twelfth year of

Pius IX's pontificate was celebrated with the issue of a medal representing the pontiff on the obverse and a running locomotive on the reverse.[8]

Secchi's observatory was thus a key site in the pontifical modernization project. Secchi, principal advisor to the pontiff on scientific matters, and his colleagues devoted much of their time directing the installation of telegraphic lines, water systems, and lighthouses. Among their most publicized achievements was the calculation of the trigonometric basis of the Appian Way for the future triangulation of the papal state. Secchi's observatory practice was directed at the scientific investigation of terrestrial and celestial physics while being a component of the pontifical project of sociotechnical reform. As illustrated by his use of telegraphic lines, Secchi systematically pushed the boundaries of his observatory toward the borders of the pontifical state: new infrastructures could turn into scientific instruments and instruments could serve the cause of discipline and centralization. The observatory of the Collegio Romano was thus the site where the urgent needs of the state were linked with the new observatory practices of physical astronomers.[9]

The inspection of artifacts from Secchi's observatory can offer further clues to the correlation between observatory practice and the struggle for survival of the pontifical state. Secchi himself designed a number of instruments, from spectroscopes to barometers, the most famous being his meteorograph for the automatic registration of meteorological parameters (1858; figure 15). This massive machine was built in collaboration with Brassart, the Roman instrument makers, to systematize and organize the constant flow of information from faraway stations and thus facilitate the study of weather, which for Secchi was itself an "immense machine." Registration could take place at an indefinite distance from the instrument, as Secchi had integrated mechanical and electrical technologies, using his expertise in telegraphic transmission. The pontifical government used this imposing piece of technology to break its diplomatic and cultural isolation. Pius IX donated around twenty thousand francs for the completion of an elegantly crafted exemplar to be sent to the Universal Exhibition of Paris in 1867 (note that fifteen hundred francs then was a good annual salary for a civil servant in Rome). The instrument won the Grand Prix and Secchi was awarded the Legion of Honor, to the discontent of Italian officials.[10]

The meteorograph's design embodied important aspects of Secchi's scientific practice. Above the church of Saint Ignatius pontifical astronomers collected, elaborated, and published an unprecedented amount of astronomical, meteorological, and magnetic data. Data were produced by

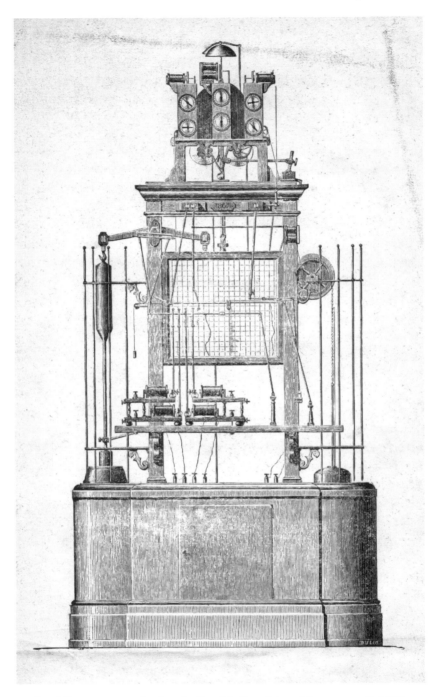

15. Secchi's meteorograph. Angelo Secchi, *Descrizione del meteorografo dell'Osservatorio del Collegio Romano* (Rome: Tip. Delle Belle Arti, 1870), frontispiece.

a capillary network of stations manned by few but reliable collaborators. The need to coordinate action with faraway collaborators was crucial to the new practice and raised issues of standardization, training, and reliability. Secchi's observational network included fellow Jesuits but also civil servants, pontifical engineers, and ecclesiastics from various orders. The implementation of a strict discipline was an essential element for the success of his distributed observatory practice. Not all pontifical institutions responded adequately: if telegraphic engineers were particularly efficient in providing Secchi with useful information, certain religious orders charged with collecting meteorological data seemed to have been much less keen on his regime. Secchi described it as being informed by the "German spirit," that is, "the patient observation and material repetition of the same thing for millions of times." Admittedly some of Secchi's students also resisted the new discipline, a failure which the Jesuit Secchi attributed to a lack of character. Instead the meteorograph's automatic registration and comparison of data offered the best model of the discipline that would govern observatory practice in the future.[11]

Secchi's meteorograph did not contain any essentially new component. Rather, like most of his new spectroscopic and meteorological instruments, it was designed to construct relations between phenomena at the very moment of their measurement. It produced and displayed parameters to emphasize possible connections between different orders of phenomena. The construction of relations between astronomical, meteorological, and magnetic phenomena was indeed at the core of Secchi's observatory practice. It should be noted that the meteorograph displayed data visually, through graphs rather than numerical tables. Secchi thought that this sort of display would facilitate the identification of relations and influences between phenomena of different kinds. The preference for the construction of visual rather than numerical representations of phenomena informed all aspects of Secchi's observational regime, as he believed in the heuristic and epistemological superiority of the visual over the abstract and the mathematical. It was in the attempt to mechanize the registration of visual information that Secchi began studying personal errors in astronomical observation. He hoped that in the future the human element could be entirely eliminated thanks to the integration of photographic and electrical technologies. Secchi had begun experimenting with collodion in 1851, taking photographs of the moon, the planets and the sun. He had great hopes for the advancement of photography as a means of objective registration. While waiting for the perfect nonhuman observer one had to do what one could, which meant strict discipline and standardization in astronomical drawing as well as meteorological registration. Indeed, the amount of effort invested by Secchi and his collaborators into producing painstakingly precise colored plates for their publications became a distinguishing feature of the observatory of the Collegio Romano.[12]

Finally, the meteorograph claimed for pontifical science a supranational, universal dimension. Not only was Secchi's observatory at the center of a telegraphic network that connected meteorological stations distributed across the pontifical territory; it was also at the center of a network of Jesuit observatories and meteorological stations distributed worldwide. Exemplars of Secchi's meteorograph were installed as far away as Shanghai, Havana, Manila, and Washington. The scope of their activity was truly global, as shown by the systematic study of terrestrial magnetism. The global dimension of pontifical scientific research mirrored the global dimension of the spiritual mission of the church, and was crucial to the construction of Rome as a universal, supranational (and thus distinctively non-Italian) city. The worldwide network of Jesuit observatories and seismological stations would bring to the twentieth century Secchi's truly *catholic* vision of scientific research.[13]

The Unified Universe

The aspects of Secchi's practice that we have sketchily outlined above were rooted in a specific conception of physical reality and the way it should be studied. In the messy rooms on the roof of Saint Ignatius there was little space for positional astronomy. The study of the celestial bodies, and of the sun in particular, was meaningful and worth pursuing only insofar as it favored the understanding of the physical relations between the various components of the creation. From this perspective, around 1850–51 Secchi began applying his experience in measuring electrical currents to investigating the physical properties of the sun. From there to his grand-scale study of magnetic variations or his cataloguing of stellar spectra, he kept bridging the gap between the Earth and the Heavens by means of his electrical, magnetic, and spectroscopic instrumentation. Even the traditional practice of optical observation had changed meaning in Secchi's observatory, where it was devoted primarily to detecting morphological types of phenomena (like solar prominences) to be related to specific physical properties, and therefore to strengthen the overall picture of a unified universe.[14]

The impurity of Secchi's scientific practices, his continuous mixing up of the Heavens and the Earth, reflected a conception of the universe as essentially homogeneous and understandable in terms of matter and motion. Secchi derived this mechanistic image of physical reality largely from experimental and philosophical currents in Catholic culture. His defense of religious dogma based on a rigidly dualistic metaphysics was also a well-known apologetic strategy. The early nineteenth century had seen various attempts to integrate recent developments in the sciences, particularly physics, with the apology of Catholic religion and the defense of theological dogma. In Rome the Accademia dei Nuovi Lincei (1801), the chair of Sacred Physics at La Sapienza University (1816), and the School of Application for Engineers (1817) were institutions designed to support a renewal of scientific studies in harmony with the restoration of religious faith and social order. Pro-science ecclesiastics operated under the assumption that science had the capacity to illustrate the superior harmony of creation and did not necessarily imply materialistic doctrines. The goal of Catholic scientists should be precisely that of breaking the ill-fated alliance between empirical research and impious philosophy that was the abuse of science made by the Enlightenment. Granted, the practical value of science was fully recognized, as well as its role in defeating scholastic metaphysics and superstition. This early concordant (*concordista*) strategy was soon taken to yet another level of sophistication by Jesuit scholarship. By giving back the Collegio Romano to the Company of Jesus in 1824, the pope aimed at reviving the Jesuit scientific tradition. Meanwhile, Galileo's works were taken off the *Index of Prohibited Books*, and the apostolic letter *Quod divina sapientia* offered full theological legitimization to experimental research. The alliance of religion and science was grounded on the tenets of a mythical Galilean experimentalism: the enemy was not science per se but rather materialist mechanism and mathematical determinism.[15]

In the 1820s one of Secchi's teachers and comrades in American exile, the Jesuit Giovan Battista Pianciani, had shown how electrical and magnetic research could lead to the discovery of a deep order and unity beyond the variety of physical phenomena. Pianciani, who would publish numerous chemical and electrical studies, rejected the notion of "electrical fluid" and the related physico-mathematical models in favor of a dynamist model. Natural processes were explained in terms of interrelated agents capable of transforming one into the other, such as heat and motion, electrical forces and magnetic forces. This Jesuit atomism and dynamism was developed in opposition to the contemporary neoscholastic resuscitation of hylomorphism and was rooted in an Augustinian vision of the *causa prima* underlying all natural phenomena.[16]

Secchi, the most authoritative ecclesiastical scientist of his age, contributed substantially to the success of this spiritualistic experimentalism. His philosophical manifesto was a book published in 1864 on the unity of the physical forces, which he wrote to breathe new life into physics teaching in Rome and which was soon translated into French and German. In it all physical laws and the structure of matter were treated in the context of a unitary representation of phenomena, the basis of which was molecular movement. Secchi criticized the realist interpretation of the force, the notions of force acting at a distance and of various and mysterious material fluids. Instead, expanding on the mechanical theory of heat, he introduced ether as the material (but imponderable) medium for all physical transformations. All natural forces could thus be shown to be simply different forms of motion of ponderable and imponderable matter. Ultimately, Secchi believed, even the constituents of ponderable matter would be reduced to nothing more than vortexes of ether.[17]

The alliance of dualistic metaphysics and what Secchi called a "healthy experimentalism" found its best expression in the scientific practice at the pontifical observatory. The connection between new dynamist physics and observatory sciences was only too obvious to Secchi. Traditional astronomy was unable to explain an increasing number of celestial phenomena precisely because it ignored the dynamist theory of the unified universe and its relation to spectroscopic techniques. In turn, the research program of astrophysics and its observational techniques could provide plenty of empirical support for Secchi's natural philosophy and were keys to his broader project of reducing all physico-chemical laws to matter and movement.

Considering this background of assumptions and expectations, it becomes clear why the news about Gustav Kirchhoff's discoveries in spectrum analysis (1859) and his map of the solar spectrum was received differently in Rome compared to other major Italian observatories. To be sure, spectroscopic expertise was not exclusive to Roman observatories. As early as 1860, in Florence, Giovan Battista Donati and the instrument maker Giovan Battista Amici had developed a slit spectroscope with collimator—which soon became the canonical spectroscope—and used it in the study and classification of a few stars. The relation found by Donati between star spectra and their color was indeed one of the starting points of Secchi's spectroscopic research. However, Donati was still working in a traditional framework, using spectroscopic instruments not to investigate the physical structure of celestial bodies but rather to calculate their position. Because of his ancillary vision of spectroscopic techniques, Donati was skeptical about introducing spectroscopy into "first-order observatories." This stance would set him apart from Secchi and his young Italian followers, especially Pietro Tacchini, the future president of the Society of Italian Spectroscopists, and the Barnabite priest Francesco Denza, the physical astronomer who would later be director of the Italian Meteorological Society.[18]

Aware of Donati's work on stars and with a Hofmann pocket direct-vision spectroscope in his hands—a gift from Jules Janssen—Secchi began studying star spectra in 1862. He immediately linked them to the material constitution of celestial bodies and the solution of "important cosmic questions." Secchi commissioned Hofmann and Merz to produce special spectroscopes with multiple prisms and an objective prism that allowed the contemporaneous vision of numerous star spectra. With these he carried out his stellar classification (1867), which included more than four thousand stars divided into four spectral types.[19]

Secchi saw his own mechanistic attempt to reduce all phenomena to matter and movement as being at odds with deterministic and materialist perspectives and regarded the attribution of self-organizing properties to matter as an unscientific speculation. Against "material and mathematical necessity" Secchi argued for the contingent nature of physical laws, very much in line with John Tyndall, whose work he often cited in his theoretical pieces. Secchi's reductionism, according to which the phenomena of heat, light, electricity, magnetism, chemistry, and even gravitation were to be understood as the movement of molecules and ether, was not seen as the triumph of matter over spirit. Quite the opposite: the admirably simple functioning of the machine of the universe revealed its teleological nature and pointed toward a superior spiritual sphere, an ordering mind, a directing principle. What lay between matter and life was an unfathomable hiatus, as was that between the unified material universe and the structure of the "great network of beings," which Secchi extended to extraterrestrial life and to beings that could be "immensely more capable than ourselves." Ultimately, existing physical laws depended on the infinite freedom of the creator. The continuous transformation of physical forces one into the other made it necessary to postulate the act of creation, conceived as the primordial transmission of movement from God to inert matter. The experimental laws of the conservation of energy could therefore be read as a sign of the continuous supernatural action in the physical world.[20]

Secchi's atomistic and dynamist theories did not go unchallenged in contemporary Catholic culture. His essay on the unity of physical forces, a defense of the new physics as well as of the new astronomy, caused much discussion among philosophers and theologians in Rome. Most notably, Secchi's scientific practice was attacked by neoscholastic philosophers who accused him of downright materialism and, after 1870, managed to isolate the pontifical observatory. The neoscholastic doctrine of hylomorphism was indeed gaining support in Catholic cultural institutions worldwide, its hegemony officially sanctioned with the papal bull *Aeternis Patris* (1879). Thus it was not surprising that Secchi should have felt surrounded by enemies in his later years. These included anticlerical politicians and intellectuals but also Roman colleagues and fellow Jesuits, like Giovanni Maria Cornoldi and the editors of the Jesuit periodical *Civiltà Cattolica*. Secchi's isolation took a striking physical dimension in 1873, when owing to the pressure of the anticlerical ministers Quintino Sella and Ruggero Bonghi, the Collegio Romano was confiscated by the Italian authorities and the Jesuits expelled. Only Secchi and his assistants above the church were left untouched and would continue to live and work in this bizarre, suspended condition. The observatory was to be eventually confiscated by the Italian authorities in 1879, after Secchi's death, thus closing down the last pontifical site for astronomical research.[21]

The collapse of the papal state as a significant territorial entity in 1870 caused a rapid decline in experimental life at Secchi's observatory. In new political and institutional conditions, Secchi's heterogeneous practices were less relevant to the Holy See than straightforward theological orthodoxy. Neoscholastic philosophy was rapidly becoming the cement of Catholic culture worldwide, thus limiting the variety of theoretical positions that had characterized scientific research at the Collegio Romano in the previous decades. But if a stateless church had lost interest in Secchi's observatory practice, new national entities began translating and endorsing them: above all the young Italian nation.

The Making of Italian Astrophysics

Between 1850 and 1870 Secchi's unified universe provided pontifical scientific culture with a most effective working framework: one that could be used to support technoscientific innovation, that updated scientific teaching and dialogue with the liveliest sectors of international physical and astronomical research, that favored the transformation of pontifical observatories into sites for elaborating and implementing techniques deemed essential for the reform of the state—all the while sustaining a metaphysical dualism that safeguarded the integrity of religious dogma by placing it beyond the limit of legitimate scientific investigation. These multiple interpretations of Secchi's experimental practice can explain the apparent paradox of a culture in which ultraconservative policy and traditionalist theology coexisted with ambitious plans of technoscientific modernization. In the same years Italy had been politically united, and in 1870 Rome had been occupied and proclaimed its capital. In the 1860s Italian politics had been taken up by the "Roman question," the eternal city being a symbol that the new nation's liberal élites could not afford to leave in the hands of priests. At stake was not only the administrative status of the city but a redefined relation between church and state as much as between reason and faith. Hence the distinctively anticlerical spirit of the Italian Risorgimento, its radical tones, and the perception of the Catholic Church as a deadly threat to the nation. Not surprisingly the age of the Risorgimento coincided with the heyday of Italian positivism, a doctrine used by the liberal élite to frame socioeconomic problems confronting the new nation and identify their possible solution. Liberal anticlerical élites that controlled the Italian government believed that technical and scientific advancement was a necessary condition for both succeeding in the nation-building process and being admitted into the restricted club of world powers. Italy's imperialistic ambitions were being discussed in European cabinets, and soon the new nation would set off on its ill-fated colonial adventures. However, with an ongoing civil war in the south and a population that by and large did not speak a word of Italian, what the liberal élites needed most was the rapid implementation of order and control under a centralized power. The natural and mathematical sciences were in particular mobilized in these processes. As proclaimed by Prime Minister Quintino Sella, in Italianized Rome the religious universalism of the church would soon be replaced by the universalism of modern science.[22]

For the government, the opportunity to be involved directly in astronomical research came with the organization of the eclipse expedition to Sicily in 1870. The emergence of a new physical astronomy in the 1860s and 1870s was marked by a series of eclipse expeditions aiming to test various hypotheses about the physical nature of solar phenomena. It has been argued that these expeditions and the imperialistic strategies of western colonial powers were intrinsically connected. In the Italian case the emergence

16. The eclipse of 1870 as drawn by Tacchini. Secchi, *Le Soleil* (Paris: Gauthier-Villars, 1875–77), 1:335.

of a prominent astrophysical community and its first scientific expeditions were constitutive elements of the very process of nation building.[23]

The expedition of 1870 was organized around Angelo Secchi's figure and work, although for obvious reasons he could be invested with no official role. It included astronomers from Padua, Florence, Rome, Naples, and Palermo, as well as technicians and photographers. A few experimental physicists were also invited, like Pietro Blaserna, the promoter of new, centralized, coordinated laboratory practices in Italian universities, based on the German model.[24] The warship *Plebiscito* sailed southward, collecting instruments along the way. Meanwhile, in Calabria military engineers and topographers were taking position along the railway line to define the exact limit of the area of totality, in constant telegraphic contact with the Sicilian expedition. The scientific program included visual and spectroscopic observations of the corona and the prominences, with Secchi in charge of the photographic aspects (figure 16). Eventually, because of cloudy weather, the data collected were scarce and not particularly enlightening. The emission line in the spectrum of the corona seen by William Harkness the year before was observed, but its position could not be measured accurately. Hydrogen and helium lines were observed in the spectra of prominences and a strong polarization was measured in the corona.[25]

But the significance of the expedition went well beyond these meager results. For one thing, this was the first Italian scientific expedition ever, the debut of a new nation on the international stage of science. When a British expedition party guided by Norman Lockyer set up tents nearby, the Italians could proudly report to Rome that they had "remained with these distinguished gentlemen in friendly and cordial relations, as it should be between citizens of learned and free nations." In constant telegraphic contact with the government, they also reported that on the day of the eclipse "the national flag was waving on the bastion of the Castle of Augusta [base of the Italian expedition]." A symbol of tyranny, the old Bourbon castle was turned into a "site of science" for "the national honor and the progress of civilization." The following day local authorities organized banquets in various Sicilian cities to celebrate the Italian, British, and American expeditions.[26]

Moreover, thanks to the expedition the new national scientific spirit had reached Sicily, the most rebellious among Italian regions. The island was still run by powerful local élites, and Palermo, its capital, had been the theater of a major anti-Italian insurrection as recently as 1866. This was an expedition to the extreme borders of the Italian territory, where the legitimacy of the government and the monarch was still visibly challenged. Under heavy protection provided by the national Italian army, scientists crossed this reportedly uncivilized island with chronographs, telescopes, and Italian flags, bringing the lights of science and exact national time. The establishment of more solid contacts with southern scientists and the revival of the Palermo observatory—once run by Giuseppe Piazzi—in the context of Italian national research had also played an important part in the planning of the expedition.

The expedition marked the beginning of the official reorganization and standardization of observatory practices among the entire Italian astronomical community. Practitioners from ten medium-sized and small observatories scattered through the peninsula were gathered for the first time. Cloudy weather was not the greatest problem that the expedition needed to overcome. Much more significant, although less emphasized in official reports, were the difficulties faced by the group in trying to work as a coherent unit. In both provisional observatories set up in Sicily, knowledge about the sun was not being produced efficiently enough. Of course, under Secchi's leadership scientists and technicians had met well in advance to discuss priorities and techniques, test and select chronometers, and exchange and calibrate instruments, many from the Collegio Romano. Amid this scientific community in the making, navy officers from the Plebiscito

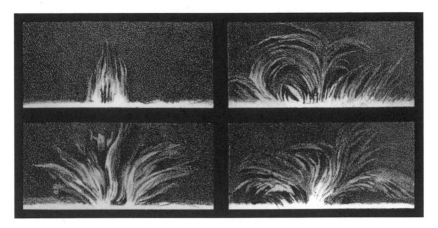

17. Representing solar prominences (Secchi and collaborators). Angelo Secchi, *Le Soleil* (Paris: Gauthier-Villars, 1875–77), vol. 2, plate H.

acted as additional assistants. Efforts were made to standardize techniques for representing and measuring solar phenomena. These included providing common answers to questions such as: How to describe a prominence? How to draw the surface of the sun? Along these lines, a whole series of micro-processes of standardization and calibration took place in the weeks before the eclipse. Pietro Tacchini and Giuseppe Lorenzoni, for instance, spent days drawing solar prominences as they appeared with a full sun. They alternated at a Merz equatorial telescope, drawing the same protuberance on the same sheet but in different colors, so that they could see "analogies and differences" between their drawings (figure 17). This exercise went on until they felt their styles had become close enough to make their representations trustworthy. In fact Secchi and his colleagues were astonished by the differences in the descriptions and representations of observers studying the same phenomena with the same instruments. They drew differently, they spoke differently, and bonds of trust were dangerously weak. Secchi was convinced that the poor quality of his photographs was only partly due to clouds. He suspected that his temporary assistants were too slow in obeying his orders and that the photographer had altered certain chemicals without informing him, quite unlike his fellow Jesuits at the Collegio Romano. Italian science, like Italy itself, was still in the making.[27]

Standardization of observatory practices continued through the year 1871, with Tacchini in Palermo and Secchi in Rome coordinating the monitoring of the solar limb. Shortly after the observatories of Padua and Naples joined the program, using spectroscopic instruments acquired on the occasion of the eclipse expedition. Toward the end of the year Secchi and Tacchini announced the creation of a national society that would effectively implement the new observational regime nationwide. Thus the Society of Italian Spectroscopists began its activities, with Tacchini as its director and the participation of five national observatories. Physicists and chemists interested in spectrographic techniques were also invited to contribute ("their works will be published along with those of the astronomers"). In line with Secchi's earlier research, the society grounded its activities on the essential connection between solar and terrestrial physics. Morphological and spectroscopic analysis had to be directed primarily to studying the physical constitution of the sun and identifying the effects of solar activity upon the earth.[28]

The society's *Memorie* (launched in 1872), edited by Tacchini, was the first journal entirely devoted to physical astronomy. The journal functioned as the site for the standardization of measuring and representational techniques. Here one finds a gradual shift away from the publication of drawings executed by one individual and the use of idiosyncratic terminology and toward more detached, objective forms of knowledge. Consider the representation of the solar limb, which was recorded daily by the members of the society. Originally all the drawings for the same day from different observatories were published. By the second year these were already replaced by a single, objective monthly sheet bearing no reference to its origin. Drawings were cut and pasted from different observatories' reports, so that each day's observation was given only one graphic representation. Every day the solar limb therefore had a unique, definitive form.

The internal reorganization process of physical astronomy was accompanied by related processes of professionalization and disciplinary policing. A comparison between reports of the solar eclipse of 1870 and the articles appearing in the *Memorie* is instructive. The reports recorded the observations of a multitude of enthusiastic nonprofessional astronomers. From aristocratic amateurs like the Sicilian Prince of Lampedusa to school and college directors, parish priests, and military officers, everyone felt entitled to take careful notes of the spectacular event and send them to the organizers of the eclipse expedition. Remarkably, the commission decided to publish—or at least refer to—many of them. Distinguished ladies seem to have been considered particularly suited to comment upon the change in the colors of the spectrum as the totality approached: "The spectacle was sublime, worthy of the Eternal Creator!," commented Angela Vanneschi, who proclaimed herself "full of marvel and respect for science and the scientists." Astronomers and dilettanti alike copiously referred to the supernatural character of the spectacle. One would look in vain for similar statements in Father Secchi's dry prose.[29]

Only a couple of years later it was out of the question for the *Memorie* to publish reports stemming from the all-too-colorful world of nonprofessional observers. Published authors included only members of the society, most notably Secchi and Tacchini, and a few Italian and foreign correspondents from well-known observatories. The issue of setting up professional boundaries implicitly emerged in Tacchini's comments on a British article about the possibility of using small telescopes to observe the chromosphere and the prominences. Tacchini dismissed with few ironical remarks the use of portable instruments, stressing the substantial qualitative difference between the observational reports of the amateur and those of the professional, the only one able to properly "see" the phenomena under study and describe them accurately. Strategies of professional boundary drawing were also apparent with respect to the divide between this emerging community of solar physicists and traditional positional astronomers. Against more cautious colleagues like Donati, Secchi and Tacchini insisted on the need to fill the great domes of the first-order observatories with spectroscopic instruments and to modify traditional telescopes so as to facilitate spectroscopic observation.[30]

The new community of physical astronomers legitimated itself also through the reconstruction and display of its genealogy. In the first number of the journal Tacchini traced the history of Italian spectroscopy back to the early works of Secchi and Respighi, showing how it spread to the observatories of Padua, Naples, and Palermo on the occasion of the eclipse of 1870. Announcing the central role of spectroscopy in modern astronomy, Tacchini pleaded for a more rational division of labor among observatories, particularly in those research areas that required continuous and coordinated spectroscopic observations. The task of drawing national maps of the solar limb was skillfully used by Tacchini as a means to implement Secchi's observational regime in the national observatories that had joined the society. Tacchini's guidelines for the observatories participating in the common effort contained extremely detailed instructions, from the correct method of representing sunspots and the standard width of the spectroscope's slit in observing prominences, to the kind of pencil and paper to be used to draw them.[31]

The expedition of 1870 and the constitution of the Society of Italian

Spectroscopists provided a new generation of patriotic astronomers led by Tacchini and Denza with the opportunity to draw resources to astrophysical and meteorological research, placing these disciplines at the center of renewed Italian astronomy. For Tacchini and his colleagues, Secchi's unified universe and his observational regime became crucial resources to link their scientific practice to the construction of the new nation and its research system. The lack of a national central observatory made it urgent for the Italian community to organize itself, raising major issues about the standardization and coordination of observational practices. This anomalous situation can explain why the new discipline of physical astronomy was able to find more institutional recognition in Italy than in other European countries. Weak resistance to the penetration of astrophysical practices into the main observatories and the relatively low cost of spectroscopic instruments favored the implementation of Secchi's observational regime, offering a working model to young Italian astronomers.

Tacchini played a key role in mediating between Secchi, the Italian astronomical community, and the Italian government. His requests for astrophysical research funding were met favorably by the minister of public instruction, the scientist-turned-politician Quintino Sella. Indeed the government was attentive to technoscientific issues, sensitive to the international image of the country, and supportive of processes of centralization and standardization.[32] Significantly, Tacchini would be entrusted with the directorship of the observatory of the Collegio Romano after Secchi's death in 1878. In the following year, at the peak of his political and scientific career, he would also assume the directorship of the newly founded Royal Central Meteorological Office, which unified and coordinated the meteorological and seismological networks. It should be noted that Tacchini entrusted the construction of meteorological and seismological instruments for the national network to Brassart, Secchi's instrument makers. Meanwhile Tacchini had also designed a national reform project for coordinating and redistributing observational tasks and funds among the national observatories, a project which received governmental approval in 1876.[33]

Interestingly enough, for a few months the Italian government had also considered the possibility of enrolling Secchi himself as the leader of the new Italian astronomy. The mathematician and politician Francesco Brioschi had discreetly contacted Secchi immediately after the occupation of Rome in September 1870, and secret negotiations continued throughout the eclipse expedition. Secchi was offered the chair of physical astronomy at the new University of Rome. Initially he accepted, on the condition that restrictions on Jesuit teaching and research be lifted and that the chair be named "physical astronomy and meteorology." The government refused to deviate from its general anticlerical line, and Jesuit institutions were not spared. Toward the end of November 1870, in the middle of the eclipse expedition, negotiations collapsed and Secchi refused to cooperate any further with Italian authorities. Since its failure to gain Secchi's cooperation, and with Tacchini well established as his authoritative heir, the Italian government began undermining Secchi's activity in a rather literal sense, as we have seen. In 1872 the Italian government formally protested his representing the Holy See at the Commission Internationale du Mètre in Paris. In 1874, on the occasion of the Italian expedition to India to observe the transit of Venus, Secchi's integration with the Italian group—led by Tacchini—was no longer an option. Secchi's collocation in the pantheon of Italian science remained problematic well after his death, as shown by the story of a planned, but never completed, commemorative monument to him. As a nineteenth-century hand wrote on the title page of a copy of *L'unità delle forze fisiche* at Padua university library: "A great mind, too bad he is a priest."[34]

Epilogue

Although the papal observatory of the Collegio Romano did not long survive the Italian occupation of Rome, its observational regime thrived in the network of nationalized Italian observatories. Secchi, who until the pontiff died remained loyal to him and who became a leader of the Roman legitimist scientists after 1870, did not play any official role in the Italian system of research. Still his observatory practice and theoretical framework became valuable resources for the emerging Italian astrophysical community. The reasons for this success are to be found in the particular conditions of astronomy in post-unification Italy, primarily the decentralized structure of the field, and in the heterogeneous nature of Secchi's practice, which could sustain the more general national priorities of control and standardization.

However, one could wonder why Secchi's unified universe was translated into a national research program in spite of its explicit metaphysical underpinnings. That is, why Secchi's unified universe prevailed rather than some materialistic version of it, more in line with the radical culture of the Risorgimento and with contemporary materialist trends in Italian scientific culture. A possible answer is that during the 1870s and 1880s the rationalist and materialist doctrines that had nourished the culture of the Risorgimento were losing ground. The positivist season was waning in Italy, along with the optimistic forecast of rapid socioeconomic development for the entire nation. While anticlericalism remained strong among the Italian élites, religious dogma did not seem the most dangerous enemy of enlightened liberal order any more. As confirmed by the circumstances of the Paris Commune, troubles for the liberal élites were now arriving from other directions. With bread riots and socialist demonstrations on the rise, the "social question" soon replaced the Roman question at the center of political discourse. The authoritarian turn of the Italian government and its repressive reaction to popular unrest would become all the more apparent through the 1880s and 1890s.[35]

Far from being an obstacle to its widespread adoption, the antimaterialist and spiritual underpinnings of Secchi's unified universe made it all the more attractive to scientific and general audiences. Jesuits in black frocks might have been out of fashion, but the fortunes of Secchi's "spiritual mechanism" were still ascendant, and not only in astronomy. His antimaterialist approach to scientific practice found very receptive and sympathetic audiences amid the increasingly conservative Italian bourgeoisie. And indeed Secchi's blending of healthy experimentalism and rigid metaphysical dualism enjoyed a long-lasting success in fin de siècle science, in Italy and beyond. Once freed from its clerical connotations, his approach could be used to legitimate technoscientific innovation and economic development while subordinating both to a superior sphere of spiritual values and unquestionable national objectives. To be sure, the metaphysical perspective of Italian élites around 1870 was grounded neither in Catholic dogma nor in the revolutionary mystique of 1848: new spiritual horizons were being opened for future generations. In the discourse of Italian nationalism an increasingly complex and pervasive mythical apparatus was emerging, centered on an idealistic understanding of the classical heritage, the redeemed fatherland, and its spiritual mission. Increasingly this language informed the technoscientific renewal and the reorganization of labor in laboratories and observatories. Newly nationalized practices and observational regimes could thus produce and give meaning to scientific values (e.g. precision) whose diffusion was deemed essential to constructing the unitary nation and realizing its spiritual destiny.[36]

Notes

I would like to thank the participants of the workshop at the Max Planck Institute in Berlin for their comments on an earlier version of this article. I am also grateful to Ileana Chinnici (Osservatorio di Palermo), Laura Gasparini (Biblioteca Panizzi, Reggio Emilia), Agnese Mandrino (Osservatorio di Brera), and Luisa Pigatto (Osservatorio di Padova) for their kind assistance.

1. Secchi to Denza, 7 February 1872, repr. in Beltrano, *Presenze scientifiche illustri al Collegio Romano*, 34.

2. Pius IX, "Syllabus Errorum," *Enchiridion delle Encicliche*, 2 (Bologna: EDB, 1996), 521–25. The plates are in Paolo Cacchiatelli and Gregorio Cleter, *Le scienze e le arti sotto il pontificato di Pio IX* (Rome: Forense, 1860–63, repr. Aureli, 1863–67; 1865, repr. Tiberina, 1870; 1874). On Pius IX's scientific studies see Giovanni Maria Mastai Ferretti [Pius IX], *Le macchine ottiche: esercizio fisico-matematico* (Firenze: Allegrini, 1809), and Pierluigi Pizzamiglio and Giorgio Tabarroni, *Pio IX e le macchine ottiche* (Bologna: Clueb, 1981). On Pius IX see Martina, *Pio IX*. On pontifical science under Pius IX see also Mazzotti, "For Science and for the Pope-King." For a historical and bibliographical survey of Jesuit observatories worldwide see Udías, *Searching the Heavens and the Earth*.

3. For biographical data and publications see Joseph Pohle, *P. Angelo Secchi, ein Lebens und Kulturbild* (Cologne: Bachem, 1883); Carlo Bricarelli, "Della vita e delle opere del P. Angelo Secchi," *Memorie della Pontificia Accademia dei Nuovi Lincei* 4 (1888): 41–105; and Abetti, *Padre Angelo Secchi, il pioniere dell'astrofisica*. See also Ileana Chinnici and Wiktor Gramatowski, "Le carte di Angelo Secchi S.J. (1818–1878) conservate presso la Pontificia Università Gregoriana: un inventario inedito rivisitato," *Nuncius* 16 (2001): 571–627.

4. Angelo Secchi, *L'Astronomia in Roma nel pontificato di Pio IX* (Rome: Della Pace, 1877), and Gaspare Ferrari, *Lorenzo Respighi: suo elogio nell'anniversario della morte* (Rome: Cuggiani, 1891).

5. The Cauchoix refractor accompanied Secchi on the pontifical eclipse expedition to Spain in 1860. On that occasion the comparison of Secchi's solar photographs with Warren De La Rue's proved that prominences were not a play of light but an actual solar phenomenon (see Aubin's chapter in this volume). See Warren De La Rue, "Comparison of Mr. De La Rue's and Padre Secchi's Eclipse Photographs," *Proceedings of the Royal Society of London* 13 (1863–64): 442–44. On Secchi's magnetic instrumentation see Secchi, *Descrizione dell'osservatorio magnetico del Collegio Romano e sunto delle osservazioni fatte nel 1859 e 1860* (Rome: Belle Arti, 1860).

6. Paolo Gavazzi, "Pio IX e la riforma degli ordini religiosi (1846–1857)," *Atti del II convegno di ricerca storica sulla figura e sull'opera di papa Pio IX* (Senigaglia: Centro Studi Pio IX, 1977), 203–42; Alberto Polverari, *Vita di Pio IX* (Vatican: Editrice Vaticana, 1986–88), 2:97; Martina, *Pio IX*, 1:463–65, 2:693–8; and Carlo Fiorentino, *La questione romana intorno al 1870* (Rome: Archivio Guido Izzi, 1997), 45–113.

7. Secchi, *Memoria sulla relazione de' fenomeni metereologici colle variazioni del magnetismo terrestre* (Rome: Belle Arti, 1864), 16. Secchi also used telegraphic lines to "observe" shooting stars. See his letter to Schiaparelli, 13 August 1866, in Secchi and Schiaparelli. *Corrispondenza*, 63.

8. The inscription on the medal reads: "Providentia P[ontificis] M[aximis] / Ferrea Via Romam Provinciis Jvngi Curavit." See Pizzamiglio and Tabarroni, *Pio IX e le macchine ottiche*, 13, and Mario Caravale and Alberto Caracciolo, *Lo Stato pontificio da Martino V a Pio IX* (Turin: Utet, 1978), 667–740.

9. Secchi, *Misura della base trigonometrica eseguita sulla via Appia per ordine del governo pontificio nel 1854–55* (Rome: Rev. Camera Apostolica, 1858). Similar connections between the Paris Observatory and railway and telegraphic networks are examined in Aubin, "The Fading Star." The cultural history of observatory science was inaugurated by Schaffer, "Astronomers Mark Time." On the connections between nineteenth-century laboratories and their sociotechnical environment see Bruno Latour, *The Pasteurization of France*, trans. Alan Sheridan and John Law (Cambridge: Harvard University Press, 1988).

10. Secchi, *Descrizione del meteorografo dell'osservatorio del Collegio Romano*, 8. Paolo Brenni has restored one of Secchi's meteorographs for the Istituto e Museo di Storia della Scienza in Florence. See Brenni, "Il meteorografo di padre Angelo Secchi."

11. On undisciplined students see Secchi to Schiaparelli, 30 June 1868, in *Corrispondenza*, 177. On undisciplined monks see Secchi to Denza, 22 February 1876, in Beltrano, *Presenze scientifiche illustri al Collegio Romano*, 35–36.

12. On the visual display of data to facilitate understanding see Secchi, *Descrizione del meteorografo dell'osservatorio del Collegio Romano*, 7. On his controversy with Rudolphe Radau see Brenni, "Il meteorografo di padre Angelo Secchi," 225–29. On his repeated claim that with analysis alone one cannot ground any physical theory see Secchi, *L'unità delle forze fisiche* (Milan: Treves, 1874 [1864]). On the complete mechanization of observation see Secchi to Schiaparelli, 8 January 1868, *Corrispondenza*, 166. On Secchi's photographic techniques and machinery see Secchi, *Fotografie lunari e degli altri corpi celesti* (Rome: Belle Arti, 1859); Secchi, *Relazione delle osservazioni fatte in Spagna durante l'eclisse totale del 18 luglio 1860* (Rome: Belle Arti, 1860); and Laura Gasparini, "Padre Angelo Secchi e l'applicazione della fotografia nelle osservazioni astronomiche," *Fotologia* 12 (1992): 35–47. See also Piero Becchetti, *La fotografia in Roma dalle origini al 1915* (Rome: Colombo, 1983).

13. See M. Sanchez Navarro-Neumann, "Os Jesuitas e a sismologia," *Broteria* 24 (1937): 141–51; Augustin Udías and William Stauder, "The Jesuit Contribution to Seismology," *Seismological Research Letters* 67 (1996): 10–19; and Udías, *Searching the Heavens and the Earth*.

14. Secchi, *Ricerche di reometria elettrica* (Rome: Belle Arti, 1850), and Secchi, *Sul modo di valutare la forza del raggiamento solare* (Rome: Belle Arti, 1851).

15. On the early concordant approach see S. Proja, "Cenni intorno alla cattedra di fisica sacra dell'archiginnasio romano," *Giornale arcadico di scienze, lettere ed arti* 74 (1837): 108. On the Jesuit approach see Enrico Vasco, *Il Ratio studiorum adattato*

ai nostri tempi (Rome: Civiltà Cattolica, 1851). On the philosophical and metaphysical underpinnings of Secchi's scientific practice and their roots in Catholic culture see Pietro Redondi, "Cultura e scienza dall'illuminismo al positivismo," *Storia d'Italia, annali 3: scienza e tecnica nella cultura e nella società dal Rinascimento a oggi,* ed. Gianni Micheli (Turin: Einaudi, 1980), 797–811. See also Redondi, "Physique et apologétique: le *Cosmos* de l'abbé Moigno et de Marc Seguin," *History and Technology* 6 (1988): 203–25.

16. See Giovan Battista Pianciani, *Istituzioni fisico-chimiche* (Rome: Puccinelli, 1833–34).

17. Secchi, *L'unità delle forze fisiche* (Milan: Treves, 1874 [1864]). See also his *Lettera su di un problema cosmologico* (Rome: Marini, 1862). Secchi's experimental physics was indebted to the work of Italian physicists of the first half of the century such as Macedonio Melloni (1798–1854), who had argued for the analogy of light and heat, and Ambrogio Fusinieri (1775–1849), whose phenomenal and antimaterialist approach had been extremely influential in the Restoration age. See Macedonio Melloni, *Mémoire sur l'identité des diverses radiations lumineuses, calorifiques et chimiques vibrées par le soleil et les sources terrestres* (Geneva: Ramboz, 1842); Ambrogio Fusinieri, *Memorie sperimentali di meccanica molecolare* (Padua: Sicca, 1844); and Ambrogio Fusineri, *Memorie sopra la luce, il calorico, la elettricità, il magnetismo, l'elettro-magnetismo ed altri oggetti* (Padua: Sicca, 1846). On Fusinieri see also Redondi, "Cultura e scienza dall'illuminismo al positivismo," 721–29.

18. On early spectrum analysis and spectroscopic instrumentation see Bennett, *The Celebrated Phenomena of Colours,* and Hearnshaw, *The Analysis of Starlight,* 51–103. On the emergence of the "new astronomy" see A. J. Meadows, *Early Solar Physics,* 20–82, and Hufbauer, *Exploring the Sun,* 42–65. On early spectroscopy in Italy see Giovan Battista Donati, *Intorno alle strie degli spettri stellari* (Florence: Galileiana [1860]); and Ileana Chinnici, "Nineteenth-Century Spectroscopic Instruments in Italian Astronomical Observatories," *Nuncius* 15 (2000): 671–80. On Amici see Edorado Proverbio and Pasquale Tucci, "Giovanni Battista Amici costruttore di telescopi e cannocchiali acromatici," *Physis* 30 (1993): 145–75; Alberto Meschiari, "Corrispondenza di Giovanni Battista Amici con Carlo Matteucci e Angelo Secchi," *Nuncius* 14 (1999): 233–61. On Denza see Giuseppe Boffito, *Scrittori barnabiti* (Florence: Olschki, 1933), 1:606–38.

19. On Donati versus Secchi see *Memorie* 1 (1872): 54–56. On Janssen in Rome see David Aubin, "Orchestrating Observatory, Laboratory, and Field." On Secchi's stellar spectroscopy and his early classification system see Secchi, "Sugli spettri prismatici della luce de' corpi celesti," *Memorie dell'Osservatorio del Collegio Romano* 2 (1860–63): 121–28; Secchi, *Le scoperte spettroscopiche in ordine alla ricerca della natura de' corpi celesti* (Rome: Belle Arti, 1865); Secchi, *Catalogo delle stelle di cui si è determinato lo spettro luminoso all'Osservatorio del Collegio Romano* (Paris: Gauthiers-Villars, 1867); Secchi, "Sugli spettri prismatici delle stelle fisse," *Memorie della Società Italiana delle Scienze* 1 (1867): 71–78; Secchi, "Sugli spettri prismatici delle stelle fisse: memoria II," *Memorie della Società Italiana delle Scienze* 2 (1869): 73–133.

20. See Secchi, *La grandezza del creato*, two lectures published posthumously in Secchi, *Lezioni elementari di fisica terrestre* (Turin-Rome: Loescher, 1879), 183–218.

21. Giacomo Martina, *Storia della Compagnia di Gesù in Italia, 1814–1983* (Brescia: Morcelliana, 2003), 47–71. On neoscholasticism see Antonio Masnovo, *Il neotomismo in Italia* (Milan: 1923); Antonio Piolanti, *Pio IX e la rinascita del tomismo* (Vatican: Lib. Vaticana, 1974); and Antonio Piolanti, *L'Accademia di Religione Cattolica: profilo della sua storia e del suo tomismo* (Vatican: Lib. Vaticana, 1977). On Cornoldi see Giovanni Maria Cornoldi, *I sistemi meccanico e dinamico circa la costituzione delle sostanze corpore, considerazioni rispetto alle scienze fisiche* (Verona: Piacentini e Franchini, 1864), and Luciano Malusa, *Neotomismo e intransigentismo cattolico* (Milan: IPL, 1989). See also Secchi's controversy with the Dominican Vincenzo Nardini in Secchi, *Intorno alla soluzione di un problema fisico-cosmologico* (Rome: Morini, 1862), and Nardini, *Risposta alla lettera del P. Angelo Secchi intorno alla soluzione di un problema fisico-cosmologico* (Rome: Cesarelli, 1862). On the confiscation of the Collegio Romano see Ruggero Bonghi, *Collegio Romano: la Biblioteca Vittorio Emanuele e i musei. Discorso inaugurale* (Rome: Barbera, 1876).

22. Alberto Caracciolo, *Roma capitale* (Rome: Riuniti, 1999); Giacomo Martina, "Roma dal 20 settembre 1870 all'11 febbraio 1929," *Storia d'Italia: annali 16: Roma, la città del papa* (Turin: Einaudi, 2000), 1059-1100; Quintino Sella, speech to the parliament, 14 March 1881, repr. in Quintino Sella, *Discorsi parlamentari* 1 (Rome: 1887–90): 273-311; Guido Quazza, *L'utopia di Quintino Sella: la politica della scienza* (Turin: ISRI, 1992); and John Davis, *Conflict and Control: Law and Order in Nineteenth-Century Italy* (London: Macmillan, 1988). On the role of scientists and particularly mathematicians in early Italian political life see Umberto Bottazzini, *Va' pensiero: immagini della matematica nell'Italia dell'Ottocento* (Bologna: Mulino, 1994). The values of political liberalism, anticlericalism, and positivism at the popular level were supported by successful publication series like *Biblioteca del popolo* (published in Milan by Sonzogno) and *La scienza del popolo* (published in Milan by Treves). On the popularization of science in Italy in this period see Paola Govoni, *Un pubblico per la scienza. La divulgazione scientifica nell'Italia in formazione* (Rome: Carocci, 2002).

23. On eclipse expeditions see Pang, *Empire and the Sun*, and David Aubin's chapter in this volume.

24. Pietro Blaserna, "Sullo stato attuale delle scienze fisiche in Italia e su alcune macchine di fisica," *L'Italia alla Esposizione Universale di Parigi nel 1867: rassegna critica descrittiva illustrata* (Paris: Racon, 1868), 70–74. On the reorganization of experimental physics in post-unitary Italy and its French and German models see Barbara Reeves, "Le tradizioni di ricerca della fisica italiana nel tardo diciannovesimo secolo," *La scienza accademica nell'Italia post-unitaria*, ed. V. Ancarani (Milan: Angeli, 1989), 53–112.

25. Cacciatore, *Rapporti sulle osservazioni dell'ecclisse totale di sole del dicembre 1870*; Secchi, "Progresso delle cognizioni solari ottenuto in occasione dell'ecclisse solare del 22 dicembre 1870," *Atti dell'Accademia Pontificia de' Nuovi Lincei* 24 (1871-72): 1-15; Pietro Tacchini, *Eclissi totali di sole del dicembre 1870, del maggio 1882 e*

1883, e dell'agosto 1886 e 1887: relazioni e note (Roma: Botta, 1888), 233-36; and Ileana Chinnici, "Eclissi totali di sole 1860-1870: la nascita della fisica solare," *Giornale di astronomia* 1 (2000): 40-45.

26. Quotes from Cacciatore, *Rapporti sulle osservazioni dell'ecclisse totale di sole del dicembre 1870,* 7-9.

27. On Tacchini and Lorenzoni see Cacciatore, *Rapporti sulle osservazioni dell'ecclisse totale di sole del dicembre 1870,* 101-2; for Secchi's complaints see p. 21. On the sociotechnical issues related to the making of reliable observers and recorders see Schaffer, "On Astronomical Drawing."

28. Pietro Tacchini, "Sulla nuova società degli spettroscopisti italiani," *Memorie della società degli spettroscopisti italiani* 1 (1872): 3-6. Against the arguments of Hervé Faye, Secchi and Tacchini defended the connection between solar and terrestrial physics in a long appendix to the second number of the *Memorie* (1873).

29. Quotes from Cacciatore, *Rapporti sulle osservazioni dell'ecclisse totale di sole del dicembre 1870,* 208. On the prince of Lampedusa see Ileana Chinnici, "Gli strumenti del Gattopardo," *Giornale di Astronomia* 1 (1997): 24-29.

30. Pietro Tacchini, "Sulla precedente memoria del prof. Lorenzoni," *Memorie della Società degli Spettroscopisti Italiani* 2 (1873): 43.

31. Tacchini, "Sulla nuova Società di Spettroscopisti Italiani"; Tacchini, "Protuberanze solari osservate contemporaneamente a Palermo, Roma e Padova nel luglio e agosto 1871," *Memorie* 1 (1872): 25-32.

32. Ileana Chinnici, "Nascita e sviluppo dell'astrofisica in Italia nella seconda metà dell'ottocento," paper presented at the XVIII Congresso di Storia della Fisica e dell'Astronomia, Como, 1998, and Giorgia Foderà Serio, "Dalla Società degli Spettroscopisti Italiani alla Società Astronomica Italiana," *L'astronomia in Italia,* ed. Fabrizio Bonoli (Naples: Arte Tipografica, 1998), 21-48.

33. On the extreme centralization and uniformity of the Italian university system compared to the French and German see Burton Clark, *Academic Power in Italy: Bureaucracy and Oligarchy in a National University System* (Chicago: University of Chicago Press, 1977); and Pier Paolo Giglioli, *Baroni e burocrati: il ceto accademico italiano* (Bologna: Mulino, 1979).

34. G. Castellani, "Nomina e rinunzia del P. Angelo Secchi a professore di astrofisica nell'Università di Roma," *Civiltà Cattolica* 95 (1944): 39-46, 159-69, 170-79, and Riccardo Finzi, "Il mancato Monumento Scientifico al Padre Angelo Secchi," *Atti e memorie della deputazione di storia patria per le antiche provincie modenesi* 6 (1971): 137-47.

35. It is worth signaling the clear shift in the image of science promoted by popular series such as La Scienza del Popolo and Biblioteca Utile, both published by Treves in Milan. The latter reprinted Secchi's *L'unità* in 1874.

36. On the fortune of Secchi's experimentalism see Redondi, "Cultura e scienza dall'illuminismo al positivismo." On the crisis of positivist culture in Italy see also Giuliano Pancaldi, *Darwin in Italy: Science across Cultural Frontiers* (Bloomington: Indiana University Press, 1991), 152-69. On the cultural climate of fin de siècle Italy and the discourse of "spiritual values" see Furio Jesi, *Cultura di destra* (Milan: Garzanti, 1993).

Chapter 3

Eclipse Politics in France and Thailand, 1868

DAVID AUBIN

Everywhere in the East Indies it is believed that when the Sun and the Moon eclipse one another, it is because some dragon, with very dark claws, stretches towards both stars wishing to grasp them. On those occasions you can see rivers covered with the heads of Indians in water up to their necks, a most devout position well adapted to defend themselves against the dragon.
—BERNARD LE BOVIER DE FONTENELLE (1686).[1]

On croit les Asiatiques plus naïfs qu'ils ne le sont.
—PROSPER MÉRIMÉE.[2]

On 18 August 1868, that is on Tuesday, the first day of the waxing moon in the tenth month of the year of the Dragon, year 2,411 of the Buddha Era, an unusual crowd gathered on the desolate beaches of the Wako district in southern Thailand (then Siam). Dozens of Europeans and Americans—or *farangs*, as the Siamese called foreigners—diplomats, traders, navy officers, and ship crew, anxiously stared at an overcast sky. There was also a handful of scientists expressly dispatched from faraway France, led by Édouard Stéphan, director of the Marseilles Observatory. At ten o'clock in the morning the king of Siam, Mongkut (later known as Rama IV), went out on the terrace of his three-storied wooden palace built for the occasion and peered through his telescope. More than a thousand Siamese from his court, including many of his wives and children and the heir apparent, as well as countless horses, cattle, and fifty elephants, had journeyed there from Bangkok, 140 miles away (figure 18). Then, to everyone's great relief, the clouds opened and the sun shone, though not as brightly as it should have at this time of the day. "It could be seen that the eclipse had already started. The fanfare therefore started the music, and the King took his bath of purification. . . . At exactly thirty-six minutes and twenty seconds after

Éclipse totale de Soleil visible à Wah-Wan, dans la presqu'île de Malaca, le 18 août 1868. à 11ʰ-45-45.

18. Borde's representation of the Bay of Wako, Thailand, at the time of the eclipse of August 1868. Courtesy of the Observatoire de Bordeaux.

eleven o'clock, the sun was in total eclipse. At that moment, it was dark as if it were nighttime, around the twilight time. Those sitting close to each other could not see nor could they recognize each other's faces."[3] At the king's signal, a cannon shot was fired. According to Buddhist mythology, an angry creature named Rahu had swallowed the sun—or grasped it, no one knew for sure—but luckily, the roar of drums and trumpets from a nearby village no doubt helped to scare him away, so that some six minutes and forty seconds later he let it go free again. Everyone rejoiced and the king gave out gifts of money to his entourage.[4]

With its long totality, the eclipse of 1868 drew the attention of European astronomers as well. For historians of astronomy, the eclipse is most significant for the discoveries made using the spectroscope.[5] Less than a decade after Robert Bunsen and Gustav Kirchhoff set the foundation of spectrum analysis and showed how to use it to determine the chemical constitution of the sun, this was the first opportunity for spectroscopes to be directed at the limb of the eclipsed sun. Several European parties dispatched to distant lands, from Aden to Indonesia, were then able to shed lingering doubts about the nature of prominences—those pinkish flames around the dark disk of the moon only visible during eclipses. They were neither effects of the earth's atmosphere nor Olympian mountains on the moon; they definitely belonged to the sun. Observing the flames through their spectroscopes, astronomers saw the emission lines rather than the dark Fraunhofer lines typical of the sun's spectrum and inferred that prominences were gigantic outbursts of incandescent gases, mainly consisting of hydrogen. As a result of this single observation, no one would have any doubt about the possibility of analyzing spectroscopically the chemical constitution of celestial bodies. Not long after, when a certain spectral line was studied more carefully and found to correspond to no known substance on earth, some felt so confident in the spectroscopic method that they attributed the unknown line to a new element unknown on earth, which they named *helium*.[6] Above all, historians of astronomy have remembered this eclipse as the occasion when the spectroscopic method for studying prominences was discovered by Jules Janssen, sent to India by the French government. (The discovery of the method was also attributed to the Englishman J. Norman Lockyer, whose observations, independent of Janssen's, relied on early reports from these eclipse expeditions.)[7]

What historians have failed to emphasize is the regime change signaled by the eclipse with respect to efforts to organize expeditions as far away as the antipode for just a few minutes of observation time—efforts that came to nothing in case of bad weather! If they occur rarely at any given place on the surface of the earth, solar eclipses are not all that uncommon. Yet from 1800 to 1868 only a handful of the forty-one eclipses visible from earth were the subject of witness reports, still fewer to precise observations by professional astronomers. A common misconception was that the eclipse of 1868 was, as Janssen wrote, "of such duration that one must go back to the time of Ancient Greece to find something similar."[8] In fact the longest eclipse of the century had occurred barely eighteen years before, on 7 August 1850. Visible in Hawaii, as far as I know it has given rise to no memorable description. Until then only three times were several concurrent major expeditions undertaken (in 1842, 1851, and 1860). These three eclipses were all visible from Europe. Whenever eclipses were visible in other parts of the world, their observation apparently did not warrant the risk of long seafaring: reports provided by navy astronomers and officers whose duties led them close enough to the penumbra usually sufficed. After 1868, on the other hand, it became almost unthinkable for an eclipse to go unobserved by trained specialists from Europe or America. A "competitive bounding ritual" propitious to the striking of friendships, eclipse expeditions tightened the bonds of an emerging international solar physics community.[9]

For different reasons, historians of Thailand have also emphasized the importance of the eclipse of 1868. Although King Mongkut died later that year from a fever contracted during his journey, the event became a building block of Thailand's national identity. During the reign of his successor and son, Chulalongkorn, the walls of the Ratchapradit Temple were decorated with depictions of the king observing the eclipse. On the hundredth anniversary of the eclipse, Mongkut was granted the posthumous title of "Father of Thai Science," and in 1982, Bangkok's bicentennial year, 18 August was declared National Science Day. The population was encouraged to pay tribute to King Mongkut each year on this day. Today in Thailand, the king's eclipse expedition now stands for the establishment of modern science, which by and large followed western norms and applied western technology, but remained respectful toward traditional belief systems. Traditionally Thai historians have regarded this process of "modernization," begun in the mid-nineteenth century, as a necessary measure taken by Mongkut and his court to save the country from being colonized by either France or Britain.[10] Viewed from Siam, the eclipse expedition of 1868 was—and remains—one of the king's shrewdest political acts.

Historians of Thai science have recently paid renewed attention to the social meaning of the solar eclipse of 1868. Thongchai Winichakul has argued that the acclimation in Thailand—not only of the technology of production, trade, and war but also of certain elements of western science such as geography and astronomy—played a role in showing the West that Siam would join the imperial world-system without having to be colonized. Concomitantly, it helped to redefine the notion of "Thainess." While Thongchai has emphasized that for Mongkut the event was an occasion to show his court astrologers that western practices were superior to theirs, Nerida Cook has explained that the king's interest in astronomy could hardly be divorced from his very real belief in astrology—or at least from his need to exert some astrological control over the various ceremonials associated with monarchy.[11] Power in Thailand was traditionally legitimized by association with the supreme sources of power, and a good way to achieve this association was through religious ideology, by which the monarch could claim privileged access to Buddhist theology and Hindu gods. In the second half of the nineteenth century Siam therefore had to reconceptualize itself in relation to a new source of supreme power, which had shifted away from defeated China and India to Europe, whose ethos no longer lay in the "cosmic" spheres but in "civilization."[12] In the following discussion I claim, however, that supreme power in Europe was also cosmic—hence the crucial status held in both societies by the body of knowledge that was both cosmic and civilized: astronomy.

The Global Politics of Solar Eclipses

But what do both views of the eclipse, the French and the Thai, have to do with one another? While insisting on the European construction of "otherness" in nonwestern societies, postcolonial studies have often preserved a measure of incommensurability in the descriptions they provide for the kind of "modernity" experienced in both societies. Science posed a challenge to traditions (especially religious ones) and therefore became a bone of contention over which local élites fought and sought to extend their control. None of this is very surprising. But that similar processes took place concurrently in France and Thailand is rarely addressed. My aim is to use the eclipse as a revelator for the several overlapping power struggles in which astronomy played a role: the cosmic clash between the West and the Far East, and within society on both sides, fights over the right to speak for the heavens and the epistemological, religious, and political consequences of exerting this right.

Solar (and lunar) eclipses indeed provided dramatic stages for the display of knowledge and power. Like cannons, steamships, and colonial bureaucracies, European countries mobilized eclipse expeditions into the service of their imperial ambitions.[13] In Thailand the sudden disappearance of the God-Sun from the tropical sky was no doubt a terrifying sight for those taken by surprise. Tales of ignorant populations frightened by an unexpected, eerie nightfall were used to draw the line between rationality and superstition. In Europe the fight against unfounded fears seemed to have been won. As the *encyclopédistes* triumphantly wrote: "Today, everybody, philosophers as well as common people, know what cause eclipses." Almost a century later the director of the Paris Observatory, François Arago proudly reported that the solar eclipse on 8 July 1842 had been observed in southern France by "twenty thousand improvised astronomers."[14] In contrast, erasing their material dependence on local knowledge, nineteenth-century European witnesses of such phenomena often felt that accounts of overseas expeditions were an appropriate place to emphasize the constructed demarcation between them and the "others," between the civilized West and the savage Orient. The mastery of rational and scientific narratives both reinforced and legitimized imperial ambitions. "Civilization," an American magazine emphasized, hinged on knowledge: a citizen who did not know his rights and duties was little better than a "Feejee Islander" unable "to manage an eclipse of the sun without burning his fingers."[15] The "management" of eclipses—accurately predicting their occurrence long before and making a display of rational, professional coolness amid chaos—superbly showed European (and American) superiority.

As rhetorical strategies of European control over local populations, these simplistic accounts prove to be, unsurprisingly, much too monolithic. In the second half of the nineteenth century learned segments of nonwestern populations had a wide arsenal at their disposal in efforts to exploit solar eclipses to their own advantage. In the face of a European menace to their sovereignty, local élites thus adopted a strategy similar to that of their aggressors. To show the value and richness of their own knowledge traditions, they attempted to channel the symbolic power of eclipses in a manner more flexible than that of westerners. In syncretistic fashion they mustered the strength of both endogenous and occidental knowledge traditions. In their view solar eclipses were an ideal terrain for seducing Europeans into believing in *both* their ability to adapt to modern science *and* the value of traditional knowledge. Such demonstrations played a key role in the defense of Thailand's political independence.

The symbolic power of eclipses was at stake not only in the imperialistic collision between Europe and Asia, but also within each continent. Among Europeans, nationalistic concerns vis-à-vis their neighbors sometimes seemed to take precedence over scientific results when planning expeditions. In France, a deeply divided astronomical community haggled bitterly over who would be in charge of the eclipse expedition of 1868, while a small minority around the king of Siam recklessly tried to navigate between modernizers in favor of acculturation and imitation and traditionalists clinging to the past. When focusing on fights about who would be allowed to claim scientific mastery over eclipses, striking parallels emerge between East and West. In both cases educated élites exploited epistemological, cosmological, and religious debates as a way to appeal to and increase their constituencies. In this view, institutional brawls in France, nationalistic contests between European states, and court intrigues in Bangkok all seem to have been geared toward claiming symbolic capital from the eclipse—to be able to speak about science and modernity increased one's legitimacy.[16] Although these debates took place in vastly different contexts characterized by huge imbalances of power and resources, the processes by which French and Thai societies were "scientized" in the last part of the nineteenth century followed similar paths.

It will come as no surprise that the European modernization of society through greater reliance on science and technology provides a useful lens through which to examine similar processes in Thailand. But, as I will suggest, the confrontation of discrepant French and Siamese experiences can also help scholars to read French archival sources in a different light that will reveal some of the dynamics at play in the European metropolis.[17] The dichotomy between West and East, between the metropolis and the empire, henceforth appears less significant. In their own ways Siam and France followed similar paths toward political prestige, strengthened central power, rationalized administration, and state-controlled science. Ultimately it is the meaning of "modernization" that is questioned here.

The Beaches of Wako

As historians know well, no events can be dated more accurately than astronomical ones. Despite the variety of sources available, there is a wide consensus concerning what happened on the beaches of Wako on the morning of 18 August 1868. Viewed from Southeast Asia, there was no question about what the main purpose of astronomy was. European navigators and Thai officials were not the least bit troubled by the contemporary emergence of astrophysics. They all agreed that the principal task of astronomy was to provide accurate measurements of position and time. So it is highly significant that while most accounts of the expedition, whether Thai or European, agree on many points, they markedly differed on two: who was responsible for choosing the observation site, and who produced a better prediction of the time of the eclipse.

Some historians—and several modern guidebooks—prefer to locate the royal camp in the luscious Khao Sam Roi Yot ("Three Hundred Peaks") National Park, but most observers gathered fifty to sixty kilometers south of there on the barren beaches of Wako. In the invitation he sent to Governor Harry St. George Ord of Singapore, King Mongkut had been quite precise: they were to convene at East Greenwich 99° 42′ and latitude North 11° 39′.[18] According to the *Dynastic Chronicles*, the king himself had "calculated on charts that the actual point of the eclipse would be at eleven degrees, forty-one minutes, forty seconds north latitude, at a spot only fifty-one geographical minutes west of Bangkok, and where the time differential would only be three minutes and thirty seconds from Bangkok time." With respect to Greenwich, this translates (taking 100° 30′ for Bangkok) to 99° 41′. When they reached the spot on 25 July, the first task that the French astronomers set out to do was to measure their position. They quickly disposed of the latitude by determining it "very exactly" as 11° 42′ 35″ N. As for the longitude, lack of time forced them to leave a residual error of about 5 seconds, and they adopted the average of $6^h 29^m 50^s$, east of the Paris meridian, which (taking 2° 20′ 14″ as the longitude of Paris) is equivalent to 99° 47′ 44″ E. According to Governor Ord, the meeting took place "in Lat: 11° 38′ N. and Long: 99° 39′ East, almost at the foot of the Mountain Kow Luan 4,236 feet high."[19]

The agreement is not bad. But who had chosen to establish a camp there? In his official chronicle of Mongkut's reign, Prince Thiphakorawong was explicit. Two years earlier the king "himself had personally calculated the coming of the eclipse. He insisted that that eclipse would definitely take place."[20] A proclamation was issued. This was a serious matter: two hundred years earlier King Narai had been ridiculed by French Jesuits' prediction of the solar eclipse of 11 November 1659 in Bangkok. Now Mongkut would outdo the French savants, who, having independently "discovered" that an eclipse could be witnessed in Siam, were allowed to come and observe it from his territory. "The French scientists came and searched about many districts, . . . but they were unable to locate the precise venue of the sun's path. It was after the Chief Minister of Military Affairs had started the construction work for the King's temporary quarters . . . that the French scientists asked if they, too, might erect their own quarters there in order to witness the eclipse. Their quarters were constructed at a point eighteen *sên* south of the King's own pavilion."[21] The French, however, were just as explicit in denying any responsibility to the Thai for the choice. In a letter to the French Minister of Navy sent from Saigon on 29 May, Governor Ohier of Cochin China reported that the gunboat *Frelon* had been sent to inspect the eastern coast of the Gulf of Siam to determine the best spot for observation. Criteria that the captain was ordered to take into account included the possibility of building a temporary wharf on the coast, harboring several ships, and probable weather conditions. Jean-Jacques Hatt,

19. Photograph of the camp, taken by the members of the French expedition to Thailand. Notice Foucault's telescope in the middle. Courtesy of the Observatoire Astronomique de Marseille-Provence.

an engineer-hydrographer, selected a spot along the coast at a latitude first evaluated at 11° 36′ and later corrected with data sent from France to 11° 42′. In his letter Ohier added: "the King of Siam is apparently intent to come with his whole court in August to the spot we have chosen [*le lieu choisi par nous*]."[22]

The contradiction highlights the underlying political tension. The station was located on an almost deserted sand beach, bordered by a jungle infested with mosquitoes and tigers (figure 19). Mongkut had gone to great effort and expense to assert his dominion on this beach: "Out of the barren sands, almost overnight, sprang a noisy encampment, a straggling line of huts and buildings, spreading along the shore for a couple of miles." But pointing out that his control over this territory was but nominal, Stéphan's reports typically mentioned the possible commercial exploitation of the region, singling out its mineral riches, were it to fall into "more resolute hands."[23]

The Eclipse in Franco-Siamese Context

When Lieutenant Hatt was received by Mongkut in the spring of 1868, he was greeted with suspicion. The king said he could not understand why the French would not observe the eclipse from their own newly acquired territory, on the southernmost tip of modern Vietnam, a territory Hatt himself had just surveyed.[24] The eastern shore of the Malacca peninsula had been singled out by the director of the Paris Observatory, Urbain Le Verrier, some fifteen months earlier, because of the protection offered by the mountains against dominant westward winds and, during the monsoon, guarantees of better weather.

At the time Mongkut could not see without considerable worry French warships crisscrossing the gulf and surveying the coasts of his kingdom.[25] On 11 August 1863, less than five years before Hatt's visit, in a move that humiliated and angered the Siamese court, Cambodia had been forced to accept French "protection." Culturally close to Thailand, Cambodia was traditionally a vassal state placed under the personal protection of the king of Siam, who appointed its kings from among the Khmer princes educated in Bangkok. Thus any foreign intervention in Cambodia was perceived as direct aggression against Siam. The government in Bangkok started to worry. "Since we are now being constantly abused by the French because we will not allow ourselves to be placed under their domination like the Cambodian," Mongkut wrote, "it is for us to decide what we are going to do; whether to swim up-river to make friends with the crocodile or to swim out to sea and hang on to the whale . . ."[26] The whale was Britain, which Mongkut saw as the lesser of two evils. If a British protectorate was necessary for the defense of Siam's interests against the French, then he would accept it.[27] In 1865 Kalahom Sisuriyawong asked D. K. Mason, the Thai consul in London, whether "the European powers, say France and England—have a mutual understanding or secret treaty as to interfering or non-interference with each other in the event of either desiring to conquer or have political influence over any native states in the East."[28] He went on: "For centuries the English have been quietly exerting their power in the East, from Ceylon to Singapore. The Dutch can act as they please in Sumatra, the Spaniards in the Philippines, and the French in Cochin China. The latter appears to have been by mutual pre-arrangement set apart for French accession. Now in the case of Siam should the French desire to acquire political influence or be inclined to take possession of it, would the British Government interpose in the matter? . . . I have reasons to think that this Kingdom, like Cochin China, has been set apart for French ambition." Writing to his ambassador in Paris, Mongkut confirmed this fear: "As regards the French, they are distinguished for their vainglorious disposition. Their Emperor, famed for his descent from a line of tigers and cobras, would, after his ascent to the Throne, seek colonies that are rich and vast. . . . These lands between Annam and Burma must appear to him to be ownerless and therefore desirable."[29] Even the French sometimes agreed with him. In 1869 Count de Beauvoir wrote that he believed the old Siamese kingdom to be a very "tempting *gâteau*" served between France and England.[30]

A few years before Hatt's visit, the French consul Gabriel Aubaret, who had gone to negotiate a treaty with Siam, noted: "The sight of a warship always confer great weight to the words of an agent."[31] In 1868 Mongkut once again bowed to the power of gunboat diplomacy and declared himself enthusiastic about the French expedition. He would provide it with temporary observatories and lodging at a total cost estimated at $100,000. But he made sure that British men-of-war would be present too. According to the Foreign Office archives, Constance Wilson reports, it was at the suggestion of the British consul in Bangkok that Mongkut mounted his own show at Wako, "ostensibly to view the eclipse." Having invited Sir Harry Ord to join him, Mongkut hoped that "the presence of Thai and British warships [would] show their unity in the face of possible French intrigue."[32] Altogether, eighteen steamships were anchored close to Wako in August 1868, between which elaborate gun salutes were exchanged. Stéphan was properly aware of the political nature of Mongkut's voyage, if not of his own: "The king . . . wanted to see by himself our true intentions and avoid the surprise of a devious seizure of his land." But he granted that the king's love for astronomy was reason enough to explain his trip.[33]

Mongkut knew that he could hardly count on British protection alone to safeguard Siam's autonomy. But direct confrontation with the French was also not an option. Even if the country could afford one hundred warships, Mongkut wrote in 1866, "we would still be unable to fight against them, because we would have to buy those very same warships and all the armaments from their countries. . . . The only weapons that will be of real use to us in the future will be our mouths and our hearts."[34] According to conventional historiography, the Siamese responded to the western threat with anxiety. But as Thongchai has argued, Thai "desire" also played an important part. A portion of the Siamese élite was truly attracted by the farangs' culture and sought to incorporate many of its aspects into their own: they became "*siwilai*" (civilized).[35]

But seduction, the flip side of desire, was present too. While the Thai were no doubt fascinated by western technology, Mongkut sought to reverse the flow in this politics of seduction. On a diplomatic level, he increased contacts with European and American heads of states and saw to it

20–21. This portrait of King Mongkut by F. Chit, photographer to His Majesty, was brought back by the members of the French expedition to Thailand in 1868. Courtesy of the Observatoire Astronomique de Marseille-Provence.

that Siam played a conspicuous part in the Paris Exhibition of 1867. There was also a more personal touch to this policy. The king, according to an anonymous foreign observer, longed "to be esteemed by foreigners abroad as one of the best men that ever drew the breath of life." The writings of Anna Leonowens, the Englishwoman who famously told of her experience as his children's governess in the royal palace in Bangkok, belabored Mongkut's lust for flattery.[36]

For seducing the West, king Mongkut was an unlikely candidate (figures 20–21). Meeting him in 1867, Count de Beauvoir found that "His Siamese Majesty . . . is perfectly ugly and looks like a monkey [*tient beaucoup du singe*]."[37] Having suffered partial facial paralysis in his late twenties or early thirties, he grinned constantly. His lower teeth had been replaced by a set made of deep-red Sapan wood. In the words of an American delegate, the king was, in the spring of 1868, "about as unprepossessing in appearance as can be imagined. His eyes were nearly closed, and he had a sort of sleepy look and drawling voice, which did not at all accord with the words he uttered. He was constantly chewing betel, . . . and the juice ran down his chin, rendering his whole appearance almost repulsive."[38] Nonetheless Mongkut inspired many novels, musicals, and movies—and was particularly successful in captivating imaginations in the West. After his death American periodicals boasted of having had him as a reader: "His late highness was a regular reader of the *Scientific American*, and it seems to us very likely that he learned more from its columns about forts, steamboats, railways, canals, and photography, than from the [*Evening*] *Post*."[39]

In the politics of seduction, science was one of the two supreme value systems of the West—the other being the Christian faith—and the only one that the Siamese could accept wholeheartedly. Mongkut organized exchanges of animals between Siam and France. The steamship *Gironde*, which the Siamese ambassador took to France, also brought animals for "French Zoographers."[40] The arrival of a lion skin and a stuffed lion from France mentioned in the *Dynastic Chronicles* shows the importance attributed to the gift. The "conquest" of the West was vital for the survival of Siam.[41] In this war of seduction Mongkut found a way to use his pet science—astronomy—as a weapon against the West.

The Many Uses of Astronomy

In 1851 Mongkut was crowned at forty-seven, after having spent more than twenty years in a Buddhist monastery. The eldest son of King Rama II, he had been passed over when his father died, and "preferring a crownless head to a headless body," had retired from the court.[42] Isolated from his princely milieu, he learned English and befriended French Catholic and American Protestant missionaries, who, Leonowens wrote, "helped to unlock to him the secrets of European vigor and advancement."[43] His eagerness to learn endeared him to foreigners. This much-praised receptivity, as Wilson has written, should be interpreted "as evidence of Thai initiative in seeking the information they wanted. They were eager to learn about mechanics, science, and medicine, the benefits of which were immediately observable, but they ignored or rejected those things for which they felt no need."[44] The Siamese élite were extremely discriminate in what they sought from western values. If they showed great interest in science and technology, they were much more skeptical toward religion, politics, art, and social mores.

In 1857 the British hosts of a Siamese delegation noted that "they seemed to have no taste for architecture, paintings, or sculpture but they . . . took great interest in scientific instruments of all kinds making many purchases of chronometers and scientific instruments and arms from the best makers and of the latest improvements."[45] As for religion, some missionaries had hoped that science could draw the Siamese to the gospel: "Under the starry mantle of astronomy, our holy religion is easily introduced among princes and province governors."[46] Others, however, recognized that while Mongkut welcomed their company, he had "no other object as he says than to acquire the English language and get hold of foreign science."[47]

Why did Mongkut, then, show so much interest in a hobby such as astronomy? There are three parts to the answer. First, astronomy was practically useful, especially as far as geodesy and mapping were concerned. When the French started to map the Mekong region, for example, it seemed unwise to Mongkut not to do the same. The Thai government hired an Englishmen to do the job, Mr. "*doojchók*."[48] On several other occasions, while visiting various part of his country Mongkut measured their longitude and latitude himself. During those trips, one reads again and again in the *Chronicles* that "the king made scientific measurements of the sun" and "of the stars." With geodesy, he wished to keep control over Siam's inscription on European maps.[49] Fortifying the capital city on the sea side, Mongkut had his "Holy Fortress of the Hill"—whose eclectic architecture encapsulated his policies, mixing Thai, western, and Chinese influences—built overlooking the town of Petchaburi and the mouth of the Chao Phraya River. From there he could observe the stars from his observatory tower while his ministers kept an eye on the naval traffic in the gulf.[50] Telescopes have many uses.

But Mongkut always had more success convincing the British than the French. If his goal in showcasing the eclipse had solely been to make a strong impression on the French scientists, one can only say that he failed miserably. For Stéphan, this encounter on the beach of the gulf of Siam symbolized the clash of two worlds. A navy officer in his party from the École Polytechnique, Lieutenant Pierre Chabirand, wrote that "the king's visit was one of the most curious events of our station on the west coast of the gulf of Siam." Going about their serious business, installing and calibrating their delicate equipment, and measuring their position, the French impatiently listened while the king, with a touching naïveté, displayed his astronomic erudition. After this meeting Mongkut seemed unpleased with the impression he had given the Frenchmen and sent a handwritten "astronomical statement" for clarification.[51] In an embarrassing confession, he explained that he had acquired his astronomical knowledge first, from Siamese sources and the Hindu treatise *Sûrya-Siddhânta*, before he "tested certain european books of astronomy and astronomical navigation geometry," enabling him to "understand better manner of astronomical calculation." But the *Sûrya-Siddhânta*, according to French eyes, was "not aimed at intelligence." "It is a collection of numerical rules, true codes indeed, that the disciple will have to learn by heart, verse after verse, and unperturbedly fix in his memory, so has to recite all the words in order, without letting any pass by. Then, by executing the arithmetic operations prescribed by each verse, he will mechanically, without mental work, be able to determine for any given time the apparent positions of the moon, the sun, and the five planets, to predict the moments of lunar and solar eclipses." Even so, the astronomer Jean-Baptiste Biot added, computing solar eclipses with the methods of the *Sûrya-Siddhânta* led to erroneous results: "As a scientific work, it is an incoherent assemblage of true or false, exact or inexact astronomic data, combined through generally imprecise numerical rules," in which observation played no part.[52]

Mongkut had no need to worry: his personal competence as an astronomer was, as far as the French were concerned, beside the point. Even if unimpressed by the king's knowledge of astronomy, westerners could not help being struck by the interest that an Asian despot showed in the most advanced of their sciences. To take part in the "scientific movement" was bound to attract attention. And eclipse expeditions had traditionally been a good way for monarchs to give the impression of contributing to western science. In 1858, for example, the emperor of Brazil, Don Pedro II, himself fond of astronomy, made sure that several parties were sent to observe an eclipse traversing the northern part of his country. In 1860, while most European astronomers converged in Spain, the viceroy of Egypt mandated his astronomer, Mahmoud-Bey, who had been trained in Paris, to observe the eclipse in Nubia. Even if no scientific data from the expedition were ever to be used in Europe, Egypt's participation in the "European scientific concert" was applauded by everyone, from members of the Academy of Sciences in Paris to trade journalists in the United States: "by European influences, the Khedive of Egypt, the King of Siam, and other semi-barbarous potentates have been induced to erect astronomical observatories and appoint salaried astronomers, among whom are many natives. . . . It is indeed a cheering sight to see the beginning of the *conquest*, which peaceful science is bound to make over the whole world, independent of nationalities, creeds, or races."[53] "Conquest" was precisely what Mongkut sought to avert even while he recognized the need to adapt to western science. But how was he to do this without selling out completely?

The third use that Mongkut made of astronomy was thus ideological. After the eclipse the king was by all accounts ecstatic. To his guests he said that the great noise made during the eclipse, the frantic beating of drums and firing of guns, was not an attempt to frighten the dragon Rahu but rather an expression of joy caused by their king's ability to predict the exact time of the event (figure 22). The court astrologers were the main losers in this event. They thought it impossible that the sun could be eclipsed by the moon and barely refrained from expressing their skepticism in the face of Mongkut's conviction. Returning to Bangkok after the trip, the king justifiably humiliated and punished them.[54]

According to many people present at Wako, the king boasted that his calculations were more precise than those of the French astronomers by two seconds. As we have seen, Mongkut's more accurate prediction was a second, important discrepancy between the Siamese and French accounts of the events of Wako. For French astronomers, however, a two-second difference would have fallen within an acceptable margin of error, since they had lacked the time to properly ascertain their longitude. As Chabirand wrote, to determine the longitude within a precision of one second was a very delicate operation requiring at least forty meridian observations of the moon: in the most favorable conditions, that would take roughly three months.[55]

Be that as it may, it is significant that this king's feat was widely reported. While his extraordinary precision was sure to impress amateurs, like Sir Harry who commented on it, Mongkut probably knew that it could be exploited on the domestic front. Indeed, the audience he had in mind when staging the expedition was most likely not the farangs but his own subjects. During the eclipse the prime minister "left his long telescope swinging on its axis and walked into the pavilion and addressed several of his wives, saying 'Will you now believe the foreigners?'"[56] As this remark shows, the eclipse was part of the king's efforts to make his people embrace western science. In the Bowonniwet temple in Bangkok, murals from this time commemorate the king's friendly relations with westerners.[57]

But Mongkut's attitude toward western science was hardly one of unproblematic, pragmatic acceptance. His emphasis on having to rely on both western and Asian astronomy books to compute the path of the eclipse is significant. His astronomical syncretism was intended not to show the West that traditional knowledge could compare to modern science as much as

22. "Savages" shooting at the moon during a lunar eclipse in Attopeu. Eugène Burnand, *Le tour du monde: nouveau journal des voyages* 38 (1879): 23.

to persuade his countrymen that they could, in Wilson's words, use Siam's "pre-modern social and political order as a weapon against the intrusion of the West." Mongkut's political struggle with his court was directed both against the excessive "occidentalophobia" (Bhumichitr) of his astrologers and the modernists' acceptance of the pure and simple domination of foreign over Siamese culture. In a fascinating article, Cook has emphasized the "paradox" according to which the monarch can be celebrated today as both the founder of Thai science and a reformer of Thai astrology. She has argued that Mongkut's experience with astronomy provided "justifications . . . both for the emergence of Siam into the modern world, and the retention of Siamese tradition without ceding defeat to competitive and hostile traditions."[58] To emphasize this point, one may turn to Mongkut's younger brother 'Itsaret, who held the honorific but powerless title of "second king of Siam." The fascination that Mongkut exerted on westerners often paled in comparison to the success enjoyed by 'Itsaret. By the time of the eclipse in 1868, he had been dead for two years, but his reputation reached Stéphan, who believed that he had devoted his life to the study of the heavens. But what foreigners admired most about the second king was his pragmatic take on science and technology. In their view, his conversation had none of Mongkut's quaintness: he "would chat with you (if you knew enough to keep up your end of the conversation) about the latest improvements in firearms or the most recent scientific discovery."[59] Moreover, he put his knowledge to good use. On 4 July 1848, for example, Dan Bradley noted in his diary: "This A.M., we saw something new on the river—a little model steamboat, not twenty feet long, with smoke-pipe, paddlewheel, all complete, steaming bravely against the tide with H.R.H. Chao Fa Noi [Prince 'Itsaret] sitting at the helm. It was the first native steamer on the Meinam [Chao Phraya River], entirely his own construction."[60]

'Itsaret's palace, "rebuilt after the model of an English nobleman's residence," held a library "more judiciously selected than that of his brother, [which] abounds in works of science, embracing the latest discoveries." He had a little cottage with glass windows and English furniture. Above the entry hung a sign reading: "Watches and Clocks Made and Repaired Here."[61] He named his second son George Washington.

Western science and technology had many faces. The two royal brothers acquired their familiarity with it through rather different channels. While Mongkut was still in the monastery, 'Itsaret was put in charge of rebuilding ancient fortifications. Through a crew of European engineers and artisans he had hired, he became acquainted with navigation, naval construction and armament, coastal and inland defense, engineering, transportation, telegraphy, and the working and casting of iron. Working with engineers, 'Itsaret was exposed to a different kind of science from the one Mongkut discovered while debating cosmologies and moralities with missionaries. Even while adopting western ways of measuring time and space, Mongkut understood the ideological use of science and of astronomy especially. In 1867 an essential book that endorsed western science and technology was published in Bangkok. It was called *Nagsu sadaeng kitchnanukit* (A book on diverse matters), and its author, a former minister of foreign affairs, was Prince Thiphakorawong, half-brother to Prime Minister Sisuriyawong and a close associate of King Mongkut. This major text, the first ever published by the Siamese without foreign assistance, used the observatory sciences (astronomy, geography, and meteorology) as linchpins for displacing traditional cosmological beliefs, which were cast off as mere superstition. In this hybrid text modern (western) astronomy became a way to reaffirm Buddhism. Western science and technology could be interwoven with old mental structures, to salvage their essential core.[62]

Achieving this required a reinvention of tradition. As a monk Mongkut had reformed Siamese Buddhism and founded a new sect, the Thammayut, which means "those who follow the law."[63] A Pali and Sanskrit scholar, Mongkut wished to rid Buddhism of Brahmanistic superstitions by recovering ancient texts and rituals. Recently this move has been interpreted either as a reaction to Christian attacks on Buddhism or as a way to explain clearly the nature of Buddhism to critical farangs.[64] Buttressing local religious traditions was for Mongkut just as important for the survival of his nation as the embracing of western technology. He made use of western technology—the printing press—and of the ideological appeal to reason to do this. While Wilson has argued that Mongkut found in tradition a resource to help his country resist the West, Bhumichitr has followed the Thai art historian Piriyah Krairiksh's suggestion that Mongkut actually invented parts of the very tradition upon which he leaned. For example, with his emphasis on free trade Mongkut constructed an idealized past that bore close resemblance to the nineteenth century Occident. In a way thereby directly inspired by European nationalism, he became one of the first theoreticians of Siamese nationalist ideology.[65]

In this light the eclipse was indeed "an event symbolic of [an] epistemological struggle."[66] But it also played an important role in the more straightforward power clashes. What was the nature of the tensions gripping the oligarchy at the time of the eclipse? Suspicion and jealousy had already soured relations between Mongkut and his younger brother, the second king.[67] But after 'Itsaret's death in 1866 a realignment in the power structure took place. In opposition to the Kalahom Sisuriyawong, Mongkut, whose health appears to have been frail even before he left for Wako, wished for his older son to succeed him, but he was only fourteen years old. During the journey the etiquette of the court slackened: "On no previous occasion had the Court been so completely revolutionized; the royal apartments were thrown open, and the ladies of the household brought prominently forward, whilst the younger members of the royal family were allowed to mix with their English visitors in the most friendly and sociable manner."[68] In this way Mongkut took the opportunity to allow Chulalongkorn to meet foreigners, participate in the workings of government, and visit French ships.

The king died on 1 October from a fever he had contracted in Wako, where "myriads of insect [had begun] to *zizzee*, as the Siamese onomatopoetically express it."[69] Although seriously ill himself, the young Chulalongkorn was appointed king, with Sisuriyawong and Thiphakorawong as his closest advisors. During his forty-two-year reign Chulalongkorn vigorously pursued his father's policy of "modernization." His main rival, the older George Washington, son of the late second king, had to be content with assuming his father's position as second king. Often taken as a sign that astrologers were right after all in seeing the eclipse as a bad omen, the trip to Wako, by securing his son's appointment, was indeed Mongkut's last political victory of enduring consequence.

Mongkut's eclipse expedition therefore has to be read as part of several overlapping political conflicts. Inside the Bangkok court the epistemological and religious implications of the eclipse became crucial chips in more mundane power struggles. In the cosmic confrontation between Southeast Asia and Western Europe, the eclipse functioned as a test case for precise mastery over astronomy. From East to West, exhibiting one's control over the event determined who was allowed to speak for the heavens, and ultimately for science, civilization, and modernity.

The Reasons of the Eclipse

"Transition periods offer a curious spectacle: . . . all is mingled, the fantastic and the real, the positive mind and supernaturalism, skepticism and superstition, imagination breeding chimeras and reason destroying them."[70] This comment was not directed at Prince Thiphakorawong's *Kitchnanukit*, in which modern observatory sciences sided with superstition to uphold Buddhism and polygamy, but at Camille Flammarion's *Pluralité des mondes habités*, first published in 1862. To outline the French cultural and political landscape that made possible the expedition of 1868, our detour through Siam will prove instructive. For the French also, the eclipse provided a major stake in various fights about science and modernity that involved epistemology, religion, and politics.

In 1868 it was by no means preordained that France should send astronomers on a journey of twenty thousand kilometers just to observe an eclipse six minutes long. The minister of education Victor Duruy had already petitioned Emperor Napoléon III for funds, but the director of the Paris Observatory, Urbain Le Verrier, started to fear that no one among the astronomers would agree to go to Siam. As late as 14 March 1868 the naval officer Ernest Fauque de Jonquières wrote: "I still believe that M. Hatt, the engineer who is the head of the astronomical and hydrographic service in Cochin China, can make useful observations of the eclipse . . . especially if a pocket or portable spectroscope were made available to him. To make good astronomical determinations, he already has the necessary instruments in his possession and the experience of this kind of observation. . . . Due to a lack of suitable instruments, he will admittedly not be able to . . . make photographic reproductions; but if he is, as I imagine, gifted in drawing, he will be able to supply images almost as good and exact [as them]."[71] At first Le Verrier had had no firmer conviction as to who should be sent to observe the eclipse: "Either we decide especially to send some professional astronomer, or the Navy would consent to ask some able officers from the station in Saigon to take charge of this work." Similarly, the astronomers Hervé Faye and Ernest Laugier, of the Bureau of Longitude, issued eight pages of instructions for naval officers who might be in the region at the time of the eclipse.[72] They detailed procedures to be respected: circles were to be drawn in advance on cardboard, around which the position of the main prominences would be sketched; and each time the observer's name, his geographic position, the instrument used, the mean time of the observation, and above all the zenith of the disk should be noted on the sketch, so that a composite painting could be drawn afterward. Lastly, the astronomers mentioned that a hand-spectroscope could be used to observe the areole around the dark lunar disk. "One will observe one of the three following phenomena: Either the spectrum of the areole will remain identical, except for its intensity, to the spectrum of the corresponding region of the sky before the eclipses, that is, to the ordinary spectrum of the Sun; either to that spectrum and another spectrum composed of a few bright and colored lines will be superposed; or the ordinary spectrum will be totally inverted, bright lines taking the place of the black lines of solar origin." Captain Rapatel and M. de Créty, both from the company Messageries Impériales, dutifully gave picturesque descriptions of the eclipse—which were never used by astronomers.[73] In Britain similar instructions, as well as pocket spectroscopes, were sent to the managing directors of the Peninsular and Oriental Steam Navigation Company, who dispatched them abroad—to little avail: Captain Rennoldson's comment, "I return the spectroscope, and am only sorry I could not make more use of it," was a rather typical response.[74] As Faye and Laugier themselves noted, the correct use of such an instrument was far from obvious: "To take advantage of this small instrument, one must familiarize oneself with the details of the spectrum that it produces when turned toward different regions of the sky during the day."[75]

By the start of 1868 the political situation in the astronomical community was explosive: acrimonious debates had erupted over such issues as a possible move of the Paris Observatory to the suburbs, the role of meteorology in the institution, Le Verrier's management of the observatory, Charles-Eugène Delaunay's lunar theory, and the relative responsibilities of the observatory and the Bureau des Longitudes.[76] On 3 February 1868, putting aside all his concerns about the future of the observatory, Le Verrier exposed his views about the coming eclipse to the Academy of Sciences. Nothing had been decided, he said, which is why he brought the issue to the table. One rather suspects that he was by then well aware that the Bureau des Longitudes had rival plans to send Janssen to India with the British and that Le Verrier's intervention was aimed at opening a new front in the debate. What better way to deflect the attacks of the committee that at the time was working on reforming the observatory's administrative structure? In this context the eclipse became, in Paris as much as in Bangkok, a prized trophy with which no one could dispense. While Faye argued that the mere hope of seeing a reverse spectrum in the prominences was ground enough to legitimate all the efforts, Le Verrier injected practical and political issues into the debate. The expedition that he wished to organize should duplicate Janssen's physical mission to "enjoy the benefit of controlled observations" by multiplying observers and stations.[77] That the shadow path should cross a French territory was reason enough, Le Verrier said, to observe the phenomenon in a neighboring country.

In the 1860s "science" was a force that everyone was trying to co-opt. Although he never traveled to see any solar eclipses, Napoléon III made sure to pay formal visits to the laboratories of some of the most prestigious savants in the capital. No one could afford to be seen as neglecting science. In February 1868 an Italian daily noted: "You will see that they will not reach the eclipse of 17 August [local European time]. The Bureau des Longitudes is sending Mr. Jansen in Cochin China with a letter of recommendation for the authorities. What a difference with the magnificent expedition organized by England. And it is from a French territory that one will be better able to observe this phenomenon without precedent. That's the state of science in France under Napoleon III."[78] When Minister Duruy read the article he penciled in red that however strained his budget might be, Janssen's mission had to take place. On 13 April 1868 at the Academy of Sciences, Le Verrier argued that sending Janssen to observe the eclipse in a place where there were so many more British was a waste of precious resources. But the minister had already decided to grant Janssen the mission.[79] Meanwhile the observatory had been reorganized and Le Verrier's position bolstered, so he decided on his own to send an official observatory mission to Siam: it would be composed of three astronomers and as many trained naval officers and would be allotted twice as much money as Janssen's. Altogether France would spend more than 75,000 francs for these two expeditions (that is, about half the annual budget of the observatory), without taking into account the 30 percent discount offered by the Messageries Impériales and the considerable expense of the navy and the colonial authorities in Saigon.

Engaged in a "scientific struggle" against other European nations, France never felt that it was competing with Siam on this level.[80] Nationalism and colonialism played a large part in mounting the expeditions, but it was the infighting among French astronomers that insured the manning of both expeditions by top-notched, if still junior, astronomers. Despite French fears, British expeditions were on the contrary staffed by officers from the navy or the Geodetic Survey, most notably Lieutenant John Herschell, who had little spare time to prepare for the special operations required for the spectroscopic observation of the eclipse. While the Dutch and the Papal States canceled their planned expeditions at the last minute, and the Germans and British mounted only halfhearted enterprises, France sent its best-trained scientists in astrospectroscopy—who performed in accordance with expectations.

The eclipse of 1868 was a key episode in the rise of direct state sponsorship of science. The French government later funded many new astronomical institutions at home. Expedition leaders in 1868 played a large part in overseeing the reorganization of metropolitan science: Stéphan's observatory in Marseilles soon became autonomous from Paris, while an Observatory for Physical Astronomy was created for Janssen in Meudon. After the Franco-Prussian war of 1870–71 the reorganization of astronomy was central to transforming the institutional landscape of French science.[81] Surrounding the power struggles about who could speak for astronomy, institutional reforms and the image of science were debated. But these considerations alone cannot fully explain why the Emperor's Council, the Legislative Chamber, and the Senate deemed eclipse expeditions important enough to be richly endowed.

The Eclipse of Reason?

"The State must support science just as [it supports] religion," Ernest Renan stated, for science, like religion, ultimately dealt with human nature.[82] But in the case of astronomy, science and religion intermingled just as tightly in the French political and cultural landscape as they did in Thailand. In the 1860s the cultural signification of astronomy has important religious and cosmographic implications in both places. Religion was serious business under Napoléon III. After the coup d'état of 1851 the church had been one of the strongest supporters of the Imperial Regime. Throughout the 1860s only the French contingent in Rome allowed the Pope to resist Italian unification. Then in 1858 a young shepherdess named Bernadette Soubirou reported an apparition of the Virgin Mary in Lourdes, an event that transformed the town into a national shrine and ushered French Catholicism into a thoroughly modern enterprise.[83] The fight against superstition was not over even in "our luminous Europe," as Flammarion reported. For example, during the partial solar eclipse of 6 March 1867 nuns in charge of a girl's school made them fall on their knees and pray hoping to forestall the malediction of the "Très-Haut."[84] From this context considerable interest for religious matters—and debates about them—ensued. As had become customary since the beginning of the century, science was one of the main tools used by liberal opponents of the French Catholic Church. But while acknowledging the nineteenth-century fascination with religion, historians have traditionally "minimize[d] the relevance of an obsessive religious quest"; by and large historians of nineteenth-century French science have been blinded by the later success of Comte's positivism.[85]

The relationship between science and religion has often been tackled from the point of view of what Geoffroy Cantor has called the "conflict thesis."[86] As in Britain, the "warfare" between science and religion was portrayed in France as a clash between two systems: "Jesuit system: the authority, the director, the Pope substituted to reason, to God" versus "progress and unity."[87] This perception is historically situated but not generally tenable. Instead complex and subtle interactions have been exhibited. For Faraday "the absolute authority of the Bible could not be undermined by science, but science, if practiced in a truly Christian way, can illuminate God's other book [Nature]."[88] Likewise, major astronomers did not hide their belief in God. On 5 June 1876 Le Verrier presented to the Academy his completed tables for Jupiter, the result of thirty-five years of work. At the same time he declared that he had relied on the spectacle of the greatest works of Creation and the thought that it reaffirmed the truths of spiritualist philosophy. In his major review of western cosmogonies, Faye was at pains to warn against the danger of seeing "the ancient demonstration of God's existence sink under the daring scaffoldings of our cosmogonic attempts."[89] As for Janssen, his private correspondence plainly shows that although he might have sometimes mistrusted the Catholic hierarchy, he was nevertheless a devout believer.

Under the autocratic Second Empire, some French liberal journalists saw the popularization of science as a way to circumvent censorship.[90] As the regime liberalized in the 1860s, religion would become one of many realms in which the consequences of censorship could be more or less freely debated.[91] Catholics themselves were bitterly divided between "Gallican" partisans of a greater autonomy vis-à-vis Rome and "ultramontains," for the pope's authority was supreme. In 1863 Ernest Renan published his notorious *Vie de Jésus*, which submitted the existence of Christ to historical, archeological, and philological scrutiny.[92] Already in *The Future of Science*, written in 1848, Renan explained that his love of science had destroyed his faith in revelation. Whereas in Britain the issues focused on technical innovation, natural history, and economics, the French debated cultural studies, mathematics, physical sciences, and astronomy. Debates raged in the *Revue des deux mondes*, for example, one of the main cultural journals of the empire, which over the decade published dozens of articles by the most influential intellectuals (Renan, Edgar Quinet, François Guizot, Paul Janet, Étienne Vacherot, Joseph Gratry, et al.). An accumulation of scientific topics rushed in to question the foundations of the Christian faith: Darwinism, spontaneous generation, geology, the discovery of prehistoric human fossils—and, I would like to emphasize here—spectrum analysis.[93]

Around 1862 Flammarion, still employed as a computer by the Paris

Observatory, made the first use of spectrum analysis in discussing metaphysical issues. Constituted of the same elements as the earth, the sun and stars possessed "elements that have given birth to organized beings."[94] From there an assertion that they really existed was a step merrily taken by Flammarion. He wished astronomy to become "philosophy's compass," the basis for a new "religion through science." From variously enthusiastic to downright hostile, the reception of the book was tremendous and launched its author's long career as one of the most successful writers to popularize astronomy. Another famous science advocate, Louis Figuier, also relied on spectrum analysis to conjecture that the sun's continuous outpouring of energy was due to the constant inflow of human souls. While in prison, the revolutionary activist Auguste Blanqui used spectrum analysis to argue that all possible fates actually happen—somewhere.[95]

In 1896 decades of cosmic debate about the relationship between science and religion seemed to be reaching a resolution: "Scientific and philosophical knowledge being always bounded, there is always beyond it a sphere open to all beliefs . . . which is the foundation of religious faith. . . . But faith must never rebel against science or put itself in contradiction with it."[96] Under the assault of scientific modernity, religion restrained some of its ambitions but preserved its autonomy. Similarly, in the Franco-Siamese conflict of 1893, whereby France took possession of Laos, King Chulalongkorn was forced to recognize the limits of his country's independence in order to preserve it.[97] Meanwhile, speculation about life on other planets has remained—to this very day—one of the strongest incentives to promote astrophysical research lavishly sponsored by governments.

Notes

1. Fontenelle, *Entretiens sur la pluralité des mondes* (1686), second soir; my translation.
2. "Asians are taken for more naïve than they are." Mérimée to Jenny Dacquin, 29 June 1861, commenting on the Siamese ambassadors' visit to Napoleon III; repr. in Mérimée, *Correspondance générale*, 10:315.
3. Thiphakorawong, *The Dynastic Chronicles*, 2:538.
4. Olga Lingberg, "King's Mongkut's Solar Eclipse," *Astronomy* 13, no. 1 (1985): 24–26; and Yvon Georgelin and Simone Arzano, "L'éclipse de soleil du 18 août 1868: Stéphan et Rayet, hôtes du roi de Siam à Wha-Tonne," *Astronomie* 113 (1999): 12–17. On Rahu see Alabaster, *The Wheel of the Law*, 217–18.
5. Among other works see Auguste Laugel, "Découvertes récentes dans le soleil," *Revue des deux mondes*, 1869, 585–602; Ferdinand Hoefer, *Histoire de l'astronomie*

depuis ses origines jusqu'à nos jours (Paris: Hachette, 1873), 544–45; Ernest Lebon, *Histoire abrégée de l'astronomie* (Paris: Gauthier-Vilars, 1899), 141–43; Clerke, *A Popular History of Astronomy*, 4th edn, 167–70; Mitchell, *Eclipses of the Sun*, 136–39; and Mark Littman and Ken Wilcox, *Totality: Eclipses of the Sun* (Honolulu: University of Hawaii Press, 1991), 66–74.

6. Lockyer, "The Story of Helium," *Nature* 53 (1896): 319–22, 342–46; W. H. Keesom, *Helium* (Amsterdam: Elsevier, 1942); Clifford W. Seibel, *Helium: Child of the Sun* (Lawrence: University Press of Kansas, 1968); and Schaffer, "Where Experiments End."

7. Besides references cited above see Seymour L. Chapin, "P. J. C. Janssen and the Advent of the Spectroscope into Astronomical Prominence," *Griffith Observer* 48 (July 1984): 2–15; and Aubin, "La métamorphose des éclipses de soleil."

8. Janssen to the Minister of Public Instruction, 15 February 1869. Archives Nationales, F17 2977. Partly repr. in *Archives de missions scientifiques et littéraires*, 2nd ser., vol. 5 (1868): 615. Eclipse predictions by Fred Espenak, NASA/GSFC.

9. Hufbauer, *Exploring the Sun*, 65. See also Pang, *Empire and the Sun*.

10. On Thailand's intellectual response to western challenges see Bhumichitr, *Phra Chomklao, roi du Siam*.

11. Thongchai, *Siam Mapped*, and Cook, "A Tale of Two City Pillars." See also David Turnbull, "Travelling Knowledge: Narratives, Assemblage, and Encounters," *Instruments, Travel and Science*, ed. Bourguet, Licoppe, and Sibum, 273–94.

12. Wilson, "State and Society in the Reign of Mongkut," 144–45; Thongchai, "The Quest for 'Siwilai,'" 533–34.

13. See Pyenson, *Civilizing Mission*.

14. "Eclipse," *L'Encyclopédie*, ed. d'Alembert and Diderot (1755), 5:293–98; F. Arago, *Astronomie populaire*, 3:583. On the eclipse of 1842 see also Theresa Levitt's contribution to this volume.

15. Anonymous, "Civilization," *Manufacturer and Builder* 1 (1869): 277.

16. On symbolic capital see Pierre Bourdieu, *Distinction: A Social Critique of the Judgment of Taste*, trans. Richard Nice (Cambridge: Harvard University Press, 1984).

17. On the juxtaposition of European and nonwestern sources and "discrepant experiences" see Said, *Culture and Imperialism*, esp. 37–40; and Dening, *The Death of William Gooch*.

18. Mongkut to Ord, signed by Henry Allabaster, dated 8 July 1868, repr. in Smith, *A Physician at the Court of Siam*, 47–48.

19. Thiphakorawong, *The Dynastic Chronicles*, 2:532; Stéphan, "Voyage de la commission française"; Chabirand, "Étude sur l'éclipse totale de soleil du 18 août 1868," 378, 381–82; Ord, "An Account of the Visit to the Late King of Siam at Hua Wan in August 1868," 118.

20. Thiphakorawong, *The Dynastic Chronicles*, 2:532.

21. Thiphakorawong, *The Dynastic Chronicles*, 2:533. Stéphan wrote that both camps were about one mile apart. Stéphan, "Voyage de la commission française," 124.

22. Ohier to the Minister of Navy, 29 May 1868. Copy in Archives Nationales, F17 3008, folder "Stéphan et Rayet (éclipse de 1868)," file "Ministère de la Marine."

23. Smith, *A Physician at the Court*, 46; Stéphan, "Voyage de la commission française," 123–25.

24. Ohier to Ministre de la Marine, 29 May 1868. See also *Cosmos*, 3rd ser., 3 (1868): 169–70.

25. See Wilson, "State and Society in the Reign of Mongkut," 497ff.; and Bhumichitr, *Phra Chomklao, roi du Siam*, 280ff. See also R. Stanley Thomson, "Siam and France, 1863–1870," *Far Eastern Review* 5 (1945): 28–46; Lawrence Palmer Briggs, "Aubaret and the Treaty of July 15, 1867 Between France and Siam," *Far Eastern Quarterly* 6 (1947): 122–38; and "The Aubaret versus Bradley Case at Bangkok, 1866–67," *Far Eastern Quarterly* 6 (1947): 262–82; and Wasana, "L'emprise thaïe sur le Cambodge."

26. Quoted in Moffat, *Mongkut, the King of Siam*, 124.

27. Mongkut to his consul in Britain, 18 December 1866; Bhumichitr, *Phra Chomklao, roi du Siam*, 348.

28. Both the previous and following quotes are from Wilson, "State and Society in the Reign of Mongkut," 392. Often translated as prime minister, the Kalahom was the most important post in the Siamese government.

29. Translation of letter to Phraya Suriyawongse Vayavadhana, Siamese ambassador to Paris, 4 March 1867, quoted in Moffat, *Mongkut, the King of Siam*, 122.

30. Beauvoir, *Java, Siam, Canton*, 306, 336.

31. Aubaret to Drouyn de Lhuys, minister of foreign affairs, 15 April 1865; quoted in Wasana, "L'Emprise thaïe sur le Cambodge et l'établissement du protectorat français," 71.

32. Wilson, "State and Society in the Reign of Mongkut," 497.

33. Stéphan, "Voyage de la commission française," 125.

34. Translation of a letter to Phraya Suriyawongse Vayavadhana, Siamese ambassador to Paris, 4 March 1867, quoted in Moffat, *Mongkut, the King of Siam*, 24; Wilson, "State and Society in the Reign of Mongkut," 394.

35. Thongchai, "The Quest for '*Siwilai*,'" 532.

36. Leonowens, *The English Governess at the Siamese Court* (1870) and *The Romance of the Siamese Harem Life* (1873). The anonymous quote is from Wilson, "State and Society in the Reign of Mongkut," 730. On seduction see Bhumichitr, *Phra Chomklao, roi du Siam*, 285–90. On the ambiguity of Siam's involvement in world's fairs see Thongchai, "The Quest for '*Siwilai*,'" 540–42; Pierre Aymar-Bression, *Histoire générale de l'Exposition Universelle de 1867: les puissances étrangères* (Paris: J. Claye, 1868), 403–12; and Étienne Gallois, *Le Royaume de Siam au Champ de Mars en 1878 et à la Cour de Versailles en 1686: deux rois de Siam* (Paris: Challamel Aîné, 1878–79).

37. Beauvoir, *Java, Siam, Canton*, 303–4.

38. Allen D. Brown, "A Visit to Bangkok," *Harper's New Monthly Magazine* 41 (1870): 359–68, at 365.

39. *Scientific American* 20 (1870), 10. Margaret Landon, *Anna and the King of Siam* (New York: John Day, 1944). The book was adapted for the musical *The King and I* by Oscar Hammerstein (1951) and for several subsequent movies. On Leonowens

see Susan Brown, "Alternatives to the Missionary Position: Anna Leonowens as Victorian Travel Writer," *Feminist Studies* 21 (1995): 587–614.

40. As Mongkut called them. Quoted in Moffat, *Mongkut, the King of Siam*, 103.

41. Thiphakorawong, *The Dynastic Chronicles*, 2:309–10; Bhumichitr, *Phra Chomklao, roi du Siam*, 399.

42. Feudge, "An Oriental Monarchy," 798.

43. Leonowens, *The English Governess*, 56.

44. Wilson, "State and Society in the Reign of Mongkut," 21–22.

45. Quoted in Aasen, *Architecture of Siam*, 147 n. 31.

46. P. H. Josson and L. Willaert, eds., *Correspondance de Ferdinand Verbiest* (Brussels: Palais des Académies, 1938), 237; quoted by Bhumichitr, *Phra Chomklao, roi du Siam*, 69.

47. The Reverend Bradley, quoted in Moffat, *Mongkut, the King of Siam*, 156.

48. Thiphakorawong, *The Dynastic Chronicles*, 2:308. This man may have been a Dutchman named Druisart. On this see Thongchai, *Siam Mapped*, 116–17, 193 n. 20.

49. Thiphakorawong, *The Dynastic Chronicles*, 2:338–41; Wilson, "State and Society in the Reign of Mongkut," 496; Cook, "A Tale of Two City Pillars," 287.

50. Bhumichitr, *Phra Chomklao, roi du Siam*, 224; Henri Mouhot, *Travels in the Central Parts of Indo-China (Siam), Cambodia and Laos, during the Years 1858, 1859, and 1860* (London: J. Murray, 1864), 2:50–57; and Aasen, *Architecture of Siam*, 135–36; for a picture and plan see 136–37. For a description of the palace see Apinan Poshyananda, *Modern Art in Thailand: Nineteenth and Twentieth Centuries* (Singapore: Oxford University Press, 1992), 5.

51. Chabirand, "Étude sur l'éclipse totale de soleil du 18 août 1868," 380. Mongkut's letter was reproduced *in extenso* by Stéphan. After the eclipse the king again tried to explain his astronomical knowledge to Stéphan, who found him "incomprehensible." Stéphan, "Voyage de la commission française," 146.

52. All quotes here are from J.-B. Biot, *Études sur l'astronomie indienne et sur l'astronomie chinoise* (Paris: A. Blanchard, 1869), 177, 187–88, 196–97.

53. Anonymous, "Progress of Astronomy among the Modern Barbarians: Paying Our Debt to the Arabs," *Manufacturer and Builder* 7 (1875): 58; my emphasis. On the Egyptian expedition see Mahmoud-Bey, *Rapport à son altesse Mahammed Saïd, vice-roi d'Égypte sur l'éclipse totale de soleil observée à Dongalah (Nubie) le 18 juillet 1860* (Paris: Mallet-Bachelier, 1861), 19.

54. Cook, "A Tale of Two City Pillars," 296–97; Alabaster, *The Wheel of the Law*, 12; and Thiphakorawong, *The Dynastic Chronicles*, 2:540–41.

55. Chabirand, "Étude sur l'éclipse totale de soleil du 18 août 1868," 383. See also Emmanuel Liais, *Traité d'astronomie appliquée à la géographie et à la navigation suivies de la géodésie pratique* (Paris: Garnier, 1867). On longitude determination see Guy Boistel's contribution to this volume.

56. Feltus, ed., *Abstract of the Journal of Rev. Dan Beach Bradley, M.D.*, 278; and William L. Bradley, *Siam Then: The Foreign Colony in Bangkok before and after Anna* (Pasadena: William Carey, 1981), 149.

57. Englishmen at a horse race, American ships bringing missionaries, German ge-

ologists searching for minerals, etc.; see Aasen, *Architecture of Siam*, plate 35, with a modern color photo of the Wat, facing p. 188.

58. Wilson, "State and Society in the Reign of Mongkut," 739; Bhumichitr, *Phra Chomklao, roi du Siam*, 179; and Cook, "A Tale of Two Pillars," 302.

59. Geo B. Bacon, "In and around Bangkok," *Scribner's Monthly* 5 (1873): 421–32, at 426. See Stéphan, "Voyage de la commission française," 125.

60. Feltus, ed., *Abstract of the Journal of Rev. Dan Beach Bradley, M.D.*, 60–61; quoted in Wilson, "State and Society in the Reign of Mongkut," 229.

61. Leonowens, *The English Governess*, 224–31; Moffat, *Mongkut, the King of Siam*, 50; and Feudge, "An Oriental Monarchy."

62. The *Kitchnanukit* is the basis for the first part of Alabaster, *The Wheels of the Law*; on this see Craig J. Reynolds, "Buddhist Cosmography in Thai History, with Special Reference to Nineteenth-Century Culture Change," *Journal of Asian Studies* 35 (1976): 203–20; Wilson, "State and Society in the Reign of Mongkut," 32–33, 226; and Thongchai, *Siam Mapped*, chapter 2.

63. Alexander B. Griswold, "King Mongkut in Perspective," *Journal of the Siam Society* 45 (1957): 1–47.

64. Bhumichitr, *Phra Chomklao, roi du Siam*, 96–98; and Arom Prapandah, "Le contact missionnaire au XIXe siècle: Mgr. Pallegoix et le Prince Mongkut (Rama IV)," *Études indochinoises: frontières et contacts dans la péninsule indochinoise (XVIe–XIXe siècles)*, Études et documents 13 (Aix en Provence: IHPOM, 1981), 23–47, 28.

65. Wilson, "State and Society in the Reign of Mongkut," 732–39; Bhumichitr, *Phra Chomklao, roi du Siam*, 102–5. Recall also Eric Hobsbawm and Terence Ranger, eds., *The Invention of Tradition* (Cambridge: Cambridge University Press, 1983).

66. Thongchai, *Siam Mapped*, 42.

67. Prince Chunla Chakkraphong, *Lords of Life: A History of the Kings of Thailand* (London: Redman, 1967), 201; Smith, *A Physician at the Court of Siam*, 158–59.

68. Ord, "An Account of the Visit to the Late King of Siam at Hua Wan in August 1868," 129.

69. Garnier d'Abain, "An Elephant Hunt in Siam," *Scribner's Monthly* 8 (1874): 223–28, at 224.

70. G. Vapereau, "La confusion des idées et des langues: philosophie, science, fantaisie, rêve et orthodoxie tout ensemble: M. C. Flammarion," *L'année littéraire et dramatique, ou Revue annuelle des principales productions de la littérature française et des traductions des œuvres les plus importantes des littératures étrangères, classées et étudiées par genre* 8 (1866): 383–90, at 383–84.

71. Note that for lack of time to prepare carefully, photography was excluded from the expedition. Jonquières to the minister of public instruction, 14 March 1868, Archives Nationales, F17 3008, folder "Stéphan et Rayet (éclipse de 1868)," file "Renseignements sur les localités."

72. Le Verrier to minister of public instruction, 15 March 1867, Archives Nationales, F17 3008. E. Laugier and H. Faye, *Instructions sommaires du Bureau des Longitudes sur l'observation en mer de l'éclipse totale du 18 août prochain (Connaissance des temps pour 1868)* (Paris: Lemercier, 1867).

73. Flammarion, *Études et lectures sur l'astronomie* (Paris: Gauthier-Villars, 1872), 3:97–98.

74. "Proceedings of the Council of the Royal Society with Reference to the Undertaking of Certain Physical Observations in India," *Proceedings of the Royal Society in London* 16 (1867–68), 283–94; Capt Charles G. Perrins, Captain D. Rennoldson, Captain Summerville Murray, and Captain Henry Welchman King, "Observations of the Total Solar Eclipse of August 18, 1868," *Proceedings of the Royal Society in London* 17 (1868–69), 125–27, quotation at 126.

75. Aubin, "Orchestrating Observatory, Laboratory, and Field."

76. Aubin, "The Fading Star of the Paris Observatory." On meterology at the Paris Observatory see Fabien Locher, "Le nombre et le temps."

77. Quoted in Stanislas Meunier, "Académie des Sciences: séance du 3 février 1869," *Cosmos*, 8 February 1868, 2, 24–25; see Le Verrier, "Éclipse totale du soleil, le 18 août 1868," *Comptes rendus des séances hebdomadaires de l'Académie des Sciences* 66 (1868): 220–23, 226–27, and Faye, "Sur les mesures prises par le Bureau des Longitudes pour l'observation physique de l'éclipse prochaine, aux Indes orientales," *Comptes rendus des séances hebdomadaires de l'Académie des Sciences* 66 (1868): 223–26.

78. *Riforma* (Florence), 12 February 1868, French trans. in Archives Nationales, F17 2977, Janssen file "4ème mission: Mission scientifique dans les Indes Anglaises (Éclipse du 18 août 1868)."

79. Stanislas Meunier, "Académie des sciences: séance du 13 avril 1868," *Cosmos*, 18 April 1868, 2:24.

80. Janssen, "Conférence faite à la Sorbonne sur l'éclipse totale du 18 août 1868," *Lectures académiques, discours* (Paris: Hachette, 1903), 299.

81. On new astronomical institutions see Maison, "La fondation et les premiers travaux de l'observatoire astronomique de Bordeaux."

82. Renan, *L'avenir de la science*, 253.

83. Ruth Harris, *Lourdes: Body and Spirit in the Secular Age* (London: Penguin, 1999).

84. Flammarion, *Astronomie populaire*, 259.

85. Eugen Weber, "Religion and Superstition in Nineteenth-Century France," *Historical Journal* 31 (1988): 399–423, at 399. See however Harry W. Paul, *The Edge of Contingency: French Catholic Reaction to Scientific Change from Darwin to Duhem* (Gainesville: University Presses of Florida, 1979); Association Française d'Histoire Religieuse Contemporaine, ed., *Christianisme et science* (Paris: Vrin, 1989); and Claude Langlois and François Laplanche, eds., *La science catholique: l'"Encyclopédie théologique" de Migne (1844–1873) entre apologétique et vulgarisation* (Paris: Le Cerf, 1992).

86. Cantor, *Michael Faraday*, 289–95; and J. H. Brooke, "Science and Religion," *Companion to the History of Modern Science*, ed. R. C. Olby, G. N. Cantor, J. R. R. Christie, and M. J. S. Hodge (London: Routledge, 1990), 763–82.

87. Renan, *L'avenir de la science*, 63.

88. Cantor, *Michael Faraday*; see also Susan Fay Cannon, *Science in Culture*.

89. Le Verrier, "Recherches astronomiques," *Comptes rendus des séances hebdomadaires*

de l'Académie des Sciences 82 (1876): 1280–81, at 1280; see Adolph Müller, "Le Verrier," *Catholic Encyclopedia* (New York: Robert Appleton, 1910), vol. 9. H. Faye, *Sur l'origine du monde: théories cosmogoniques des anciens et des modernes* (Paris: Gauthier-Villars, 1884), 8.

90. See e.g. Victor Meunier, *Science et démocratie* (Paris: G. Baillières, 1865–66).

91. Ludovic Vitet, "La science et la foi," *Revue des deux mondes* 57 (1865): 680–706, at 681.

92. Allan Pitt, "The Cultural Impact of Science in France: Ernest Renan and the *Vie de Jésus*," *Historical Journal* 43 (2000): 79–101.

93. Petit, "L'esprit de la science anglaise et les français"; John Farley and Gerald L. Geison, "Science, Politics, and Spontaneous Generation in Nineteenth-Century France: The Pasteur-Pouchet Debate," *Bulletin of the History of Medecine* 48 (1974): 161–98.

94. Camille Flammarion, *Pluralité des mondes habités* (Paris: Flammarion, 1862), 125.

95. Louis Figuier, *Le lendemain de la mort, ou la vie future selon la science*, 11th edn (Paris: Hachette, 1904 [1871]); Louis-Auguste Blanqui, *L'éternité par les astres* (Geneva: Slatkine, 1996). On the plurality of worlds see Crowe, *The Extraterrestrial Life Debate*.

96. Alfred Fouillée, *Le mouvement idéaliste et la réaction contre la science positive* (Paris: Félix Alcan, 1896), lvi, lx.

97. Patrick Tuck, *The French Wolf and the Siamese Lamb: The French Threat to Siamese Independence, 1858–1907* (Bangkok: White Lotus, 1995).

Chapter 4

Keeping the Books at Paramatta Observatory

SIMON SCHAFFER

I had an opportunity of advancing the interests of Science by founding an Observatory at Paramatta and I carried out two assistants at my own expence and the instruments requisite. It was soon put into a state of active operation. It may now be considered the Greenwich of the Southern Hemisphere.
—THOMAS MAKDOUGALL BRISBANE (1842).[1]

The buildings of Paramatta Observatory are in a very sad state of repair. The white ant has been most destructive, and, as the surrounding ground is full of them, it would be fruitless to attempt a repair. The building was originally of a very inferior description, being only intended as a private establishment and not calculated to last beyond a few years.
—JAMES DUNLOP (1847).[2]

Meridian Astronomy in the Penal Colony

What techniques, exactly, made somewhere count as an observatory? Consider the "curious-looking structure," a flat-roofed stone building hastily built in a park just southwest of Government House in Paramatta, New South Wales, between late 1821 and early 1822. Its small domes and rudimentary neoclassical façade were designedly characteristic of early-nineteenth-century European astronomical establishments. The penal colony's new governor, a battle-scarred Scottish veteran named Thomas Brisbane (figure 23), installed fine books, telescopes, and clocks. He also brought able assistants, the mathematician Carl Rümker from Hamburg and the Scottish handloom weaver James Dunlop. In June 1822 staff members were the first to recover Encke's Comet; they later detected several others, and tracked planetary and lunar paths. Rümker got a silver medal from London and a congratulatory land grant from Brisbane. The gov-

23. The site of Paramatta Observatory in the mid-twentieth century. The obelisk and the remains of the pier of the observatory's transit instrument are both evident. Photograph by Frank Hurley. National Library of Australia, Hurley negative collection PIC FH/7436.

ernor assigned him a convict gang to clear the property, which Rümker named "Stargard" after his German birthplace and profession. The observatory gave standard time to the colony and in 1828 provided government surveyors with their initial meridian and at least one clock. By then Dunlop had catalogued over 620 nebulae and double stars. The places of more than 7,300 southern stars seen from Paramatta were published at Greenwich in a large catalogue in 1835. Considerable work was done on meteorology and geomagnetism. Pendulum beats were measured to help determine the earth's shape. Brisbane reported that "the whole will be given to the world as the choicest of Australasian production, which is most singularly well situated both as to position and climate for an observatory." In 1828 the Royal Astronomical Society awarded Brisbane and Dunlop its gold medal. The society's president John Herschel, expert on double stars and nebulae, soon to make himself doyen of southern hemisphere astronomy from the Cape of Good Hope, then lauded Paramatta's place: "the astronomers of Europe may view with something approaching to envy the lot of these their more fortunate brethren."[3]

But despite metropolitan rewards and solid labor, controversy wracked this observatory's brief career. Under insect, climatic, and human attack, buildings and instruments decayed, and the staff bickered and quit. London astronomers fought bitterly about the standard of the Paramatta accounts. In June 1847 commissioners reported that "the floor and partitions of the building seem to be entirely destroyed by the white ant." Many of the observing books had also been eaten. Surviving stock was inventoried, and within a year the last observer Dunlop was dead and the observatory shut. New South Wales did not host another for at least a decade. In 1880 an obelisk was erected to mark the site of the Paramatta transit instrument—in the wrong place (figure 24). The cometographer John Tebbutt, whose own observatory had also been consumed by termites, then eulogized: "Often as I have stood there in the quiet solitude of the place have I called up to memory the patient toiling of those three representatives of the stars, Brisbane, Rümker and Dunlop. Nothing but the piers of the transit instrument remained to testify to the astronomical character of the ruins. The whole affair was fast sinking into oblivion."[4]

Observatory techniques contested oblivion. Paramatta observatory's short life tells of these techniques and contests. Brisbane's claim to have built a southern Greenwich reveals how meridian astronomy could make observatories count. Despite naval and mercantile astronomers' previous work in the south Atlantic and the Cape of Good Hope, European observatory managers had not yet extended full meridian surveys below the equator. Rümker lectured Brisbane's new Philosophical Society of Australasia on "a new field richly sown with unknown stars, which to register and class is his pleasant duty." At imperial headquarters, as in many observatories of the period, vast numbers of stars were catalogued on a grid by measures of celestial longitude performed with a transit instrument and a clock, and of celestial latitude with a mural circle. Meridian maps were manufactured from these accounts. Brisbane shipped out fine-ruled bound volumes, in which times and angles entered on slates each night were to be copied the following day, and a pair of lithographic presses to print astronomical charts. Accounts would then be reduced by identifying equations to standardize data, such as aberration, refraction, personality, and instrument error. Greenwich's stern boss, George Airy, rated reduction above mere observation: "an idiot with a few days' practice may observe very well." He told Brisbane how the system worked: "I should suspend observations for a time and employ the whole energy of the head and assistant of the establishment in bringing up the past work." In a memorandum on the

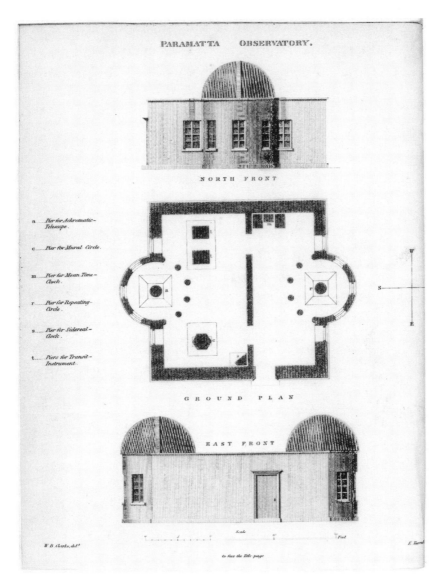

24. The plan of Paramatta Observatory, founded 1821–22. The mural circle (c) and the piers of the transit instrument (t) are at the eastern and western sides of the building; the mean-time clock (m) is just north of the transit instrument. Source: William Richardson, *A Catalogue of 7385 Stars Chiefly in the Southern Hemisphere* (1835), frontispiece. By permission of the Syndics of Cambridge University Library.

Paramatta observatory Herschel wrote that "the reduction and printing of the observations is a matter quite as important as the making them, it being fully understood at present that unreduced observations are scarcely worth transmitting home."[5] Reduced accounts were referred to an idealized epoch to make observations chronologically coincident. Brisbane regarded reduction as "work more of labor than intellect," but knew that it mattered. Paramatta observations "were all made with such arrangements as would enable computers to reduce them with least trouble." Collective celestial surveys might then be transformed into something like the uncanny vision of a single eye.[6]

Historians have found observatories' huge star lists depressingly sterile and oddly utopian. Yet meridian techniques directed observatory layout, labor, and hardware. They put observatories on the map and could make or destroy reputations. Paramatta's status depended on meridian accounts mauled by insects and astronomers. Rümker admitted that "business of more immediate importance accumulated so soon upon my hands, that I at first procrastinated the copying, left lacunas and at last, finding myself too much in arrear, entirely neglected it." These observatory techniques were performances for audiences elsewhere. Heinrich Schumacher, head of the prestigious new Altona observatory, complained to his London colleague Francis Baily about this shoddy clerkship: "The observations were written by Mr. Rümker in the observatory the moment he made them on single sheets. I told him that I looked upon this habit of his as a very bad one. I never permit it in my observatory to my assistants nor do it myself." European astronomers were decisive judges of Paramatta's techniques, especially the actuarial expert Baily, who cultivated an "archaeology of practical astronomy" which rescued past star catalogues "from oblivion." Baily labored too to restore the accounts and repute of the nation's greatest past meridian astronomer, John Flamsteed, launching fights on astronomers' morals and technique. Herschel praised this astronomical accountancy as "the sure pledge that our own works if really worthy will not be suffered to perish by time and neglect."[7]

So observatory sciences needed inventory investment. Practitioners tried to turn mutable phenomena into commodities in global networks whose nodes were privileged sites of accumulation. Distributed resources were concentrated at these sites, then put to work elsewhere. Local concentration and effective distribution were especially decisive in the Australian penal colony. Its population of about thirty thousand Europeans, half of them convicts, was the subject of precise if often remote discipline. Free migrants—such as the Paramatta astronomers—were known colloquially as "sterling," contrasted with emancipists who had served their term and with the supposedly debased "currency," the growing number of native-born settlers. The cash analogy was deliberate. Settler society understood and mocked power's metrological hierarchies. Brisbane's regime sought to overhaul New South Wales money, just as he committed himself to administrative standards throughout the colony. Temporal and spatial order was telling there. While Aboriginal culture was associated with atavistic timelessness, thus viciously disempowered, transported convicts were "doing time," a phrase coined in Sydney to capture the schedules of penal society. Both assigned labor and Aboriginal expertise proved crucial for the spatial ordering of the nascent colony. Convicts built the observatory and worked elsewhere as mechanics and artisans. Astronomically equipped surveyors such as Matthew Flinders and Thomas Mitchell relied on Aboriginal guides in coastal and inland surveys. The Aboriginal leader Bungaree, Flinders' guide in his circumnavigation of the continent, proved crucial as go-between in ceremonial arrivals at Sydney harbor. Brisbane reportedly honored him with a uniform. Special places were built to train inmates and natives to spend time well: military barracks, factories, asylums, and hospitals were all established in Sydney and Paramatta. Brisbane sought to "try the experiment of the white and black Natives of this Colony imbibing their earliest intellectual and religious ideas under a common roof."[8]

Brisbane hoped that precision accountancy would flourish under his observatory's roof too. It would have been right "to have established from the commencement of the Colony a system of order and regulations when acted upon, but here it was requisite to do more by remodeling all the established regulations: the more transactions were mystified the less easy to establish the iniquity." Brisbane had an evangelical loathing of mystique. The colony was to submit to "the philanthropist, the political economist and the philosopher." He presided over a pastoral boom, sought to turn the colony into a net exporter of goods and importer of conscientious labor, overhauled property and convict assignment based on local surveys, and established agricultural training and profit incentives for farmers. "The convict barracks of New South Wales remind me of the monasteries of Spain," he told the veteran colonial secretary Earl Bathurst. "They contain a population of consumers who produce nothing." Accounts of Brisbane's ambitious measures, accumulated against the odds, flowed from Paramatta to both the Astronomical Society and the Colonial Office.[9]

Brisbane and many of his colonial colleagues were pupils of Edinburgh masters in exact science, economic reform, and evangelical piety. Promotion of "agrarian patriotism" owed much to Scottish accounts of colonial economy. Enlightened sciences could help global government. At university the future governor sat at the feet of eminent political economists, preachers, and astronomers, and "would leave gay society to make his observations" at a nearby instrument maker's workshop. In an oft-told anecdote of spiritual conversion, Brisbane described his early realization of precision astronomy's virtues while facing shipwreck on a colonial military expedition. His accuracy allegedly saved the vessel: "Providential interference saved us. I reflected that in the course of my life I might often be placed in a similar situation in a ship where the reckoning was not accurately kept."[10] Spiritual and economic accounts reminded Brisbane how to run the world from a distance. "If Galileo laboured successfully in the bringing the most distant heavenly bodies near to us, our countryman Watt has no less immortalized himself by bringing the most distant countries near to us." Steam engineering accelerated British productivity and allowed British cotton to undercut Indian markets: similar economics would work in Australia. In Brisbane's spiritual diary, the quick tallying in a chapel pew of the future population of the British Empire took him to reflections on the inhabitants of other planets, thence to divine grace. Critics saw a fatal gap between the governor's observatory work and his administrative duties, but Brisbane saw the intimacy of accounts of heavenly government, colonial administration, and individual salvation.[11]

Paramatta observatory had to be assimilated to this disciplinary regime to prevent its assimilation by parasites and wasters. According to Herschel's memo on the observatory, "a new country peopled by a most energetic race of men," emancipists, and currency would soon degenerate into "endless litigation" without firm boundaries. A reformed observatory would help beat the bounds. Brisbane backed a trigonometric survey when the Colonial Office suggested a meridian arc be laid down in New South Wales in 1824.[12] The economic meaning of secure observatory techniques was obvious. Herschel saw stellar surveys as "one of those great masses of scientific capital laid up as a permanent and accumulating fund." Here astronomical work was fantasized as capitalist enterprise, with prudent ledgers, patient accountants, disciplined observers, well-oiled machinery, and precision values as sources of profit. Simon Werrett shows elsewhere in this book how serf-based feudal systems provided resources for the tsarist political technology of Struve's Pulkovo. In similar manner, colonial observatories were elements in imperial systems of government and in the imagination of what empire might mean. "I trust you will make conquests glorious for science in the new world to which you are going," the president of the Royal Society Humphry Davy wrote a fortnight before Brisbane's departure for New South Wales. In 1828, lauding Brisbane's "distinction worthy of a British governor," Herschel echoed Davy's aggressively pacific sentiments: "our first triumphs in those fair climes have been the peaceful ones of science, and the treasures they have transmitted to us are imperishable records of useful knowledge, speedily to be returned with interest, to the improvement of their condition and their elevation in the scale of nations."[13]

Colonial projects organized numbers in arrays which displayed and made ways of governing. One Scottish astronomer praised Brisbane because "through his munificence British science was made co-extensive with British dominion." Precise celestial knowledge had long been a tool of colonial power, a sign of that power's legitimacy, and a rationale for its exercise.[14] The colony in New South Wales had a history entangled with astronomical enterprise. William Dawes, astronomer on the First Fleet in 1788, set up a short-lived observing station overlooking the new settlement. Royal Society committeemen, led by the astronomer royal, recalled past British naval expertise in astronomy as part of their campaign to get funds for new survey and longitude projects in the South Seas. Exchanges between the Polynesian priest-navigator Tupaia and British mariners were crucial in enabling their journeys, then used to judge Polynesian cultural development by the standards of European astronomical history. These judgments offered precedents for violent denial of rights to Aborigines in the *terra nullius* of Australia by European surveyors who reckoned that its inhabitants lacked accurate notions of space and time.[15] These expeditions got their meaning from accounts of astronomy's past and modern imperium. Such accounts were a passion of the master historian of early-nineteenth-century observatory sciences, Alexander von Humboldt, whose public enterprises matter so much in this book. Humboldt's stories linked geopolitics with geophysics through a subtle account of the classical progress of the astronomical sciences in Europe and the New World. His understanding of Mayan and Aztec astronomical knowledge helped to place these territories in narratives of socioeconomic development, while his understanding of European observatory sciences helped place the Americas in narratives of physical development. Humboldt's contemporaries pursued similar programs: in principle, histories and maps could be produced worldwide.[16]

Astronomy became a "pattern science" for nineteenth-century tales of scientific and political development. "Epoch," "period," "revolution," and "zenith" became historians' terms. Some wrote of an "astronomical conception of society."[17] Observatory techniques were used to exemplify utilitarian governance and rationalist politics. Charles Dickens's dark satire of panoptic calculation imagined "an astronomical observatory made without any windows" where "the astronomer within should arrange the starry universe solely by pen, ink and paper." The fantasy hit home. "Astronomical bookkeeping" was indeed a slogan of early-nineteenth-century observatory sciences. The scientific servicemen and fiscal experts in London's new Astronomical Society, of which Brisbane was vice-president and Dunlop and Rümker medalists, rather liked the phrase. Herschel lauded the new reduction technique of Friedrich Bessel, a former accountant from Bremen and the society's gold medalist in 1829, as "the perfection of astronomical book-keeping," for "it renders the observation book itself available as a catalogue." Principles of competitive precision were at work. Remote observatories could begin adjudicating between rival European outfits. Brisbane's team measured longitudes by observing moon-culminating stars with their transit instrument, a German method that Rümker helped to introduce. These data were dispatched to London for Baily to have distributed by the Astronomical Society throughout the colonial system. Bessel spotted differences between polar distance tables manufactured in Britain and those from the German lands, so Brisbane was commissioned to run checks with the Paramatta mural circle. The Australian results of early 1824 settled the dispute in Britain's favor.[18] When Herschel proposed reform of the newfound southern constellations, his friend Thomas Maclear from the Cape Colony agreed that "the present assemblages are a hindrance in observatory book-keeping." Some saw the overhaul as "a Reform Bill for the stars, improving the representation of the skies, arranging the boundaries according to their star population."[19]

Savants cultivated surveys, censuses, and reviews to manifest good order. Measures of seconds pendulum lengths worldwide would fix the earth's shape and reform national metrology. Surveys of magnetic force and direction would reveal laws of compass behavior. Censuses of double stars would provide measures of stellar parallax, proper motions, and the universality of gravitation. With the right hardware astronomical rule might even reach distant nebulae and star clusters. The new observatory sciences depended on visionary maps of a civilized world. Their data were incorporated in reformist almanacs by metropolitan outfits such as the Society for the Diffusion of Useful Knowledge (SDUK) to teach number, weight, and measure.

In New South Wales, where print was effectively a government monopoly, settlers used almanacs to cope with the continent's unfamiliar weather, seasons, and geography. Against disreputable currency interest in astrological lore and sensational gossip, the colonial regime encouraged newfangled statistical calendars and weather surveys.[20] Brisbane set up the colony's first meteorological review from Paramatta, extending west of the Blue Mountains and south to Van Diemen's Land. Davy had already explained that coastal climate surveys would reveal whether the Australian interior was desert or covered with an inland sea: "of course you will carry out with you sets of barometers, thermometers and hygrometers." John Leslie, a controversial natural philosophy professor from Edinburgh, offered Brisbane his own meteorological kit: "a general survey would immortalize you."[21]

Under the admittedly vague notion of Humboldtian science, historians have understood such projects as parts of a newly fashioned global physics, and the increased production of precision hardware as crucial for resuscitating the exact sciences in Regency Britain and emulating other European projects. While in postwar Paris with the British occupying forces Brisbane was ordered by the Duke of Wellington to pursue this metrological work. This was a key moment for French survey work in Australia. In 1816 the scientific results of Nicolas Baudin's voyage to what he had baptized "Terre Napoléon" were at last published under the editorship of the naval expert Louis de Freycinet. Freycinet himself soon left again in 1817 on a three-year expedition designed to collect geomagnetic, geodetic, and natural-history material from the Pacific and Australia. The expedition's draftsman, Jacques Arago, reported back to his brother, the eminent savant François Arago, who in turn advised Brisbane on which pendulums he should take to Australia. On his departure from Paris in June 1818, Brisbane asked his commander in chief, the Duke of York, for the post of governor of New South Wales, because "an accurate determination of the Figure of the Earth" in "an island the high latitude of which renders it peculiarly fitted for the purpose" would "not only [be] highly interesting to science but may be beneficial to mankind." Wellington backed Brisbane's appeal with praise of his military timekeeping. Brisbane's men were enlisted in all these initiatives: geodesy, geomagnetism, meteorology, and astrometry.[22]

Many public observatories with similar layout and hardware were founded in these few years, at Edinburgh (1818), the Cape (1821), Cambridge (1823), and elsewhere. Madras Observatory was overhauled for the Indian trigonometric survey. Brisbane took advantage of his good connections to help administer the meridian observatory at Edinburgh and advise on hiring at the Cape. Other establishments in northern Europe, especially Bessel's Königsberg and Struve's Pulkovo, provided rivals and models for the network. Rümker's contacts with observatory managers such as Bessel and Schumacher linked him with dominant institutions of this new astronomy.[23] Brisbane's techniques were more Francophile. "In Astronomy," he lectured, "there is little left that has not received a finishing hand," a self-confident sentiment typical of the Parisian milieu he frequented when he joined the grateful Institut after intervening to stop Allied military marauders. In a manner akin to Humboldt's global use of Parisian hardware and numbers, Brisbane intended to produce a southern meridian catalogue to replace previous French astronomers' incomplete surveys. Just before Brisbane left Europe in 1821, the aged science administrator Jean-Baptiste Delambre read his plans to the Académie des Sciences and the Bureau des Longitudes. What French astronomers had done, Delambre reassured him, "proves there's nowhere a steady, able and diligent observer can't find very advantageous opportunities." Delambre suggested that Brisbane coordinate his surveys with those in British India. Delambre's successor at the Observatoire, Alexis Bouvard, regretted that he could not join Brisbane's meridian project in person.[24]

Colonial enterprise nourished these projects. Herschel believed that "a perfect knowledge of the astronomy of the southern hemisphere is becoming daily an object of greater practical interest now that civilization and intercourse are rapidly spreading through those distant regions—that our colonies are rising into importance." Dunlop presented his own work not only as "a valuable treasure for the present generation to possess, but an invaluable inheritance for them to transmit to posterity."[25] The mix of permanence, scope, and precision helped give meridian work its exemplary place. Sober Greenwich tradition favored separate measures of right ascensions using a transit instrument and a pendulum clock, and of zenith distances with a mural circle. The contrast was with equatorial astronomy, which devoted entire sessions to a select celestial object such as a comet or binary star, and with Herschel's natural history of the heavens, which required huge reflectors for nebular surveys. Thus readers consulting "Observatory" in the cheap SDUK astronomy handbook were peremptorily referred to the entry on "transit instrument." This entry thus granted the status of observatory science only to enterprises matching meridian work. In 1812–16 the arrogant Edward Troughton set the standard with new commissions for Greenwich, a ten-foot transit instrument, and a mural circle six feet in diameter. Other makers, notably Thomas Jones, who also made pendulums for geodesists in London, mimicked Troughton's enterprise.[26]

Paramatta's stock was designed to make these projects into a permanent part of the penal colony's science. For right ascensions the team brought astronomical clocks from London and Paris and a sixty-five-inch Troughton transit instrument, two hundred pounds in weight. Even before the Greenwich mural circle was finished in 1812, Brisbane had already obtained for his Largs observatory a Troughton mural circle two feet in diameter equipped with four eyepieces, the costliest instrument shipped to Australia. Brisbane witnessed the construction of a similar mural circle at the Paris Observatoire. Paramatta's layout, with securely separate sites for transit and mural instruments, clocks installed centrally, and the neoclassical design favored by meridian observers, testified to Brisbane's aims (figure 25). Transit observations by the received "eye and ear" method demanded that the observer listen to the clock while watching a star's image cross the wires in the eyepiece, checking the pendulum beats as each wire was passed. Workers at the mural circle and the transit instrument had to be kept apart so that the noise of the clock would not disturb this crucial ritual. At some observatories transit observations were made before dinner, "lest over-anxiety interfere with the salutary function of digestion." Columns on Brisbane's printed forms were then supposed to discipline data entry. Space was left for pressure, temperature, refraction, clock-rate, and each microscope reading. But the meridian system was a judgmental ideal—as the workers at Paramatta found out the hard way, it was tough to realize. It relied on local improvisation never easily accounted. Entries could not easily be made at the eyepiece, a basic puzzle for observatory accountancy. Rümker explained that "the observer who is more settled and less agitated, [even] if he can write down his observations without leaving his Transit, cannot do this there satisfactorily in his book with ink, at all events he will do it more slovenly, particularly if attending mural circle and transit at the same time." Two expert observers were needed, working simultaneously on each observation, then entering data the next day into the books. The surplus value of an observatory was proverbially proportional to the number of staff above two. In his memo on Paramatta, Herschel urged that "no astronomer can possibly do his duty as a public servant without an assistant who should be an excellent and practised astronomical computist."[27]

Staff and machinery had to behave, but did not. Rümker was away from

25. Sir Thomas Brisbane, governor of New South Wales, by Augustus Earle (1825). The portrait, in which military honors and astronomical objects are both apparent, was commissioned on Brisbane's departure from the colony. Earle, stepbrother of the eminent astronomer William Smyth, decorated Government House for Brisbane's farewell banquet; he later joined HMS *Beagle* as supernumerary artist. By permission of Government House, Sydney.

the observatory between June 1823 and May 1826, throwing the demanding work with slates, books, micrometers, and clocks onto Dunlop alone. In any case the transit instrument was judged defective, so from mid-1825 Dunlop fixed the mural circle in the meridian plane to attempt registration of both transit times and altitudes, entering the data in one set of books. The result was not a success. When Rümker returned to Paramatta in 1826, he complained that "as the journeyman is taken away I have to count the time from a clock in the next room and must interrupt the observations to attend the transit and mural circle." Rümker, homeless, had to sleep next to his instruments. Astronomers had to make do and mend in erratically unpredictable ways.[28] A single instrument might solve the problem, allowing a solitary meridian observer to register both transit times and zenith distances. Troughton had already built such a four-foot meridian circle for the amateur Stephen Groombridge at Blackheath in 1806. By 1817 Groombridge had used the innovative instrument to survey fifty thousand circumpolar stars, a precedent for Brisbane's team. As with the Cape and Paramatta meridian catalogues, reduction was slowly completed from Greenwich. New meridian circles became popular in the German lands under the aegis of the eminent workshops of Reichenbach and Repsold. Their client Rümker was keen to see the Australians get such a device. In London in June 1829 he sought vainly to purchase the Groombridge instrument from its owner, James South. When Rümker was replaced by Dunlop, he in turn commissioned from Thomas Jones a new meridian circle, forty-two inches in diameter, equipped with spirit levels to fix the index arms and a telescope thirty inches long. Jones was building a similar meridian circle for Oxford University. Dunlop's purchase reached him in Australia in April 1835. But its spirit level, filled with a liquid which boiled at low temperatures, burst in antipodean heat. The index circle proved badly graduated. Dunlop never used the instrument. It was "unhandy," if not "quite useless."[29]

When European instruments and regulations erred, in observatories as in courtrooms, locals had to fix them by ready reckoning. In 1832 Dunlop grumbled to Brisbane that currency workmen had wrecked their sidereal clock, made in London, by taking it down with saw and chisel. "I adjusted all the other pallets to a great degree of nicety. She has performed excellently ever since. I got the Government clock-maker to clean Breguet but would not let him touch [the sidereal clock]." Instruments were like precious livestock, every vagary requiring careful husbandry. In his contribution to this book, John Tresch points out the seeming paradox of local spontaneity and global coordination within the Humboldtian regime of instruments. Observatories cultivated an image of perfected instruments and of a well-aligned workforce run from metropolitan workshops and colonial offices by machine-like delegates regulated by error analysis and reduction. Airy argued that "the principle of division of labour" should govern them. To Herschel, "every astronomical observatory which publishes its observations becomes a nucleus for the formation around it of a school of exact practice."[30] But in the meantime, behind their workplaces' façades, meridian astronomers tried converting messy artfulness into imposing castles in the sky.

Controversy brought backstage work to public scrutiny in print. Brisbane was easily identified with his instruments. According to a story published in one of Charles Dickens's best-selling magazines, the superb go-between and Aboriginal leader Bungaree would mimic the governor simply by holding a telescope to his eye, to the vast amusement of spectators. In 1824 a newspaper in London gossiped that "Sir Thomas spends the greater part of his time in the Observatory or shooting Parrots. Lately he has been deprived of his principal Assistant Mr Rumker, who, report saith, hath returned to his farm either from anger or disgust." The journalist was well informed. It was said that Rümker quit because he loathed shooting birds; that during a lunar eclipse Brisbane's accuracy "was questioned and his character perhaps at stake"; that Rümker absented himself once he acquired his estate in Stargard; that he had published in European journals without permission; that he protested when Brisbane credited Dunlop with the heroic recovery of Encke's comet. Certainly in July 1823 a journal in Edinburgh reported Rümker's recapture of the comet, but after a few months the journal retracted its praise on Brisbane's say-so, patriotically lauding "our modest countryman" Dunlop rather than the "Germanic diligence" of his colleague.[31] On 16 June 1823 Rümker rode away from Paramatta, taking "all the observations up to this period and also the catalogues of stars" to his farm in Stargard thirty-five miles way. Four months later Brisbane sent a magistrate named Donald McLeod with Dunlop to recover the books, unsuccessfully. The governor soon wrote a formal complaint to Bathurst in London demanding that no more land be assigned to Rümker.[32]

If it was a long ride to Stargard, it took half a year for messages to travel to Britain. The delay mattered for the observatory's output and for colonial administration. Brisbane's regime was obsessed by timetables in the new disciplinary institutions they managed. The colonial secretary eventually answered Brisbane's dispatch in July 1824, just before the London newspapers, keen on leaks from Whitehall, carried news of the Paramatta fight. Bathurst's wayward proconsul was told that it was his own fault if Rümker had gone awry and within a few months Bathurst ordered Brisbane's recall. Meanwhile, Brisbane explained away the land grant and defended his administration: "Mr Rumker rendered himself obnoxious to myself and family . . . I found him devoted alone to his own objects and pursuits, [he] consequently became no further use to me."[33] From mid-1825, during the final months of Brisbane's regime, relations with Rümker collapsed.

Dunlop ran homely surveys and Brisbane sent observations of comets and planets, but Paramatta's meridian data seemed lost. Brisbane ordered the magistrate with a constable to Stargard, again without success. In October a police clerk, Samuel Salmon, arrived at the farm to copy transit accounts. Rümker said that "no person but himself could do anything" with the mural circle books of zenith distances. "Nobody could satisfactorily reduce these observations but he that was acquainted with the imperfections of the Transit and the circumstances attending the observations, that could not be sufficiently described in Remarks." Books could not capture observers' expertise. Furthermore, Rümker's habits had wrecked Brisbane's accountancy. "I made use of the vast quantity of spare paper in the printed forms of Sir Thomas Brisbane's books for my calculations, whereby these books were too much disfigured." Salmon's solitary copy of one set of transit accounts traveled home in December with Brisbane, who handed them to the Board of Longitude.[34] From 1826 the Rümker affair was entangled in polemics of metropolitan reformers obsessed by administrative accounting. They wanted to know how, and why, the Paramatta books had been kept.

As David Aubin urges in his study of eclipse expeditions, seemingly remote astronomical enterprises often mattered to politics back home. New South Wales certainly offered tools that Londoners could use in debates about how to spend time properly. In 1832 early issues of the reformist *Penny Magazine* scolded metropolitan printed trash by comparing the precision virtues of an Australian "almanac, not only undeformed by the ridiculous astrology of the English almanacs, but containing much valuable information on the agriculture and statistics of this fine region."[35] Tales of Paramatta made equally apt weapons for London reformers: a public establishment whose accounts had gone astray; insubordinate labor and incompetent managers; purloined astronomical commodities; the laxity of the Board of Longitude and the Royal Society. Reformers such as James South, Charles Babbage, and Francis Baily sought to seize judgmental authority from what they saw as the old corruption of the Royal Society and the Board of Longitude, the Treasury, and the Admiralty. Even the timorous Dunlop told Baily that "measures should be taken to expose the system of jobbing and peculation which appears to have been carried on under color of the patronial garb of science."[36] Within the Royal Society the election as president in late 1827 of a pliant Tory MP, Davies Gilbert, and the consequent expulsion of Babbage and South from its council, inaugurated three years of vivid politicking culminating with the close election of November 1830, when the reformers' candidate, John Herschel, was narrowly defeated. The conduct of Edward Sabine, the society's secretary, was made scandalous. Herschel, South, and Baily denounced the Board of Longitude's inefficiencies as an almanac maker and scientific patron, and campaigned for new global physics surveys which might escape from the board's placemen. The Astronomical Society became a reform lobby, labeled "banditti" by the Tory establishment. Its aims got pugnacious publicity in South's pamphlets and in Babbage's incendiary *Reflections on the Decline of Science in England*, which hit the press in May 1830 just as the Paramatta controversy, which Babbage's book brandished, reached its vicious climax.[37]

Londoners' scrutiny of Paramatta accounts can be reconstructed, because Baily conscientiously stashed the relevant papers which crossed his desk at Tavistock Place. He treated the Rümker case just as he later prosecuted the Royal Society in the name of the ideal astronomical accountant, John Flamsteed. Baily mastered archives to establish intellectual property and convict astronomy's patrons. "An astronomer is no lawyer," Rümker vainly protested. He spotted Baily's self-appointed role as Brisbane's legal counsel. South and Babbage, with Herschel's judicious support, were the show-trial's impresarios. It did not impress Rümker's persecutors that he was a German-trained analyst made government astronomer, assigned to survey a meridian arc in the colony and offered a salary commensurate with that of other colonial astronomers. They took these rewards as crude jobbery. So when Rümker's latest reports reached the Royal Society in early 1827, Herschel as secretary at once asked Brisbane, secure on his new Scottish borders estate, whether Rümker had the right to publish. Since he had been Brisbane's employee, these data belonged to Rümker's boss. Subordination, Herschel expostulated, "is implied in the very meaning of the words 'assistant in an observatory.'"[38]

Brisbane's interests were clear. He wanted Dunlop back in Paramatta and its accounts reduced and published as testimony to his administration's virtues. These interests nourished London reformers' plans. In June 1827 Herschel resigned the position of secretary in protest against the Royal Society's lassitude. By year's end Gilbert was president and Sabine secretary. Paramatta's accounts became a test case between dissidents and the strengthened conservative regime. On 31 January 1828 the Royal Society's council agreed to have the Paramatta meridian observations reduced and published at the expense of the Board of Longitude. The astronomers showed typically precise timing. Exactly one week later Herschel honored Dunlop and Brisbane at the Astronomical Society. South sent their medals to Scotland. But no one in Britain yet had the meridian accounts to reduce and publish. So Brisbane came down to London in May 1828, met his Astronomical Society allies, and got the Royal Society to ask the Colonial Office to recover the books from Rümker and appoint Dunlop as Paramatta astronomer. Brisbane's former companion-in-arms, the new colonial secretary George Murray, raised the thorny issue of cash. The government abolished the Board of Longitude, making Sabine leader of a new Admiralty advisory committee. Though South expostulated in the press about government waste and the virtues of Brisbane's observatory, resources for the Paramatta program dried up.[39]

News of wrangling in London about Rümker's job and the Paramatta accounts reached the papers in Sydney at the end of 1828. Rümker decided that he had to be on the spot: "my object in coming home is partly to publish my observations and to make some arrangements with Government for a better astronomical establishment at Paramatta . . . and also to efface a prejudicious opinion which I imagined to exist against me." Just before he left, Rümker at last got the Colonial Office's orders to return the books; he spent his voyage making three volumes of fair copies from his rough notes, and handed them to a bureaucrat in Downing Street when he reached London in June 1829. These three precious volumes of meridian data, with each star's mural circle and transit readings entered in matching columns, were passed via Sabine at the Royal Society to an assistant in Greenwich, William Richardson, for reduction. Whitehall, the Royal Observatory, and the Royal Society all agreed to process Rümker's meridian data and give Paramatta better hardware and housing. As a sign of goodwill, the *Philosophical Transactions of the Royal Society* devoted a special 150-page supplement, funded by the government, to Rümker's impressive observations of comets, planets, and pendulum lengths: "a future volume," the author promised, "will contain a catalogue of the stars of the southern hemisphere."[40]

But this was hubris. The conspirators decided to use Rümker's misdeeds to prevent his triumphant return to Australia and to attack the conservatives. The principles of meridian astronomy played into their hands. Managers owned observatories' output, which must be reduced from original accounts. Rümker had slighted Brisbane by printing data without his approval and subverted accountancy by giving Richardson copy books, not raw data. All this had been subsidized by Sabine and the Colonial Office. So the Royal Society was coerced into ordering Rümker to hand over not just the three copy volumes but his rough notes too. Rümker worried: "they can easily be scattered about and lost." "I must beg you will lose none of the leaves." Worse was to come as that winter approached, a time of nationwide economic and social crisis. South wrote to the newspapers attacking the Royal Society's treatment of Brisbane, while Babbage began composing a book-length polemic lambasting Sabine, Gilbert, and their coterie. After the new year in 1830 the pair got Brisbane and Dunlop to document the scandal of the Paramatta meridian accounts and its political meaning. "It does not add much to the reputation of the Colonial Department," Dunlop told South. Babbage warned Brisbane that "this injury done to science" might prevent anyone from emulating "the splendid example" he had "afforded to the Government" of its "numerous colonies."[41]

Once Baily's forensic skills were mobilized in January 1830, the issue was not in doubt. He went to the Royal Society printer to get manuscripts of Rümker's cometary data and obtained Rümker's rough notes from South. The computer in Greenwich, Richardson, forwarded the three copy books of meridian observations: "avoid bringing me into any scrap, but I hope justice will be done." Richardson was given a medal by the Astronomical Society, and the prosecutors now had the books. Brisbane was encouraged to write formally to Sabine asking where all these papers had come from, whether they were strictly "originals," and what the role of government had been. While Rümker was away visiting Schumacher at Altona and Bessel in Königsberg, the Royal Society's secretary temporized. Sabine explained to Brisbane that the society could not tell in Rümker's absence whether these were original accounts, that Rümker's handwriting was "exceedingly difficult to decypher," but that his "purpose has obviously been the public advantage in bringing together in one point that which is best seen in one view." This reply seemed absurdly vague and fatally revealing: "I really am quite astounded at that answer and almost believe I am in a dream," Baily told Brisbane. Staring at the papers on his desk, Baily was horrified. Rümker's rough notes were almost illegible, written when he was "evidently exhausted from some other work since none of them seemed to be original observations," while the fair copies were books that prompted Baily to remark, "no astronomer ever saw such before." Rümker must still have the originals somewhere in his bags, and should be stopped from returning to Paramatta "till he has given a satisfactory account of these mysterious and discreditable proceedings."[42]

Mystery and discredit were potent words in the astronomers' lexicon. The experts on Rümker's case knew how to exploit them. In May 1830 they were more active than ever. South started campaigning for a royal charter for the Astronomical Society so it could match Sabine's outfit, while Babbage released his *Decline of Science*, with venomous remarks on Paramatta's troubles and Sabine's crimes. Babbage pilloried the Royal Society's falsification of its records, implying its kinship with Rümker's misdeeds. Baily took the fight to court. With Brisbane's help he got the Colonial Office to arrest Rümker on his return to London from Germany, then persuaded the Royal Society to set up a tribunal with himself as counsel. Baily told Gilbert that Rümker mistakenly claimed "a legal right" to records, some of which were in any case "partially destroyed." Rümker protested "against such a violation of private property" and declared his loyalty to Brisbane. He got a reference from his host in Altona, Schumacher, who frankly told Baily: that "though sometimes inconsiderate" he had "ever found him honest," and that in any case Rümker had a job waiting for him in nearby Hamburg. Baily's committee moved in for the kill. On 5 June 1830, over the names of the president Davies Gilbert, the hydrographer Francis Beaufort, and Baily himself, they reported that the original Paramatta books had been kept and then wrecked by Rümker. Accounts by all those involved were published by the Colonial Office. Rümker was promptly dismissed from his colonial post and left for Hamburg. He complained still of "uncontrolled dictatorship" and the "quackery which spreads distemper into Society and even sends its pestiferous influence into the southern hemisphere."[43]

The influence was felt in both hemispheres. The Hamburg Senate did not offer the refugee Rümker a post at its observatory nor a high salary. Instead it proposed that he run the local navigation school, with whose acting director Rümker soon staged a classroom fistfight. He spent his spare time denouncing Baily's mathematics and South's politics. South's career of well-publicized violence would continue for years in the astonishing battle he waged, with Babbage's support, against Troughton's skills and repute, ending with the ritual destruction of a prized telescope. In November 1830 the Tory government at last fell from power on the issue of reform, taking Murray and his colonial administrators down with them. South issued another pamphlet against the regime's spendthrift publication of bad accounts and its endorsement of Rümker, "whose self-condemnation is before the public." The same traumatic week, Baily told Gilbert that the Royal Society must print a notice of Rümker's sins. Within a few days Gilbert resigned and civil war broke out with Herschel and his allies.[44]

Herschel prudently left for the Cape in early 1834, taking his family's great reflector for new southern surveys. He started by recounting positions in the Paramatta meridian catalogue, compiled from relics of the Australian books by Richardson, then sent south for scrutiny. The result was a disaster. Richardson sneered that "the places of stars in this Catalogue cannot be supposed as correct as those determined in great national observatories." Paramatta transit times seemed several seconds less than comparable observation made by the East India Company at St Helena. Herschel amply confirmed this damning judgment: "These great discrepancies are alarming." He groaned when Dunlop got a French prize: "I wish the awarders would come here and look for some of his nebulae and double stars." By 1836 he decided that the Paramatta transit data were hopeless. "The errors follow no traceable law. Henceforth I feel disposed to dismiss the epithet The Brisbane Catalogue as connected with this remarkable astronomical record."[45]

Brisbane and Dunlop got the bad news from Herschel's Cape recounts. Over a sultry Christmastide in 1837 Herschel hosted the New South Wales governor-elect George Gipps on his way to Sydney via the Cape. He spent Boxing Day dinner giving Gipps his prescription for Paramatta's ills. Dunlop's transit device should be sent home for repair. "No zeal, energy or attainment on the part of an observer unprovided with the chef d'œuvres of workmanship and the extent of national support" could compete with metropolitan outfits. So Paramatta should stop fundamental meridian work. Instead it must become a metrological center, the "zero point of Australian, Chinese and South Pacific Geography" and "the nucleus of a regular and scientific trigonometrical survey of British Australia." Imperial weight and length standards should be preserved there and copies given to colonial magistrates and surveyors. Better, it might become what Herschel dreamed of, a Humboldtian observatory working on tides, geomagnetism, and meteorology, even perhaps on his new friend Charles Darwin's theories of coral reefs. "May it be hoped that the department of *physical observation* will either have Observatories of its own or be connected with astronomical ones."[46] The real Paramatta never quite made it. When he got to Sydney, Gipps wrote to Brisbane that Dunlop was near death. Dunlop lingered, struggling with failing hardware and offering resources to local surveyors and scientific passers-by. He was eulogized as "an instance of original genius lost in this moral desert for want of kindred minds."[47]

Herschel made his imaginary Paramatta a model of how global physics could become proper observatory science. When he got back from Africa to the imperial capital, he touted this plan in Admiralty instructions for naval surveyors and in preaching the crusade for global geomagnetic measures. Especially eloquent was the manifesto he wrote with Sabine in 1842 for a dedicated physical observatory at Kew. When characteristically botched by the Royal Society, Kew Observatory was salvaged by the British Association for Advancement of Science (BAAS). The same year Brisbane set up his own magnetic observatory; one of his employees took over Kew a decade later. The ambitious Kew standards institution, storage facility, and test site for "optical astronomy" became an important node of networks of calculation and vigilance, along which innovative workplaces of electrotechnology and astrophysics dispatched their exacting observational accounts.[48]

However reliant on the integrity of these long-range networks and local labor processes, observatory sciences such as those at Paramatta or Kew idealized their autonomy. Though images of isolation completely misrepresent observatory life, that life—like other colonial projects—maintained self-sufficiency as the aim. Brisbane's New South Wales and his observatory might become profitably productive, rather than wastefully parasitic, if they could escape dependence, no longer needing distant subsidies and supplies. The seclusion of well-equipped, well-managed workforces seemed a condition of their reliability. Spokesmen lauded observatories' self-registering equipment as nuclei for the precision values of self-help. But their chances depended on others elsewhere: patrons, intermediaries, support staff, customers. When Paramatta shut in 1848 experts worried that its books, "though of value to an observatory, would fetch little more than as waste paper" if auctioned locally; "there are instruments that would scarcely realize *here* the value of their weight in metal." Without support systems, observatories transmuted into scrap.[49]

In early-nineteenth-century observatory sciences, pursued in remote bivouacs, on storm-tossed ships, or in ruinous shacks, this puzzle of distant resources and authorities mattered most in accounting. Good accounts needed judicious recounting—and astronomers had long memories. Many veterans of the Paramatta episode learned the moral calculus by which accounts were settled. When Airy became head of Greenwich in late 1835 he at once sought Richardson's dismissal for collusion in falsifying star catalogues. The Admiralty demurred. But a decade later Richardson stood trial at the Old Bailey for "the wilful murder of his incest child." Though acquitted, he was instantly fired from Airy's Observatory. Even when the New South Wales government proposed a new observatory in Sydney, Airy wrote: "the discreditable mismanagement of that of Parramatta will effectually prevent me from making such a recommendation." Others experienced equally telling judgments. In 1857 Rümker retired to Lisbon, where he planned the belated reduction of his southern meridian work. He contacted the astronomer William Smyth, first met four decades earlier on military service in the Mediterranean. Rümker wanted Smyth's help in reducing the Australian product. "It is a delicate matter to deal with," Smyth waspishly noted, "for Ruemker was generally esteemed a loose observer, though a clever eliminator, of his observations, and moreover there were bickerings as to the proprietorship of the whole."[50]

So the Paramatta fight was not forgotten, its place in history important but unstable. When the BAAS meeting at Glasgow contemplated making Brisbane its president, his comrade Roderick Murchison, veteran of the Peninsular War and doyen of imperial geologists, rejected Brisbane's candidacy because his "observations have been corrected to a great extent," and "though a good worthy man he is so thoroughly inefficient in leading a great body." But when Brisbane lay dying in late 1859, Murchison "as a Soldier who long ago turned his sword into a hammer reflected with pride that the Army should have produced in one its bravest and most skilful leaders so proficient an Astronomer as Sir Thomas Brisbane." For Murchison, Brisbane, and their colleagues, rhetorical connections between imperial, military, and scientific projects were serious matters both of fieldwork and publicity.[51] Scandals could therefore be forgiven, absorbed by military and imperial iconography, or turned into amiable jokes in the streets of Sydney. The moral of Paramatta's books is that some performances with instruments and paper could become genuine observatory work or else turn into woeful fraud when judged so by appropriate others away from the scene. Paramatta Observatory had important remote audiences, linked to its world of account books, telescopes, land grants, and white ants by the tenuous but potent networks of colonial power.

Notes

I thank Will Ashworth, Julian Holland and David Turnbull for their invaluable help. I gratefully acknowledge access to the following libraries: Mitchell Library, State Library of New South Wales; Northern Ireland Public Record Office; National Library of Australia; New South Wales State Records; Royal Astronomical Society; Royal Greenwich Observatory archives, Cambridge; Royal Society. This work was aided by a Research Leave Fellowship from the Arts and Humanities Research Council.

1. *Report of the Proceedings at the Grand Dinner at Largs in Honour of Sir Thomas Makdougall Brisbane of General Sir Thomas Makdougall Brisbane of Brisbane and Makerstoun* (Kelso: Johnston, 1842), 10 (draft at National Library of Australia, Parkes, MSS 4036/1/48, repr. in Brisbane, *Reminiscences*, 89–93).

2. Dunlop to the Board of Visitors of the Paramatta Observatory, 18 August 1847, in Service, *Thir notandums*, 197.

3. Service, *Thir notandums*, 139–42; Brisbane to Barnard (24 February 1823), Northern Ireland Public Record Office, MS D207/67/58; and John Herschel, *Essays from the Edinburgh and Quarterly Reviews* (London: Longmans, 1857), 489–503. There is now a very thorough study of the observatory's career in Shirley Saunders, "Sir Thomas Brisbane's Legacy to Colonial Science: Colonial Astronomy at the Parramatta Observatory, 1822–1848," *Historical Records of Australian Science* 15 (2004): 177–209.

4. Service, *Thir notandums*, 214; H. C. Russell, "Astronomical and Meteorological Workers in New South Wales, 1778-1860," *Australasian Association for the Advancement of Science, Reports* 1 (1888): 45–94, appendices G–L; and Haynes, ed., *Explorers of the Southern Sky*, 44–47. Remnants of the equipment are at Sydney Observatory. Charles Pickett and Nick Lomb, *Observer and Observed: A Pictorial History of Sydney Observatory* (Sydney: Powerhouse, 2001), 25.

5. Charles Rümker, "On the Astronomy of the Southern Hemisphere," *Geographical Memoirs on New South Wales by Various Hands*, ed. Barron Field (London: John Murray, 1825), 267; Airy to Wood, 22 March 1847, Royal Greenwich Observatory Archives, Cambridge University Library, MS 6.2/259–64; Airy to Brisbane, 9 December 1844, Mitchell Library, State Library of New South Wales, MS 1191/1/177; "A Statement of Facts Relative to the Conduct of Mr. Rumker," Royal Astronomical Society Archives, MSS Baily 5 fol. 16; and Herschel to Gipps, 26 December 1837, Royal Society Archives, HS 19.72.

6. Brisbane to Herschel, 26 March 1828, Royal Society Archives, MSS HS 5.102.

7. Dewhirst, "Meridian Astronomy in the Private and University Observatories of the United Kingdom"; Rümker, *On the Most Effectual Means of Encouraging Scientific Undertakings* (Hamburg: Nestler, 1831), 4; Schumacher to Baily, 9 March 1830, Royal Astronomical Society Archives, MSS Baily 5 fol. 70; Herschel, *Essays from the Edinburgh and London Quarterly Reviews with Addresses and Other Pieces*, 597–98; Ashworth, "'Labour Harder Than Thrashing.'"

8. Graeme Davison, *The Unforgiving Minute: How Australia Learned to Tell the Time* (Oxford: Oxford University Press, 1993), 16-28; J. Brook and J. L. Kohen, *The Parramatta Native Institution and the Black Town* (Kensington: NSW University Press, 1991), 180; Keith Vincent Smith, *King Bungaree* (Kenthurst, New South Wales: Kangaroo, 1992), 26–65; and D. W. A. Baker, *The Civilised Surveyor: Thomas Mitchell and the Australian Aborigines* (Melbourne: Melbourne University Press, 1997), 72–77.

9. Brisbane to Barnard, 24 February 1823, Northern Ireland Public Record Office, D207/67/58; J. J. Eddy, *Britain and the Australian Colonies, 1818–1831* (Oxford: Clarendon, 1969), 126–33; New South Wales State Records, Kingswood, Colonial Secretary MSS 4/1753, p.160 (5 July 1822); Brisbane to Bruce, 28 March 1822, Mitchell Library, State Library of New South Wales, MSS A1627; and Brisbane to Bathurst, 29 November 1823, *Historical Records of Australia* 11 (1917): 181.

10. T. N. Clarke, A. D. Morrison-Low, and A. D. C. Simpson, *Brass and Glass* (Edinburgh: National Museums of Scotland, 1989), 26, and Brisbane, Lecture at Glasgow Astronomical Institution, 16 December 1836, Mitchell Library, State Library of New South Wales, MSS 1191/1, 521-47. For "agrarian patriotism" see C. A. Bayly, *Imperial Meridian* (London: Longman, 1989), 121-25, and Drayton, *Nature's Government*, 90-92.

11. Brisbane, *Reminiscences of General Sir Thomas Makdougall Brisbane of Brisbane and Makerstoun*, 13-4, 47, 89-93; Brisbane, Lecture at Glasgow Astronomical Institution, 16 December 1836, Mitchell Library, State Library of New South Wales, MSS 1191/1, 521-47; and Brisbane, "Sacred thoughts," Mitchell Library, State Library of New South Wales, MSS 419/2, 1 (September 1823). Compare Brisbane to Barnard, 24 February 1823, Northern Ireland Public Record Office, D207/67/58: "my internal monitor fully acquits me."

12. Herschel to Gipps, 26 December 1837, Royal Society Archives, HS 19.72; Bathurst to Brisbane, 23 January 1824, *Historical Records of Australia* 11 (1917): 201-2; and Brisbane to Rümker, 9 August 1824, Mitchell Library, State Library of New South Wales, A1559/2/62.

13. Davy to Brisbane, 3 May 1821, Mitchell Library, State Library of New South Wales, MSS 419/1/29-31, and Herschel, *Essays from the Edinburgh and London Quarterly Reviews with Addresses and Other Pieces*, 498-99.

14. John Pringle Nichol, *Views of the Architecture of the Heavens*, 3rd edn (Edinburgh: Tait, 1839), 81. For survey colonialism elsewhere see Patricia Seed, *Ceremonies of Possession in Europe's Conquest of the New World* (Cambridge: Cambridge University Press, 1995); Bernard Cohn, *Colonialism and Its Forms of Knowledge* (Princeton: Princeton University Press, 1996), 3-11; and Drayton, *Nature's Government*, 90-91. Dubious claims that "there is no evidence that physics or astronomy played any significant role in British imperial policy or colonial rule" and that "those who achieved distinction overseas in the exact sciences did so by staying beyond the reach of the imperialist octopus that moved men across oceans to satisfy metropolitan appetites" were made respectively by Paolo Palladino and Michael Worboys, "Science and Imperialism," *Isis* 84 (1993): 91-102, at 101, and Lewis Pyenson, "Why Science May Serve Political Ends: Cultural Imperialism and the Mission to Civilize," *Berichte zur Wissenschaftsgeschichte* 13 (1990): 69-81, at 77.

15. Haynes, *Explorers of the Southern Sky*, 21-36; Bernard Smith, *Imagining the Pacific: In the Wake of the Cook Voyages* (New Haven: Yale University Press, 1992), 135-72; and Dening, *Performances*, 215-21.

16. Dettelbach, "Global Physics and Aesthetic Empire."

17. For example, Whewell to Jones, 21 August and 19 December 1834, in Isaac Todhunter, *William Whewell* (London: Macmillan, 1876), 2:186, 193; Ian Hacking, *The Taming of Chance* (Cambridge: Cambridge University Press, 1990), 131.

18. Dickens, *Hard Times*, 99; Herschel, *Essays from the Edinburgh and London Quarterly Reviews with Addresses and Other Pieces*, 510; Francis Baily, "On the Method of Determining the Difference of Meridians by the Culmination of the Moon," *Memoirs of the Astronomical Society* 2 (1824): 1-26; and John Brinkley, "Results of

Computations on Astronomical Observations Made at Paramatta," *Memoirs of the Astronomical Society* 2 (1825): 106–28.

19. Maclear to Herschel, 13 December 1837, in Brian Warner and Nancy Warner, *Maclear and Herschel* (Cape Town: Balkema, 1984), 207, and Whewell to Somerville, 29 August 1838, in Mrs Stair Douglas, *Life of William Whewell* (London: Kegan Paul, 1881), 194.

20. Maureen Perkins, "Almanacs and Astrological Belief in Colonial Australia," *Journal of Religious History* 17 (1993): 465–74; Sybil Jack, "Cultural Transmission: Science and Society to 1850," *Australian Science in the Making*, ed. R. W. Home (Cambridge: Cambridge University Press, 1988), 45–66.

21. Davy to Brisbane, 3 May 1821, Mitchell Library, State Library of New South Wales, MSS 419/1/29–31; Leslie to Brisbane, 29 May 1822, National Library of Australia, Parkes, MSS 4036/11/8; Thomas Brisbane, "Meteorological Observations at Paramatta," *Edinburgh Philosophical Journal* 11 (1824): 119–20, and "Meteorological Tables Kept in 1822 at Macquarie Harbour and Hobart's Town," *Edinburgh Journal of Science* 2 (1825): 75–77. Brisbane's meteorological journal on *Royal George* (1821) is at Royal Society Archives, Meteorological Archives, 56; his later meteorological and geomagnetic notes are at National Library of Australia, Parkes, MSS 7852/1.

22. Cannon, *Science in Culture*, 73–82; David Philip Miller, "The Revival of the Physical Sciences in Britain 1815–1840," *Osiris* 2 (1986): 107–34, at 120–21; Brisbane, *Reminiscences of General Sir Thomas Makdougall Brisbane of Brisbane and Makerstoun*, 33–35; and Brisbane to Duke of York, 8 June 1818, Mitchell Library, State Library of New South Wales, MSS 419/1, 2–3. See Arago to Brisbane, 11 May 1816, National Library of Australia, Parkes, 4036/11/4; for French surveys of Australia see John Dunmore, *French Explorers in the Pacific* (Oxford: Oxford University Press, 1969).

23. Dewhirst, "Meridian Astronomy," 148; H. A. Brück and M. T. Brück, *The Peripatetic Astronomer* (Bristol: Hilger, 1988), 22; and Brisbane to Maclear, 8 July 1833, in Warner and Warner, *Maclear and Herschel*, 5.

24. Brisbane, Lecture at Glasgow Astronomical Institution, 16 December 1836, Mitchell Library, State Library of New South Wales, MSS 1191/1, 547; Delambre to Brisbane, 2 March 1821, and Bouvard to Brisbane, 21 April 1821, Mitchell Library, State Library of New South Wales, MSS 419/1, 21–22, 25–27. For Brisbane and Laplace see Nicollet to Brisbane, 1 February 1819, Mitchell Library, State Library of New South Wales, MSS 1191/1, 95, 2.

25. Herschel, *Essays from the Edinburgh and London Quarterly Reviews with Addresses and Other Pieces*, 491; Dunlop to Herschel, 27 April 1828, Royal Society Archives, HS 6.515; and James Dunlop, "A Catalogue of Nebulae and Clusters of Stars in the Southern Hemisphere," *Philosophical Transactions* 118 (1828): 113–51, at 150–51.

26. Bennett, *The Divided Circle*, 169–74.

27. Maclear to Herschel, 14 April 1835, in Warner and Warner, *Maclear and Herschel*, 84; Rümker, *On the Most Effectual Means of Encouraging Scientific Undertakings*, 32; and Herschel to Gipps, 26 December 1837, Royal Society Archives, HS 19.72.

28. Dunlop's mural circle book, Royal Astronomical Society Archives, MSS Paramatta 5.1, p. 4 (16 June 1823); Service, *Thir notandums*, 143–46; Richardson, *A Catalogue of 7385 Stars Chiefly in the Southern Hemisphere*, vii; Charles Rümker, "Observations Made at the Paramatta Observatory," *Memoirs of the Astronomical Society* 3 (1829): 370–85, at 370; and Darling to Murray, 29 August 1828, *Historical Records of Australia* 14 (1920): 355.

29. Bennett, *The Divided Circle*, 174–77; Rümker, *On the Most Effectual Means of Encouraging Scientific Undertakings*, 6–7; and Russell, "Astronomical and Meteorological Workers," appendices G, J.

30. Service, *Thir notandums*, 171; Ashworth, "John Herschel, George Airy and the Roaming Eye of the State," 161–64; Airy to Harcourt, 5 September 1832, in *Gentlemen of Science: Early Correspondence of the BAAS*, ed. Jack Morrell and Arnold Thackray (London: Royal Historical Society, 1984), 152; and Herschel, *Essays from the Edinburgh and London Quarterly Reviews with Addresses and Other Pieces*, 640. Compare the founding statement of the Philosophical Society of Australasia (June 1821): "our ignorance arises in a great measure from the want of some nucleus which might gather around it the many valuable facts that are floating about." *Royal Society of New South Wales, Journal* 55 (1921): lxvii–viii.

31. "Bungaree, King of the Blacks," *All the Year Round* 1, no. 4 (1859): 77–83, at 79; Brisbane to Bathurst, 23 May 1825, *Historical Records of Australia* 11 (1917): 606–14; George Bergman, "Christian Carl Ludwig Rümker," 254–55; [David Brewster,] "Comet of Encke Re-Discovered in New South Wales" and "History of the Rediscovery of Encke's Comet," *Edinburgh Philosophical Journal* 9 (1823): 193, 391.

32. Royal Astronomical Society Archives, MSS Paramatta 5.1, p. 4 (16 June 1823); "Statement of Facts Relative to the Conduct of Mr. Rumker," Royal Astronomical Society Archives, MSS Baily 5 fol. 16; Dunlop to Baily, 2 February 1830, Royal Astronomical Society Archives, MSS Baily 5 fol. 82; and Brisbane to Bathurst, 15 November 1823, *Historical Records of Australia* 11 (1917): 154.

33. Geoffrey Blainey, *The Tyranny of Distance: How Distance Shaped Australia's History* (Melbourne: Macmillan, 1975), 56; Thomas Brisbane and Charles Rümker, "Observations Made at Paramatta and Sydney," *Transactions of the Royal Society of Edinburgh* 10 (1826): 112–16; Brisbane, "Observations Made in the Years 1823–24 at Paramatta," *Memoirs of the Astronomical Society* 2 (1826): 65–83; Bathurst to Brisbane, 15 November 1823, *Historical Records of Australia* 11 (1917): 304–5; Brisbane to Rümker, 9 August 1824, Mitchell Library, State Library of New South Wales, MSS A1559/2/62; and Brisbane to Bathurst, 13 and 23 May 1825, *Historical Records of Australia* 11 (1917): 480, 606–14.

34. "Observations Made at Paramatta in New South Wales by Sir Thomas Brisbane to Which Are Annexed Observations Made by Charles Rumker at Stargard," *Memoirs of the Astronomical Society* 2 (1826): 277–84; Dunlop to Baily, 2 February 1830, MSS Royal Astronomical Society Archives, Baily 5 fol. 82; New South Wales State Records, Kingswood, Colonial Secretary MSS 4/1844A/719, p. 419 (4 October 1825); "Statement of Facts Relative to the Conduct of Mr Rumker," Royal Astronomical Society Archives, MSS Baily 5 fols. 18–19; Rümker to Brisbane, 26 June

1829, and to Baily, 1 July 1829, MS Royal Astronomical Society Archives, Baily 5, fols. 29–30, 46–47.

35. Perkins, *Visions of the Future*, 192–93.

36. Ashworth, "The Calculating Eye," 410–15; and Dunlop to Baily, 20 May 1830, Royal Astronomical Society Archives, MSS Baily 5 fol. 1.

37. Charles Babbage, *Reflections on the Decline of Science in England* (London: Fellowes, 1830), 61, and David Philip Miller, "The Royal Society of London, 1800–1835" (Ph.D. diss., University of Pennsylvania, 1981), 297–372, at 324. For "banditti" see Ashworth, "The Calculating Eye," 430–34.

38. Ashworth, "'Labour Harder Than Thrashing,'" 201–5; Rümker, *On the Most Effectual Means of Encouraging Scientific Undertakings*, 29; Bergman, "Christian Carl Ludwig Rümker," 263; and Herschel to Brisbane, 15 March 1827, Royal Society Archives, MSS HS 5.101. In 1825 Brisbane painstakingly noted occasions when Rümker published data without crediting him: National Library of Australia, Parkes, MS 7852/2.

39. Brisbane to Herschel, 26 March 1828, Royal Society Archives, MSS HS 5.102; Royal Society Archives, MSS Council Minutes 10 (1822–28), 322–24, 326–27; Herschel, *Essays from the Edinburgh and London Quarterly Reviews with Addresses and Other Pieces*, 498; Herschel to Baily, 12 May 1828, and Young to Herschel, 13 May 1828, Royal Society Archives, HS 3.79 and 18.339; Brisbane's copy of Royal Society Archives, Council resolution, 22 May 1828, National Library of Australia, Parkes, MSS 4036/17/1; Brisbane to Murray, 9 June 1828, and Murray to Brisbane, 3 July 1828, Royal Astronomical Society Archives, MSS Baily 5 fols. 31, 58; Herschel to Gilbert, 2 September 1828, Royal Society Archives, HS 8.116; and Royal Astronomical Society Archives, Papers 2.2, Council Minutes, 11 November 1828.

40. Rümker to Brisbane, 26 June 1829, Royal Astronomical Society Archives, MSS Baily 5 fols. 29–30; Bergman, "Christian Carl Ludwig Rümker," 263–67; Baily memorandum, Royal Astronomical Society Archives, MSS Baily 5 fols. 20–23; Rümker to Brisbane, 1 July 1829, Royal Astronomical Society Archives, MSS Baily 5 fol. 45; Twiss to Brisbane, 9 July 1829, Royal Astronomical Society Archives, MSS Baily 5 fol. 58; Rümker to Richardson, 14 July 1829, Royal Astronomical Society Archives, MSS Baily 5 fol. 43; Twiss to Darling, 21 December 1829, *Historical Records of Australia* 15 (1921): 298; Rümker, "Astronomical Observations Made at the Observatory at Paramatta," *Philosophical Transactions* (1829), part 3, 1–152, at iv–v.

41. Rümker to Richardson, 15 December 1829, Royal Astronomical Society Archives, MSS Baily 5 fol. 39; Rümker, *On the Most Effectual Means of Encouraging Scientific Undertakings*, 8–10; Babbage to Herschel, 15 and 18 December 1829, Royal Society Archives, HS 2.241–2; Brisbane to South, 4 January and 19 January 1830, Royal Astronomical Society Archives, MSS Baily 5 fols. 78 and 80; Dunlop to South, 5 January 1830, Royal Astronomical Society Archives, MSS Baily 5 fol. 84; and Babbage to Brisbane, 14 January 1830, Mitchell Library, State Library of New South Wales, MSS 419/1/137.

42. Baily to Brisbane, 28 January and 16 February 1830, Royal Astronomical Society

Archives, MSS Baily 5 fols. 93 and 97; Richardson to Baily, 13 February 1830, Royal Astronomical Society Archives, MSS Baily 5 fol. 61; Brisbane to Sabine, 18 January and 21 February 1830, Royal Astronomical Society Archives, MSS Baily 5 fols. 24 and 34 (= Royal Society Archives, MSS Domestic 4.74–77); Sabine to Brisbane, 6 February and 13 March 1830, Royal Astronomical Society Archives, MSS Baily 5 fols. 26 and 33 and National Library of Australia, Parkes, MSS 4036/11/27; Brisbane to Baily, 11 February 1830, and Baily to Brisbane, 13 February 1830, Royal Astronomical Society Archives, MSS Baily 5 fols. 57 and 95. For Richardson's medal see *Monthly Notes of the Astronomical Society* 1 (1830): 165.

43. J. L. E. Dreyer and H. H. Turner, *History of the Royal Astronomical Society* (London: Wheldon and Wesley, 1923), 50–52; Miller, "The Royal Society of London," 337–40; Brisbane to Murray, 28 February 1830, Royal Astronomical Society Archives, MSS Baily 5 fols. 68–69 (= Royal Society Archives, MSS Domestic 4.78); Brisbane to Murray, 2 May 1830, Royal Society Archives, MSS Domestic 4.81; Rümker to Baily and Baily to Brisbane, 17 May 1830, Royal Astronomical Society Archives, MSS Baily 5 fols. 72 and 100; Baily to Gilbert, 20 May 1830, Royal Astronomical Society Archives, MSS Baily 5 fol. 55; Rümker, *On the Most Effectual Means of Encouraging Scientific Undertakings*, 16, 59; and Schumacher to Baily, 9 March 1830, Royal Astronomical Society Archives, MSS Baily 5 fol. 70. Baily's draft of his report is Royal Astronomical Society Archives, MSS Baily 5 fols. 6–14; Brisbane's copy is Mitchell Library, State Library of New South Wales, MSS 419/2/6; Rümker's copy is in *On the Most Effectual Means of Encouraging Scientific Undertakings*, 18–28.

44. Bergman, "Christian Carl Ludwig Rümker," 272, 275–76; Rümker, *On the Most Effectual Means of Encouraging Scientific Undertakings*, 33–34; Michael Hoskin, "Astronomers at War: South versus Sheepshanks," *Journal of the History of Astronomy* 20 (1989): 175–212; Brisbane to Gilbert, 17 November 1830, Royal Society Archives, MSS Domestic 4.103; Baily to Gilbert, 5 November 1830, Royal Astronomical Society Archives, MSS Baily 5 fol. 106; and Miller, "The Royal Society of London," 350–61.

45. Richardson, *Catalogue*, v–vi, 260–71; Beaufort to Herschel, 10 May 1832, Royal Society Archives, MSS HS 3.327; and Warner and Warner, *Maclear and Herschel*, 40, 84, 113, 135, 141. Herschel's working copies of the Brisbane Catalogue are listed in Isabella Herschel, *Catalogue of the Herschel Library*, ed. Sydney Ross (Troy: Editor Books, 2001), 81.

46. Warner and Warner, *Maclear and Herschel*, 180, 204; Herschel to Dunlop, 15 December 1837, UT H/L 0145; David S. Evans et al., eds., *Herschel at the Cape* (Austin: University of Texas Press, 1969), 67, 333; and Herschel to Gipps, 26 December 1837, Royal Society Archives, HS 19.72.

47. Service, *Thir notandums*, 189, 201–3; James Dunlop, "Observations by the Transit and Mural Circle of Mars and Vesta," *Memoirs of the Royal Astronomical Society* 8 (1835): 227–29; and Mitchell lecture (1855), Mitchell Library, State Library of New South Wales, MSS A295/3, p. 604.

48. Ashworth, "Herschel, Airy and the State," 173; Herschel, *Essays from the Edinburgh*

and London Quarterly Reviews with Addresses and Other Pieces, 63–141 (June 1840); Robert Scott, "The History of the Kew Observatory," *Proceedings of the Royal Society of London* 39 (1886): 37–86, at 47–52; Brisbane, *Reminiscences of General Sir Thomas Makdougall Brisbane of Brisbane and Makerstoun*, 68–70; and Graeme Gooday, "Precision Measurement and the Genesis of Teaching Laboratories in Victorian Britain," *British Journal for the History of Science* 23 (1990): 25–51.

49. Phillip King to Colonial Secretary, 21 December 1848, in Russell, "Astronomical and Meteorological Workers in New South Wales," appendix N.

50. Allan Chapman, "Private Research and Public Duty: George Biddell Airy and the Search for Neptune," *Journal of the History of Astronomy* 19 (1988): 121–39, at 126 and 137; Airy to Parker, 9 November 1849, in Wayne Orchiston, "Illuminating Incidents in Antipodean Astronomy: P. P. King and the Founding of Sydney Observatory," *Vistas in Astronomy* 32 (1989): 285–301, at 291; Bergman, "Christian Carl Ludwig Rümker," 282; and Smyth to Lee, 19 December 1861, Royal Astronomical Society Archives, MSS Paramatta 1.

51. Murchison to Harcourt, 4 May 1839, in Morrell and Thackray, *Gentlemen of Science*, 312; Murchison to Sacker, 18 October 1859, Mitchell Library, State Library of New South Wales, MSS 1191/1/A(i)/263; Brisbane, *Reminiscences of General Sir Thomas Makdougall Brisbane of Brisbane and Makerstoun*, 80–81; James Secord, "King of Siluria: Roderick Murchison and the Imperial Theme in Geology," *Victorian Studies* 25 (1981): 413–42.

Chapter 5

Training Seafarers in Astronomy: Methods,

Naval Schools, and Naval Observatories

in Eighteenth- and Nineteenth-Century France

GUY BOISTEL

TRANSLATED BY DAVID AUBIN

AND CHARLOTTE BIGG

> Si la Marine a besoin des astronomes, les astronomes ont besoin du
> ministre de la Marine.
> —JÉRÔME LALANDE (1803).[1]

In astronomy the eighteenth century was dominated by the quest for a
method to determine longitude at sea. A great number of scientific studies
of the time were devoted to theoretical and technological innovations that
would help seafarers and astronomers in this respect. To this end celestial
mechanics (especially lunar-motion theory) was developed, octants, sextants, and naval chronometers were successively improved, the theory of
error analysis was refined, and attempts were made to simplify methods of
nautical astronomy. This search undertaken in the early eighteenth century
was pursued without interruption over the following century. Contrary to
what is usually assumed in longitude stories, which emphasize the role of
John Harrison's clocks, I contend that there is no significant rupture between the eighteenth and nineteenth centuries in the techniques of nautical
astronomy, but rather, as this chapter will show, a striking continuity in the
problems and solutions debated among astronomers and seafarers.

This chapter surveys the vast panorama of astronomy and navigation
in France, beginning in 1667 when the building of the Royal Observatory
in Paris commenced, to 1882 when its director, Admiral Ernest Mouchez,
organized the telegraphic distribution of time to French merchant seaports.
I shall discuss the different initiatives to disseminate astronomical practices

among seafarers. Only at the end of the nineteenth century did having accurate timekeepers emerge as the most practical and affordable method for determining longitude at sea. Until this time astronomical methods—such as those based on the measurement of lunar positions—were, I argue, preferred by the rising French scientific élite. But the lunar method required extensive computing, and its practical use at sea was hampered by the navigators' poor level of mathematical training throughout the period covered here. As a result, from 1765 the scientific training of naval officers became a pressing issue for the French state. To address it, successive generations of French navy ministers chose to rely on astronomers, a decision that would play an important part in their professionalization.

Within nautical astronomy theoretical and practical considerations went hand in hand with social reforms. Astronomers' involvement in navigational issues therefore went beyond teaching, writing textbooks, designing instruments, and testing potential candidates: they were drawn into debates about the organization of training and research institutions for nautical astronomy. Was their proper role to dispense extensive scientific training to seafarers, or should they endeavor to simplify their methods and make them more accessible? While there was no great rupture in this period either in the debates that pitted astronomers against seafarers nor in the astronomical methods used in navigation, crucial developments can nevertheless be identified. As the use of marine clocks became widespread and transformed seafarers' procedures, naval observatories and naval training schools had to be adapted accordingly. I briefly discuss the importance of navigation for the history of astronomy before turning to the debates about seafarer training and then to the creation of naval observatories. In conclusion I ask whether one can truly speak of a transfer of techniques from the observatory to the ship.

Astronomy and Navigation: An Old Story

When the great royal observatories were established in Paris and Greenwich in 1664–75, the task assigned to astronomers was clearly stated: to solve the vital problem of determining longitude at sea. In Britain the astronomer royal was instructed "to apply himself with the most exact care and diligence to Rectifying the Tables of the Motions of the Heavens and the Places of the Fixed Stars, in order to find out the so much desired Longitude at Sea, for perfecting the Art of Navigation."[2] Indeed the nation that would uncover the "secret of longitude," to use the eighteenth-century

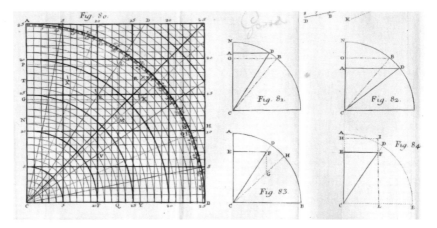

26. Seamen reported data obtained by gross reckoning on the sinical quadrant: the estimated speed of the ship and the rhumb line (i.e. a steady course steered with respect to meridian lines). Regular graduations allowed them to deduct the sine of the various angles used in the basic nautical computations detailed in all British and French navigation treatises at the end of the eighteenth century. François-Vincent Dulague, *Leçons de navigation* (Rouen, 1784), fig. 80.

expression, would be sure to dominate commercial routes while keeping its crews safe from sudden shipwreck caused by maritime maps' dramatic lack of precision.

Until about 1750 latitude and longitude were determined on ships with separate procedures. Astronomical observations were exclusively used for determining latitude. The height of the celestial North Pole over the horizon was usually measured by a rudimentary instrument called a "Jacob's staff," and then increasingly with the octant, an instrument requiring more delicate manipulation. In this period naval officers, astronomers, and instrument makers designed various types of octants. Their precision was largely a function of their price and maker. The sextant, invented in 1757 by the British naval officer John Campbell, increased the precision of latitude determinations.

Longitude was determined by dead reckoning. The ship's route was approximated by measuring its speed using a loch, a rope with regularly spaced knots that drifted freely behind the ship. The difference in longitude proper depended on wind drift. It was approximately determined using the sinical quadrant (figure 26) on the basis of the difference in latitude between points of departure and arrival, the estimated speed of the ship, and observations of wind direction.

At the start of the eighteenth century colonial kingdoms promulgated acts promising important rewards for the discovery of a more reliable method that would unveil the "secret of longitude." Decreed by Queen Anne on 3 July 1714, the British Longitude Act was the most famous: it "provid[ed] a publick reward for such person or persons as shall discover the longitude at sea." The act also aimed at rationalizing existing methods, none of which were excluded from the outset. To examine proposals an expert committee was set up, the Board of Longitude, led by Sir Isaac Newton. In France many an inventor's fantasy was fueled by the vague promises of magnificent rewards made by the Regent Philippe d'Orléans in 1716. The Royal Academy of Science decided in 1722 to use a bequest from Count Rouillé de Meslay to fund a biannual prize for navigation, which unfortunately had little effect on nautical astronomical research.[3]

A difference in longitude equals (the measurement of) the difference in time between a standard meridian (Paris, Greenwich, or any other meridian passing through an observatory) and the meridian on which the ship is located. The main difficulty therefore was to know the time at the "home" meridian (the standard) at the very moment when observations were made at sea. In theory this required no more than an accurate timekeeper, which gave the time at the port of departure. But this solution remained impractical well into the nineteenth century. Not until John Harrison claimed the British Prize in 1765 with his improved clocks did the Paris Academy of Sciences finally overrule its negative view of naval clock making and encourage trial voyages to develop naval chronometers.[4]

The efforts to determine the shape of the earth in the early eighteenth century must also be understood in connection with astronomers' and seafarers' wish to improve navigation. Behind the debate between the old Cartesians and the young Newtonians about the new physics lay the desire to know the shape of the earth for navigational purposes. Pierre-Louis Moreau de Maupertuis's expedition to Swedish Lapland (1736–37), for instance, was publicized in four books in which astronomical navigation figured largely: *Élémens de géographie* (1740), *Discours sur la parallaxe de la Lune* (1741), *Astronomie nautique* (1743), and *Traité de loxodromie tracée sur la véritable surface de la Mer* (1748).[5]

A direct consequence of this expedition was to put astronomers and members of the Royal Academy more firmly in charge of improving navigation. Shortly after Maupertuis's return from Lapland, the minister of the navy, Pierre-Louis Moreau de Maurepas, made him "responsible for the improvement of navigation" or "of the Navy in all its forms" (*préposé au perfectionnement de la navigation or de la Marine sous toutes ses formes).[6]* In 1745 Pierre Bouguer succeeded Maupertuis. He had begun his career as a professor of hydrography before acquiring fame as a mathematician and above all as a member of the Peru expedition, the second geodetic investigation to determine the shape of the earth, sponsored by the Academy of Sciences (1733–44). Bouguer's authoritative rejection of chronometer-based longitude determination held sway at the academy until Harrison. After Bouguer's death his position was split between the mathematician Alexis Clairaut and the astronomer Pierre-Charles Le Monnier, both experts on lunar tables and their nautical uses.[7] Put in charge of perfecting the navy, they worked without interference from their peers under the direct control of the minister. In the eighteenth century improving nautical astronomy was considered a task not for naval officers but for the scientific élite: royal astronomers and members of the Royal Academy of Sciences.

At that time the only reliable method for determining longitude relied on the predicted timing of astronomical phenomena such as the eclipses of Jupiter's satellites, or on the angular distances between the bright edge of the moon and selected zodiacal stars. Astronomers increasingly argued that the moon was the only natural clock that could be used regularly at sea. In 1749, after he had perfected the octant, the French West Indies Company officer Jean-Baptiste d'Après de Mannevillette had been the first naval officer to apply the lunar distance method at sea. His voyage to the Cape of Good Hope (1750–54) with Nicolas-Louis de Lacaille did much to rekindle practical interest in the lunar distance method.

The practical implementation of the method was also made possible by important theoretical advances. Since perturbation of the lunar orbit by the sun was very significant, the moon's motion was one of the most arduous problems of celestial mechanics. The accurate prediction of lunar motions required an approximate solution to the infamously intractable three-body problem. Even the contribution of Edmond Halley and other astronomers could not sufficiently improve the precision of lunar tables for reliable longitude determinations at sea: to determine a difference of longitude of half a degree (that is, with an error of fifty kilometers on the equator), astronomers had to provide the ecliptic longitude of the moon with a precision of one minute. The mathematical analysis of the perturbations of the lunar orbit by Clairaut, Leonhard Euler, and Jean Le Rond d'Alembert, as well as by the German astronomer Tobias Mayer, between 1749 and 1754 offered the first approximate solution to the three-body problem.[8] On this basis Clairaut and Mayer significantly improved the precision of lunar tables. In

1754 Lacaille produced the first nautical almanac. Giving lunar distances every three hours, it provided the model for Nevil Maskelyne's *Nautical Almanac*, first published in London at the end of 1766.[9]

After the scientific voyage of the frigate *Flore*, sponsored by the academy to test naval timekeepers (1771–72), Jean-Charles de Borda codified the lunar distance method for practical use. He specified the order in which astronomical observations were to be made and described precisely how to measure heights above the horizon and angular distances using nautical instruments. His scheme explained the procedure to follow in the lengthy spherical trigonometric computations required for determining longitude differences by comparing observations with lunar tables (figure 27). He also introduced his famous repeating circle, an instrument that would be more successfully applied to geodetic terrestrial operations than to naval observations in the nineteenth century.[10]

The Scientific Training of Seafarers

HOW TO "PUT THE LONGITUDE METHOD WITHIN THE REACH OF ORDINARY NAVIGATORS"

The success of the trial runs at sea by the team on board the *Flore* in 1771–72 was portentous. Both Borda's standardized lunar distance method and the accuracy of naval timekeepers tested on board proved satisfactory. Naval chronometers were at long last deemed accurate and stable enough to be used at sea. From then on most astronomers and naval officers believed the lunar method to be a perfect solution to the longitude problem. But one should not conclude from this that Borda solved the problem and that the method was immediately integrated in sea captains' daily routines. Time-keepers were not immediately put to general use: technical requirements, notably thermal stability, were still wanting; and perhaps more significantly, accurate timekeepers long remained rarities unaffordable to most seafarers, especially merchant ship commanders.

The theoretical issue having been satisfactorily solved, the debate among specialists shifted to the practical feasibility of this method, requiring long series of logarithmic computations to reduce observations made at sea by means of spherical trigonometry.[11] Was this method a preserve of the élite? French naval historians have drawn attention to the ineffectual reforms in the scientific training of *gardes de la marine*, or naval cadets, decided and implemented by the navy ministers Étienne-François de Choiseul de Stainville in 1765 and Charles-Eugène de la Croix de Castries in 1786–87.[12]

Éléments du calcul.	Réduction de la distance.

Éléments du calcul.

Latitude . . . . . 16°. 10
Heure approchée . 5ʰ. o
Long. estimée . . 27 . o
Heure de Paris . . 6ʰ. 48
Demi-diam. ☉ . . 15′. 56
Demi-diam. ☾ . { 15 . 52
Aug. du ½ diam. { o . 11
Demi-diam. corr. 15′. 43″
Parall. hor. ☾ . . 56 . 55

Dis. obs. ☉ ☾ 116°. 8′. 50″
Demi-diam. ☉ o . 15 . 56
Demi-diam. ☾ o . 15 . 43
_____
116 . 40 . 29
Déviat. . . . — o . 20
_____
116 . 40 . 9
Err. du miroir. — o . 26
_____
Dist. ap. ☉ ☾ 116° . 39 . 43

Haut. obs. ☉ 18 . 40 . 55
Dép. de l'hor. — 4 . 3
_____
18 . 36 . 52
Demi diam. ☉ + 15 . 56
_____
Haut. app. ☉ 18 . 52 . 50
Réf. — par . — 2 . 37
_____
18 . 50 . 13
Corr. therm. + o . 6
Corr. bar . + o . 1
_____
Haut. ap. ☉ 18 . 50 . 20

Haut. obs. ☾ 44 . 15 . 25
Dép. de l'hor. — 4 . 3
_____
44 . 11 . 22
Demi-diam. ☾ + 15 . 43
_____
Haut. app. ☾ 44 . 27 . 10
Par. — réf. { + 59 . 1
{ + o . 39
Ther. et bar. + o . 2
_____
Haut. vr. ☾ 45 . 6 . 52

Réduction de la distance.

Dist. app. ☉ ☾ 116°. 59′. o″
Haut. app. ☉ . 18 . 52 . 50   com. cos. o . 0240192
Haut. app. ☾ . 44 . 27 . 10   com. cos. o . 1464065
_____
Somme   179 . 59 . o
Demi-somme. 89 . 59 . —   cos. 6 . 1626961
Moins la dist. 26 . 39 . 50   cos. 9 . 9511907
Haut. vraie ☉ { 18 . 50 . 20   cos. 9 . 9760886
Haut. vraie ☾ { 45 . 6 . 52   cos. 9 . 8486158
_____
Somme. 63 . 57 . 12   Somme 36 . 1090169
    Demi-somme 18 . 0545084
Demi-somme. 31 . 58 . 36   cos. { 9 . 9285509   8 . 1259775 . sin. A
     cos. A { 9 . 9999611   A = o°. 45′. 57″
_____
     Somme 9 . 9284920
C'est le sin. de la demi-distance . . . . 58°. o′ . 54″
         Distance . . . 116 . 1 . 48
     Quantité restituée . . . . — o . o . 43
_____
DISTANCE RÉDUITE . . . . 116 . 2 . 31   Différ. log. logi.
Dist. prises dans les tables. { 1ʳᵉ à 6ʰ 115 . 39 . 5   o°. 25′. 26″ 8854
{ 2ᵐᵉ à 9ʰ 117 . 9 . 9   1 . 30 . 4 3007
_____
     Différ. 5847
   Ce qui répond à . . . oʰ. 46′. 50″
   Heure de la premiere distance . . . 6 . o . o
_____
HEURE DE PARIS . . . 6 . 46 . 50

Calcul de l'heure du vaisseau.	Déclinaison.

Calcul de l'heure du vaisseau.

Hauteur ☉ . 18°. 50′. 20″
Latitude . . 16 . 10 . o com. cos. o . 0175226
Dist. pol. . 20 . o com. sin. o . 0124757
_____
Somme 111 . 20 . 20
Dem. som. 55 . 40 . 10   cos. 9 . 7512535
Moins haut. 36 . 49 . 50   sin. 9 . 7777554
_____
    Somme 19 . 5590050
    Demi-somme 9 . 7795015
C'est le sin. du demi-angle hor. 37°. o′ . 14″
       Multiplié par 8
_____
HEURE du VAISSEAU 4ʰ. 56′. 1″. 52‴
HEURE DE PARIS 6 . 46 . 50
_____
Différence en temps 1 . 50 . 48
LONG. A L'OUEST DE PARIS 27° . 42′

Déclinaison.

Décl. le 26 15°. 54′. 31″ʰᵒʳ.
Et le 27 13 . 53 . 40
Dif. en 24ʰ o . 19 . 9
Part. proportionnelles.
Pour 6ʰ { o . 4 . 47
Pour 45′ { o . o . 36
Pour 2′ { o . o . 2
_____
Pour 6ʰ 47′ o . 5 . 25
Déclin. . 13 . 39 . 56
Dist. pol. 76 . 20 . o

Calcul des observations de longitude faites par un seul observateur.

27. Borda's original standardized sheet for logarithmic spherical trigonometric computations used in the lunar-distance method. The Dépôt de la Marine later printed this sheet and distributed it to sailing ships. J.-C. de Borda, *Description et usage du cercle de réflexion avec différentes méthodes pour calculer les observations nautiques* (Paris: Firmin Didot, 1802), 64.

Beyond scientific issues, a social struggle pitted proponents of recruiting naval cadets primarily from the nobility (such as Borda and Louis-Antoine de Bougainville) against those who believed that the navy should also rely on talented men who were neither noble nor gardes de la marine already.

Within the Academy of Sciences there was no consensus. Noticing as early as 1754 that most seafarers lacked the scientific training needed to carry out the lunar method, Lacaille argued that this method should be adapted and made practical for navigators (put "within the reach of ordinary seafarers").[13] In Nantes the hydrography professor Pierre Lévêque, an ardent follower of Borda, castigated merchant commanders as "vile routineers" who neglected astronomical observation in favor of "gross reckoning" of the ship's route.[14] About to be made examiner of the royal hydrography schools and of the navy, Lévêque wrote the first published presentation of Borda's method. His *Guide du navigateur*, published in Nantes in 1778, cleared the way for a new generation of navigation manuals and treatises.[15]

Would this be enough to change practices and minds? Ten years later a future minister of the navy, Pierre Claret de Fleurieu, produced a sorry assessment of the practical state of astronomical navigation: "The English have employed their computers to [compose] tables that make up an enormous volume in which ready-made computations are to be found. Merchant-ship masters [in France] are unable to use them; they fear the immensity of these computations, through which they have to sift, the proper rules of proportion, the plus or minus signs, the erratas, &c. . . . but when they are shown Borda's method, they cry out [']this is impossible. If knowledge of this is required to become a Captain, we shall never succeed. Taking the moon in our hand or performing this computation is equally difficult for us.[']"[16]

But in his letter to the minister Fleurieu skirted the fundamental issue: the organization of the naval personnel's scientific training. Subscribing to a widespread view, he castigated the laziness ("molle routine") of French mariners and appealed to them to "discard apathy and devote themselves zealously to the small amount of supplementary work imposed by the observation of distances or the use of naval timekeepers."[17]

Yet the debate among scholars had never been so intense. In 1788 Abbot Guillaume Thomas Raynal bequeathed to the Paris Academy of Sciences the funds for a prize "having practical navigation as its principal object."[18] Two years later the Academy proposed a topic that echoed Lacaille's call of 1754: "to find, for the reduction of the apparent distance between two celestial bodies to a true distance, a sure and rigorous method that nonethe-

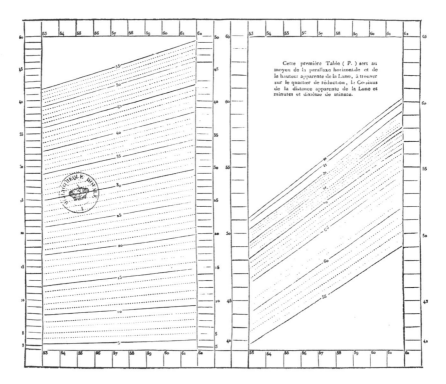

28. The graphic method introduced by Alexis Rochon was based on the use of two graphs and a sinical quadrant (see fig. 26). Rochon's method allowed seafarers to compute graphically the corrections to be applied to the observed lunar distance, e.g. to take into account the horizontal parallax of the moon. *Mémoire sur l'astronomie nautique* (1798), read on the 1st Ventôse an 6 (Imprimerie de Prault).

less in practice requires only simple computation and [is] within the reach of the greatest number of seafarers."[19] In his *Abrégé de navigation*, published in 1793, Jérôme Lalande presented the first alternative to the lunar distance method since 1773. Lalande based his method on solar observations, hoping to obtain the navy's support for the publication of his *Tables horaires* of the celestial position of the sun. The timetables simplified computations considerably and were geared toward the needs of the average seafarer. In 1798 the Abbot Alexis Rochon published two texts introducing a simplification of the lunar distance method (figure 28). The first was printed in Brest and quoted Lacaille in its title: *Exposition d'une méthode facile, et à portée du commun des navigateurs, pour résoudre les utiles problèmes de la latitude et de la longitude (ca. 1798)*. In his *Mémoire sur l'astronomie nautique*, presented to the Institut de France on the 1 Ventôse of year VI (19 February 1798), Rochon wrote: "It is no doubt saddening to think that the art of reaching down to the level of most men is not without difficulties; it is a sad truth that first-rank savants perhaps do not feel as vividly as less educated men. I have seen savants justly famous for the extent of their knowledge, who were not always as useful as they should have been, had they better judged, known more about the influence of a neglected education on the great majority of men."[20]

To simplify and popularize was the astronomer's task, according to Lalande, echoing Rochon: "Tables have simplified the remaining aspects of longitude calculations . . . One part however greatly lengthens [works with] the longitude method and prevents many seafarers from engaging in these studies: if they persist in neglecting these observations at the risk of their fortunes and lives, it is the astronomers' duty to smooth difficulties and to remind them of their vital interests."[21] But in 1798 Borda and Lévêque expressed the opposite opinion when they condemned the use of graphic methods, which had "the drawback of having men, only too inclined to it, becoming used to a process in some way automatic." Elsewhere Borda concluded that the best way for navigators to avoid the difficulties and inconveniences of computing was to be properly taught how to compute. Borda's and Lévêque's expectations were high and their opinion of seafarers low: "It is about time that seafarers ceased looking at the mathematical and physical sciences as having no practical use in navigation and its progress. Without the help of science, the Navy would still be in infancy."[22]

For most mathematicians, navy examiners, and savant-officers of the French navy it was up to seafarers to rise to the exigencies and competence required by new navigational methods, and not up to mathematicians to simplify solutions and contrivances in order to circumvent direct computation.[23]

NAVAL SCHOOLS AND NAUTICAL ASTRONOMY
IN NINETEENTH-CENTURY FRANCE

During the French Revolution and the First Empire successive governments did not concern themselves much with the navy. A law passed on 15 May 1791 even purely and simply abolished the navy corps and marine garde corps. Instead it set up hydrography schools in seaports, free of charge for students. But it sidelined the issue of officers' training for the state navy.[24] Admittedly the rank of midshipman was created in 1791, but with little practical effect. On 27 September 1810 the Emperor Napoleon decreed the foundation of two on-board schools (*écoles embarquées*) based in

29. Practical training in the use of sextants on the on-board school *Borda*. [Pech de Cadel,] *Histoire de l'école navale et des institutions qui l'ont précédée, par un ancien officier* (Paris: Maison Quantin, 1889), 284.

Brest and Toulon that became operational at the end of 1811 (figure 29). After the fall of the emperor King Louis XVIII abolished both imperial schools and on 31 January 1816 established the Collège Royal de la Marine. This school was located inland in Angoulême, chosen for its central location near the military seaport of Rochefort. Although both Louis XVIII and his successor Charles X supported the naval college, it quickly proved inadequate for the needs of the navy. In the 1820s Louis XVIII's minister of the navy, François-Joseph Gratet du Bouchage, made the following assessment: "For young people intending to serve in the Navy, practical and theoretical training cannot be conducted simultaneously. The first demands a special college, and the second expressly designed ships."[25] As a result the school ship *Orion* was outfitted for the task, but remained in duty for only three years. Inaugurated in January 1818, the Collège Royal was active until 1 November 1830, when it was abolished by King Louis-Philippe. Going back to Napoleon's idea, the new king decreed that the college be replaced with an on-board naval school registered in Brest. The laws of 20 April 1832 and 4 May 1833 spelled out the regulations of the school. Originally set at twenty months, the duration of training was extended to two years in 1833.

Governmental efforts to organize seafarers' scientific training notwithstanding, the authors of naval treatises continued to voice the same concerns as their forerunners. They seemed equally perplexed as to what merchant-ship commanders should be taught. Should they provide complete mathematical proofs? How far should they simplify the presentation of astronomical methods of navigation? For the astronomer and physicist Jean-Baptiste Biot, whose second edition of the *Traité d'astronomie physique* included additions on nautical astronomy written by Édouard de Rossel (1811), there was no doubt: "men who have displayed the most ingenuity in the sciences are the most capable, I would almost say the only men capable of giving their teaching the right direction."[26] This bravado concluded a report on the *Traité de navigation* submitted for the academy's approval in 1808 by Professor Jean-Baptiste du Bourguet of Dieppe. In the name of a committee composed of Rochon, Bougainville, and Delambre, Biot judged the methods described by the author as follows:

A large number of authors from various nations have published treatises on this topic. Most were content to give only strictly necessary notions of astronomy and computation and restricted themselves to giving practical details rather than developing demonstrations and the gist of the methods. Mr. Du Bourguet's plan was both more instructive and more extensive: he carefully provides proofs . . . of all the rules that must guide the navigator; he looks for direct solutions to all problems in nautical astronomy and, when the nature of the problem is such that this is not possible, instead of resorting as is usually done to successive estimates and approximations, he looks to differential calculus for corrections required by the assumptions he was forced to make at the beginning of the computation. This way of proceeding is more mathematical and more satisfying for the mind. The basic idea of this work therefore deserves navigators' gratefulness and savants' approval.[27]

Clearly little had changed since 1743, when Maupertuis published his *Naval Astronomy [Astronomie nautique]*, a treatise for the shipwrecked entirely based on calculus![28] In the early nineteenth century merchant-ship commanders seemed, like Maupertuis's shipwrecked, lost in spherical trigonometry computations. In 1818 Professor Pierre Caillet, from the Public School of Hydrography in Paimbœuf (near Saint-Nazaire in the Loire estuary), wrote a *Manuel du navigateur* that stated: "Of all the computations that are done at sea, the most difficult is that of longitude using the distance of the moon to the sun and stars. Thus, even when they are convinced of its usefulness, many seafarers, either because they fear its length or for lack of time, sail at a given latitude and estimate longitude by reckoning . . . this results in great uncertainties, unexpected landings at night, and the fear of having bypassed the destination when hundreds of leagues are still to be covered."[29] According to a professor of hydrography from Nantes, the situation had not improved much by 1840: "All computations used for navigation are based on a mathematical knowledge that most merchant naval officers, not to speak of those intending to become coasting pilots, do not possess. This knowledge can only be acquired through special, time-consuming studies; most naval personnel begin navigating at an early age, and cannot master the theories that lead to a knowledge of the perfect science of navigation . . . The goal of this book would be achieved if it heartens the naval officer who fears reading overly scientific books and considers as beyond his reach the computations that are for him a daily necessity."[30] On 18 May 1877 the astronomer Antoine Yvon Villarceau and Lieutenant Henri-Julien de Magnac presented to the academy their book on "new navigation" and its methods.[31] In this method (also known as the Sumner, or Marcq Saint-Hilaire, method) the line of position played a central role, allowing for the first time a simultaneous measurement of latitude and longitude. That year a long polemic on the book unfolded. In their discussion Antoine Yvon Villarceau from the Paris Observatory and Admiral Mouchez used terms that echoed those exchanged by naval officers and astronomers a century earlier.[32] In the 1780s new navigation had meant the delicate measurement of lunar distances. A century later the term was applied to computing thermal compensation for naval timekeepers. While Yvon Villarceau and his followers were in favor of complex numerical methods (some of them probabilistic), Mouchez and other naval officers wanted simple, direct, and efficient algorithms using graphic methods. Rejecting the use of the Taylor series advocated by Yvon Villarceau as "long and arduous [computations that are] impracticable and useless for seafarers," Mouchez concluded: "These speculative studies are a little too subtle, of no interest other than theoretical, and similar to the nautical astronomical studies of scores of professors and savants barely familiar with the needs of the Navy: they can be of no practical use whatsoever. We need simpler, surer, and quicker procedures to guide a ship."[33]

Like Borda in his time, Yvon Villarceau wished to rationalize the methods used at sea and to draw a clear line of demarcation between the "old" and the "new" in navigation. Mouchez's reaction was sharp: "I have always protested in the most formal way against the many treatises written by authors foreign to the Navy. On this matter, there is too great a difference between theory and practice for professors and astronomers who have never sailed to pretend doing anything else than to produce possible variations on trigonometric combinations, and, if they wish to introduce some new solution, they most often run the risk of coming up with problems devoid of practical usefulness and of the type that old mathematicians used to call 'mathematical recreations.' . . . In a word, it is not on land that one learns the naval profession, even from the point of view of nautical astronomy."[34] Forced to acknowledge that he was no specialist of nautical astronomy, Yvon Villarceau modestly said that his only wish was to respond to the request of the Ministries of Navy and Public Instruction. Like Maurepas choosing Maupertuis as his aide for naval affairs (*préposé de la marine*) in 1739 to address problems in astronomical navigation, the nineteenth-century French state preferred to rely on savants sitting in Paris, at the Academy of Sciences.[35]

The Rise and Fall of Naval Observatories in France

CALIBRATING MARINE CLOCKS AND
TRAINING MERCHANT-SHIP CAPTAINS

Seafarers' increasing disaffection of astronomical methods may also have stemmed from the rapid development of marine clock making and the wider distribution of timekeepers in both the military and civilian navies. Until the 1860s the lunar-distance method was almost exclusively used on board military ships, and only for calibrating and regulating marine chronometers.[36]

But having a marine timekeeper at hand did not necessarily mean being able to use it. By the beginning of the nineteenth century clock making had greatly improved in France. At two exhibitions of "Products of French Industry" in 1802 and 1806, the official naval clockmaker Louis Berthoud, as well as his later successor Louis Breguet, displayed naval clocks and timekeepers of unprecedented precision. Their average daily deviation was below two seconds.[37] Tested in the worst sea conditions, Breguet's chronometers differed from each other by less than one minute over a period of six months. French naval timekeepers now fulfilled the requirements set for the prize of £10,000 offered by the British: an error smaller than two minutes in six months. The success of the timing method for direct determinations of longitude at sea seemed granted. After 1814 all navigation manuals followed the model provided by the naval officer Édouard de Rossel in the new edition of Étienne Bézout's *Traité de navigation* and included chapters on a "method to calibrate a marine timekeeper," "determination of longitudes by means of naval timekeepers," and a "method for correcting longitudes obtained from naval timekeepers."[38]

Upon a visit to Lorient in 1822 the minister Aimé-Marie Gaspard de Clermont-Tonnerre noted that docked ships had no way of calibrating their chronometers and ordered that an observatory be built there.[39] Who remembers the Lorient Observatory today? Historians, notably Olivier Sauzerau, have recently drawn attention to the forgotten history of the Nantes naval observatory and its director Frédéric Huette, an optician to the navy. Sauzereau's micro-historical approach has partly lifted the veil on the history of French naval observatories and hydrography schools in the first half of the nineteenth century.[40] The founding of naval observatories was closely related to the diffusion of marine chronometers aboard military and civilian ships.

A survey of the French navy archives allows us to draw a preliminary picture of the early diffusion of chronometers. In 1825 no more than eighteen chronometers could be counted in the merchant navy; ten years later their number had risen to forty-three. The Royal Navy had thirty-four chronometers on board its ships in 1815, forty-three in 1817, and fifty-one in 1818–20. By 1832 altogether 143 naval chronometers equipped the French Navy. This however seems blatantly insufficient: the official protocol recommended that three naval timekeepers equip each ship for a safe sailing.[41]

After the success of Berthoud and Breguet at the exhibitions of 1802 and 1806, several private initiatives investigated practical means of calibrating naval timekeepers. One of the most significant seemed to have been Charles Guépratte's. As a professor of hydrography in contact with Berthoud, Guépratte organized a service for synchronizing clocks in Brest. Between 1811 and 1816 Berthoud and, after his death in 1813, Breguet, together with Guépratte and the navy depot director François-Étienne de Rosily, established a network for distributing time and taking care of the delicate marine chronometers. Several seaports, including Rochefort, Brest, and Toulon, were equipped with naval observatories, where astronomers took charge of synchronizing clocks before they were sent to sea. In 1816 Guépratte was named director of the Brest Observatory, where he remained until 1852.[42]

Initially the time distribution service relied on the goodwill of a few devoted men. But it quickly became necessary to organize and regulate more permanently the networks of time distribution and clock maintenance. From 1815 chronometers were delivered only by an official order from the navy minister.[43] On 1 September 1816 the minister, François de Jaucourt, circulated a set of instructions for marine clocks (*Instructions auxquelles devront se conformer les Officiers chargés des montres marines*). Distributed to all seaports, these instructions were included in Guépratte's *Problèmes d'astronomie nautique et de navigation*.[44] Officers were reminded of their duty in the upkeep of chronometers for cartographic needs. For the first time in France, standard tables were printed for testing the accuracy of naval chronometers. The seaport astronomer merely needed to read observations reported on these tables to check the drift of on-board timekeepers. Starting in 1819 or 1820, standard labels were attached to clock cases to keep track of successive cleanings and oil changes.[45]

On 7 August 1825 Charles X's royal decree made official the measures taken since 1811 and organized the schools of hydrography. This law determined their number and regulations, as well as the salary of hydrography professors and naval examiners.[46] Schools and professors were organized into four classes. In the first-class schools at Le Havre, Saint-Malo, Nantes, Bordeaux, and Marseilles, "a space devoted to the establishment of an observatory should also be provided so that seafarers can attend to astronomical observations."[47] For most of the century these regulations ruled over the daily life of as many as forty-four hydrography schools and their forty-four professors.

As Sauzereau has shown, life in the hydrography schools and attendant observatories was harsh. The professors' modest requests, even for such basic instruments as sextants and octants, could meet with the city's objection. While some Parisian journalists denounced the laziness of merchant-ship masters and their refusal to adopt new methods, hydrography examiners complained of Parisian intrigues mocking hydrography professors. Little by little in the following decades, the lives of the hydrography faculty deteriorated.[48]

By 1879 the time distribution service was in a dire state. The director of the Paris Observatory, Ernest Mouchez, recalled that in all five French military seaports a *lieutenant de vaisseau* was in charge of calibrating and synchronizing clocks. The navy depot in Paris lent chronometers after submitting them to standard testing at different temperatures in drying cabinets. But according to Mouchez such facilities were nonexistent in commercial seaports, where as a rule the calibration and synchronization of marine timekeepers was left to a local clockmaker. Underequipped and unable to perform astronomical observations, they most often resorted to a clock whose accuracy no one could vouch for.[49] Supporting Mouchez's assessment, the astronomer Hervé Faye expressed the wish that commercial seaports be treated with as much care as military ones. Looking with envy to Britain, Germany, and Norway, he recommended that special observatories be set up and directed by officers equipped with excellent instrumentation.[50]

By the late nineteenth century the laws and regulations adopted in 1825–30 for the establishment of naval observatories in the French trade ports had clearly become unsuited to the needs of navigation. Local authorities seem to have played a determining role in shaping each observatory. Nantes is a striking exception in the midst of general neglect. The Nantes school and observatory were only able to perform their duties thanks to the exceptional commitment of their director Huette, who had secured the support of the local chamber of commerce. Although the history of nineteenth-century French hydrography schools remains largely unexplored, a cursory look at the situation abroad reveals paths not taken.

In Britain and in the United States the history of time distribution in commercial seaports and of the synchronization of naval timekeepers is closely related to the establishment of time balls. In 1818 the Royal Navy captain Robert Wauchope drew a "Plan for Ascertaining the Rates of Chronometers by an Instantaneous Signal," by which a ball sliding along a long mast atop an observatory fell at a given speed and at a fixed time (noon in Britain, 1:00 p.m. in the United States). This visual signal enabled naval officers to synchronize their timekeepers without leaving the ship. In England the first operational time ball was put in service in 1829 in Portsmouth, where the Naval Academy was located. A second one was built on top of the Greenwich Observatory in 1833. Liverpool and Edinburgh were equipped in 1836. Wauchope submitted his scheme to the French and American ambassadors when they visited London and Liverpool in 1830. Three months later the U.S. Naval Repository and the U.S. Naval Observatory were established in Washington; the first American time ball went into service in 1845.[51]

In France the calibration, maintenance, and distribution of naval chronometers had from early on been assigned to the navy depot, and while political considerations had ruled over the establishment of naval observatories attached to the various hydrography schools in the main trading

30. Engraving of the naval observatory in the Montsouris Park showing shacks brought back from the Transit of Venus expedition to Saint-Paul Island (8 December 1874), as well as wooden sheds built with the help of the architect for the City of Paris and set up in Montsouris at the end of 1875. *Annuaire du Bureau des Longitudes* (1876), 456.

seaports of the country, many northern European states, like Britain and the United States, had opted for a more efficient, centralized organization. In the mid-nineteenth century observatories staffed by competent astronomers were set up in a few selected seaports (Hamburg and Kiel in Germany, Trondheim and Bergen in Norway). From these central institutions time was distributed along the burgeoning telegraphic network. This was the model that Mouchez and Faye wished for France to follow.

THE OBSERVATORY OF THE BUREAU OF LONGITUDES

The failure of French naval observatories to assist merchant seafarers gave rise to several private initiatives. A perfect example is the establishment under Mouchez's leadership of the Observatoire de la Marine at Montsouris, south of Paris, also known as the Bureau of Longitudes Observatory. In the course of his numerous hydrographical missions around the globe, Mouchez perfected the astronomical determination of geographical coordinates, put forward stricter routines for using theodolites, and adapted terrestrial astronomical instruments for sailing conditions (e.g. the short meridian telescope modified by Jean Brünner in 1850). Capitalizing on this experience, he led one of the French expeditions to observe the transit of Venus on 9 December 1874 (figure 30). The excellent observations that he brought back despite trying conditions won him the acclaim of the Academy of Sciences. As a result, on 19 July 1875 the academy elected him to its astronomy section.[52] He had already been elected to the geography and navigation section of the Bureau of Longitudes in July 1872.

During his hydrographical campaigns Mouchez had already considered including astronomical instruments used in observatories into the standard equipment of seafarers concerned with cartography and hydrography. For this reason he was keen to "make more astronomical observations than are usually done during ordinary voyages at sea."[53] In 1875 the time had come for these projects to materialize, and after his election to the academy they were quickly set in motion. Though at the time a mere captain, Mouchez persuaded the Bureau of Longitudes to establish, with the support of the ministry of the navy, a small observatory in Montsouris Park, near the meteorological observatories set up by Charles Sainte-Claire Deville in 1868. Despite the difficulties facing the Bureau of Longitudes at the time, Mouchez's observatory opened on 4 October 1875.[54]

As Mouchez explained, this observatory was to "foster the taste for astronomical observation and to facilitate its study." It would welcome "not only officers from the Navy and the General Staff, but also everyone willing to be instructed in the practice of observation, whether to prepare for some voyage of exploration or to devote themselves more exclusively to the study of astronomy."[55] Public declarations in favor of opening the observatory to a large audience notwithstanding, this was above all an observatory dedicated to training naval officers. A complete training program for seafarers was created, with the aim of providing officers with a complementary education: "The Montsouris Observatory is precisely founded on this idea: it fills a gap in the education provided by the special schools and gives officers the possibility of acquiring further training in some very useful branches of the sciences."[56]

The courses offered to implement this program included practical astronomy (taught by Maurice Lœwy and Faye), hydrography (Mouchez), terrestrial magnetism (Edme Marié-Davy), electrical telegraphy (Louis Breguet), meteorology, *physique du globe*, and photography (Alfred Cornu and Charles Angot).[57] Mouchez and Loewy carried out the regular work in the observatory, together with the best experts of the capital. The *Annales de l'Observatoire de Montsouris* regularly published the observations recorded by officers transiting through the observatory. Cohorts of six officers were trained every other year, three of which remained to partake in the teaching and training of the new recruits. The observatory was equipped with instruments of high quality, the very ones made for Mouchez's transit of Venus expedition to Saint-Paul Island: two equatorials (8 and 6 inches aperture), their micrometers for extrameridional observations; several telescopes, at least one equipped with a photographic camera (daguerreotype and collodion); transit instruments; portable meridian circles; and one theodolite.

Mouchez was promoted at the beginning of 1878 to rear admiral and then, on 27 June, to director of the Paris Observatory, a position that he occupied until his death in 1892. The establishment of the Montsouris Observatory enabled the Bureau of Longitude to recover its autonomy and justify its continued existence. The choice of Mouchez as head of the Paris Observatory might also be interpreted as the expression of a political will to repair the scientific and human damages caused by Urbain-Jean Le Verrier's and Charles Delaunay's directorships since 1854. Mouchez seems to have been the consensus candidate, able to lessen tensions and give new impulse to the scientific activity of the observatory.[58]

Conclusion: Technical Transfers from the Observatory to the Navy?

The instruments used by seafarers and astronomers for measuring the height of stars have traditionally been the same. Ultimately those equipping naval ships evolved very little after the end of the eighteenth century. Even today the fundamental instrument—when GPS cannot be used—remains the sextant, invented in 1757. Seafarers' daily routines were focused on the immediate needs of navigation, and above all the determination of longitudes for perfecting cartography and improving the safety of navigation. Astronomers' instruments, on the other hand, have undergone fundamental changes. The move from refractors to reflectors entailed modifications of research programs and the introduction of new techniques throughout the nineteenth century, such as spectroscopy and photography.[59]

Rather than speak of a transfer of techniques and practices from the observatory to the ship, one should perhaps insist on the constant adaptation of observatory techniques to the needs of navigation. This is the sense in which Mouchez transformed several observation instruments to cater to seafarers' specific needs, such as his portable meridian spyglass, designed in collaboration with Brunner in 1850, or his portable imitation of the Jacob staff to measure the height of the sun at the solstice under all latitudes: "There are two kinds of instruments which one may use to measure the height of stars: those that are fixed and used in observatories and those that are portable, such as the reflection instruments used by seafarers. Even shrunk to the smallest possible size, the former are too cumbersome, and are in need of too stable a setting, to be used in voyages of exploration; the latter, on the contrary, are small and in need of no prior installation, and are the right instruments for such voyages."[60] Moreover, in France traditional observatories showed very little interest in the problem of distributing time to merchant seaports. This lack of interest was especially striking at the large observatories in two major naval cities. In Bordeaux, the director Georges Rayet was more interested in photography and spectroscopy than in synchronizing the city's clocks. Likewise, at the Marseilles Observatory, which had been relocated further from the seacoast in 1862 to the Longchamp Plateau, the large 80-cm Foucault telescope dominated Édouard Stephan's research agenda and drew his attention away from observations relevant to nautical uses.[61] In 1879, when the opportunity of establishing chronometric observatories for testing and calibrating marine clocks was discussed, Faye answered Mouchez's objection over cost: "The costs would be low in Marseille and Bordeaux, where we already have civilian observatories. Their directors, MM. Stephan and Rayet, would gladly perform these tasks, provided the State would ask them to do so and would pay for a few hot boxes, cupboards, logbooks and an additional auxiliary. It would be otherwise, I must say, in Le Havre, Saint-Nazaire, Nantes, Dunkerque, La Rochelle, Bayonne, etc., for in these ports, everything would have to be organized from scratch, resources and personnel."[62]

That someone like Faye, who paid close attention to the needs of nautical astronomy, could be unaware that in Nantes there was already a working, well-endowed naval observatory can be taken as symptomatic of the failure of French efforts to establish naval observatories and official chronometric workshops to meet the needs of the navy. By the end of the nineteenth century, a methodological renewal (the line of position method, later radio navigation) definitely rendered obsolete the system dreamed up in the 1830s.

Notes

1. "The Navy may need astronomers, but astronomers need the Navy minister." Lalande, *Bibliographie astronomique*, 702.
2. Maunder, *The Royal Observatory, Greenwich*; Howse, *Greenwich Time and the Discovery of Longitude*, 19–44; Forbes, *The Birth of Scientific Navigation*; Raymonde Barthalot, "The Story of Paris Observatory," *Sky & Telescope* 59, no. 2 (February 1980): 100–107; and Débarbat, Grillot, and Lévy, *L'Observatoire de Paris*.
3. This chapter is largely based on my dissertation. See Boistel, "L'Astronomie nautique au XVIIIe siècle en France."

4. Jim Bennett, "The Travels and Trials of Mr Harrison's Timekeeper," *Instruments, Travel, and Science*, ed. Bourguet, Licoppe, and Sibum, 75–95; and Andrewes, ed., *The Quest for Longitude*.

5. Initially due to be published in 1742, this last memoir was read at the Academy of Sciences on 30 May 1742 and published in *Histoire de l'Académie Royale des Sciences pour l'année 1744*, Mémoires (1748), 462–74.

6. On this position see Boistel, "Pierre-Louis Moreau de Maupertuis: un inattendu préposé au perfectionnement de la navigation (1739–1745)," *Annales 2003 de la Société d'histoire et d'Archéologie de l'Arrondissement de Saint-Malo* (2004), 241–61.

7. Guy Boistel, "Au delà du problème des trois corps: Alexis Clairaut et ses tables de la Lune à vocation nautique (1751–1765)," *Actes du congrès de la Société Française d'Histoire des Sciences et des Techniques, Poitiers, 20–22 mai 2004*, ed. A. Bonnefoy and B. Joly, Cahiers d'histoire et de philosophie des sciences, hors-série (Paris: SFHST, 2006), 20–29.

8. René Taton and Curtis Wilson, "The Problem of Perturbation Analytically Treated: Euler, Clairaut, d'Alembert," *The General History of Astronomy*, ed. Hoskin, 2B:89–107.

9. See Boistel, "L'Astronomie nautique au XVIIIe siècle en France," 2:123–73, 3:332–82. See also Donald Harry Sadler, "Lunar Distances and the Nautical Almanac," *Vistas in Astronomy* 20 (1976): 113–21.

10. Jean-Charles de Borda, Jean-René-Antoine Verdun de la Crenne, and Alexandre-Gui Pingré, "Opérations faites tant à bord de la frégate du roi *La Flore* qu'en différens ports ou rades d'Europe, d'Afrique et d'Amérique pour la vérification des instrumens & des méthodes relatives à la détermination des Longitudes sur mer, & à d'autres objets concernant la navigation," *Histoire de l'Académie royale des sciences 1773* (Paris, 1777), Mémoires, 258–334. See also *Voyage fait par ordre du Roi, en 1771 & 1772, par MM. Verdun, de Borda et Pingré à bord de la Flore* (Paris, 1778). On the use of the Borda circle in geodesy see Alder *The Measure of All Things*, and the contributions by Sven Wildmalm and Martina Schiavon in this volume.

11. See Boistel, "Les longitudes en mer au XVIIIe siècle sous le regard critique du père Pezenas."

12. Michel Vergé-Franceschi, *La Marine française au XVIIIe siècle: les espaces maritimes: guerre, administration, exploration* (Paris: SEDES, 1996), 370–75.

13. Boistel, "L'Astronomie nautique au XVIIIe siècle en France," 441–500.

14. Pierre Lévêque to Alexandre-Gui Pingré, 12 January 1775, Bibliothèque Sainte-Geneviève, Paris, MS 2251, fol. 24–25.

15. Pierre Lévêque, *Le guide du navigateur, ou traité de la pratique des observations et des calculs nécessaires au navigateur* (Nantes: Despilly, 1778). See also Vincent-François Dulague, *Leçons de navigation* (Rouen: J. Racine, 1784; Rouen: Besongne, 1787).

16. "Les anglais ont employé leurs calculateurs à des tables qui forment un volume d'une grosseur énorme où l'on trouve les calculs tout faits. Les capitaines marchands ne peuvent en faire usage; ils s'effraient à l'aspect de l'immensité de ces calculs, où il faut chercher celui qui leur convient, quelques règles de proportion qu'il y a à faire, les signes plus et moins des erratas &c. . . . mais si on leur montre

cette méthode de Borda, ils s'écrient c'est la chose impossible. S'il faut savoir cela pour être Capitaine, jamais nous ne le serons. Prendre la Lune avec la main ou faire ce calcul, c'est la même difficulté pour nous." Pierre Claret de Fleurieu to Mgr. Le comte de la Luzerne, ca. 1787–88, Archives Nationales, Marine, G96, fol. 43–44.

17. Pierre Claret de Fleurieu to the Navy Minister [César-Henri de la Luzerne], 4 March 1792, Archives Nationales, Marine, G96, fol. 126r°.

18. *Procès-verbaux de l'Académie Royale des Sciences*, 9 May 1788, fol. 109r°; 28 May, fol. 132r°; 7 June, fol. 157v°.

19. Lalande, *Bibliographie astronomique*, 614. See also Jean-Charles de Borda and Pierre Lévêque, "Rapport sur le mémoire et la carte trigonométrique présentés par le citoyen Maingon, lieutenant de Vaisseau," *Procès-verbaux de l'Académie des Sciences de l'Institut de France*, séance du 11 Vendémiaire an VII (2 October 1798), 1:465–73.

20. Rochon, *Mémoire sur l'astronomie nautique et particulièrement sur l'utilité des méthodes graphiques pour le calcul de la longitude à la mer, par les distances de la Lune au Soleil & aux étoiles* (Paris: Prault, 1798), 2.

21. Jérôme Lalande, *Histoire générale des mathématiques par M. Montucla* (Paris: H. Agasse, 1799), 4:581.

22. All quotes from Borda and Lévêque, "Rapport sur le mémoire et la carte trigonométrique," 472–73. Borda also expressed his opinion about seafarers in a letter to the navy minister (Georges René Pléville le Pelley), 4 March 1798, Archives Nationales, Marine, G96, fol. 55.

23. Boistel, "De quelle précision a-t-on réellement besoin en mer?"

24. See A. Rouyer, *L'École Navale pendant la Révolution et l'Empire (1789–1815)* (typewritten, 1954), Service Historique de la Marine, Vincennes, 3-S-1107. See also Anon., *Histoire de l'École Navale et des institutions qui l'ont précédée*, 145.

25. *Histoire de l'École Navale et des institutions qui l'ont précédée*, 161.

26. Jean-Baptiste Biot, "Compte rendu sur le traité de navigation de Jean-Baptiste du Bourguet," 224–25. See Jean-Baptiste du Bourguet, *Traité de navigation* (Paris: Veuve Courcier, 1808).

27. Biot, "Compte rendu sur le traité de navigation de Jean-Baptiste du Bourguet," 218–19.

28. Pierre-Louis Moreau de Maupertuis, *Astronomie nautique, ou Élémens d'astronomie tant pour un observatoire fixe, que pour un observatoire mobile* (Paris: Impr. Royale, 1743), foreword.

29. Pierre Caillet, *Manuel du navigateur* (Nantes: Forest, 1818), foreword.

30. Frédéric H[uette], *Calculs de navigation à l'usage des officiers de la marine marchande et des capitaines au cabotage, par H., professeur d'hydrographie à Nantes* (Nantes: Suireau, 1840), foreword.

31. Antoine Yvon Villarceau and Henri-Julien Aved de Magnac, *Théorie et pratique de la nouvelle navigation* (Paris: Gauthier-Villars, 1877).

32. See *Comptes-rendus de l'Académie des Sciences* 84–85 (1877), and esp. Mouchez, "Observations relatives à l'ouvrage présenté à l'Académie par M. Yvon Villarceau,"

1207–11, and A. Yvon Villarceau, "Réponse préliminaire aux observations présentées par M. Mouchez, au sujet de l'ouvrage concernant la Nouvelle navigation, dont les trente-cinq premières feuilles ont été déposées sur le Bureau, dans la séance du 14 mai," *Comptes-rendus de l'Académie des Sciences* 84 (1877): 1251–56. On the "new navigation" see Alfred Ledieu, *Nouvelles méthodes de navigation: étude critique* (Paris: Dunod, 1877), and Michel Vanvaerenbergh and Peter Ifland, *Line of Position Navigation: Sumner and Saint-Hilaire, the Two Pillars of Modern Celestial Navigation* (Bloomington: Unlimited, 2003).

33. Mouchez, "Observations relatives à l'ouvrage présenté à l'Académie par M. Yvon Villarceau," 1210.

34. Ernest Mouchez, "Deuxième note relative à la Nouvelle navigation de M. Yvon Villarceau," *Comptes-rendus de l'Académie des Sciences* 84 (1877): 1352–53.

35. See Boistel, "De quelle précision a-t-on réellement besoin en mer?"

36. Boistel, "De quelle précision a-t-on réellement besoin en mer?"

37. Louis-Marie Bajot, "Horlogerie astronomique," *Annales maritimes et coloniales*, 1820, 141–49.

38. Édouard de Rossel, *Traité de navigation par Bézout: nouvelle édition* (Paris: Veuve Courcier, 1814).

39. Frédéric Marguet, *Histoire générale de la navigation du XVᵉ au XXᵉ siècle* (Paris: Société d'Éditions Géographiques, Maritimes et Coloniales, 1931), 260.

40. Sauzereau, *Nantes au temps de ses observatoires.*

41. See for example Louis Pagel, *La latitude par les hauteurs hors du méridien, méthode facile et courte pour déterminer la position de l'observateur par les hauteurs; aperçu sur les distances lunaires* (Paris: Impr. Royale, 1847), 268. The purchase records are taken from Marguet, *Histoire générale de la navigation du XVᵉ au XXᵉ siècle*, 258.

42. Guépratte, *Dictionnaire de biographie française* (Paris: Letouzey et Âné, 1983–85), 16:1453.

43. Marguet, *Histoire générale de la navigation du XVᵉ au XXᵉ siècle*, 258.

44. Charles Guépratte, *Problèmes d'astronomie nautique et de navigation* (Brest: Lefournier et Deperiers, 1823), 206.

45. See Louis-Marie Bajot, "Huile pour les chronomètres [de Marine]," *Annales maritimes et coloniales* 2 (1830): 171–72.

46. "Règlement qui détermine le nombre et la répartition des écoles d'hydrographie ainsi que le traitement des examinateurs de la Marine et des professeurs desdites écoles," *Bulletin officiel de la Marine* 2 (1879): 39–50, at 47–50.

47. Art. 4, quoted in Sauzereau, *Nantes au temps de ses observatoires*, 43. Military seaports such as Cherbourg, Lorient, Toulon, Brest, and Rochefort were assigned to the second class of hydrography schools.

48. Sauzereau, *Nantes au temps de ses observatoires*, 92–96.

49. Ernest Mouchez, "Envoi de l'heure de l'Observatoire de Paris aux ports de commerce pour le réglage des chronomètres," *Comptes-rendus de l'Académie des Sciences* 88 (1879): 1227–28.

50. Hervé Faye, "Observatoires chronométriques pour la marine marchande," *Comptes-rendus de l'Académie des Sciences* 88 (1879): 1143–47; and "Remarques à

l'occasion d'une Note de M. l'amiral Mouchez," *Comptes-rendus de l'Académie des Sciences* 88 (1879): 291–93.

51. On time balls see Ian R. Bartky and Steven J. Dick, "The First Time Balls," *Journal for the History of Astronomy* 12 (1971): 155–64; H. W. Dickinson, "The Portsmouth Naval Academy, 1733–1806," *Mariner's Mirror* 89, no. 1 (2003): 17–30; Steven J. Dick, "How the U.S. Naval Observatory Began, 1830–65," *Sky and Telescope* 60 (1980): 466–71; Bartky, *Selling the True Time*.

52. Mouchez, *Notice sur les travaux scientifiques*, Archives of the Academy of Sciences, Paris, and Mouchez, "Observation du passage de Vénus effectuée à l'île Saint-Paul." On the French Transit of Venus expeditions see Aubin, "Un passage de Vénus en politique"; Christophe Marlot, *Les Passages de Vénus: histoire et observation d'un phénomène astronomique* (Paris: Vuibert, 2004); Aubin, ed., "L'événement astronomique du siècle?"

53. Mouchez, *Notice sur les travaux scientifiques*, Archives of the Academy of Sciences, Paris, 3–4.

54. Ernest Mouchez, "Observatoire du Bureau des longitudes à Montsouris," *Comptes-rendus de l'Académie des Sciences* 81 (1875): 545–46; Mouchez, "Création d'un observatoire astronomique d'étude." See Guy Boistel, "Instruire les marins avec les moyens du bord: l'observatoire de Montsouris," *Les Génies de la science* 28 (2006): 28–33.

55. Ernest Mouchez, "Création de l'Observatoire de Montsouris," *Revue maritime et coloniale* 54 (1877): 510.

56. Ibid., 513.

57. In fact not all professors answered Mouchez's call. For an assessment of courses taught after ten years see Anonymous, "L'observatoire de la Marine à Montsouris," *Magasin pittoresque*, 2nd ser., 3 (1885): 30–31.

58. On controversies surrounding the Bureau of Longitudes see Eugène Yung and Émile Alglave, "À quoi sert le Bureau des Longitudes?," *Revue scientifique* 10 (1872): 481–83. On the virulent critiques voiced by Paul Bert in front of the National Assembly on 9 December 1872 see Guillaume Bigourdan, "La réorganisation du Bureau des longitudes, en 1854 et 1862," *Comptes-rendus du congrès des sociétés savantes à la Sorbonne en 1929* (Paris: Impr. Nationale, 1931), 23–34, and "Le Bureau des longitudes, chap X: suite des travaux du Bureau des Longitudes; attaques de Paul Bert, réponse de Faye, réorganisation de l'Observatoire de Paris (1872) et fondation d'observatoires en province: la réforme de 1874," *Annuaire du Bureau des longitudes*, 1933, A65–72. On the Paris Observatory in this period see Aubin, "The Fading Star of the Paris Observatory in the Nineteenth Century." On Mouchez's scientific work there, especially his involvement in the Carte du Ciel project, see Derek Jones, "The Scientific Value of the *Carte du Ciel*," *Astronomy and Geophysics* 41, no. 5 (2000): 16–20; Bigg, "Photography and the Labour History of Astronomy"; and Chinnici, *La Carte du Ciel*.

59. Chapman, "The Astronomical Revolution."

60. Ernest Mouchez, "Instrument portatif pour la détermination des itinéraires et des

positions géographiques dans les voyages d'exploration par terre," *Comptes-rendus de l'Académie des Sciences* 86 (1878): 268.

61. On Bordeaux see Maison, "La fondation et les premiers travaux de l'observatoire astronomique de Bordeaux." On Foucault's telescope see Tobin, *The Life and Science of Léon Foucault*.

62. H. Faye, "Remarques à l'occasion d'une Note de M. l'Amiral Mouchez," *Comptes-rendus de l'Académie des Sciences* 88 (1879): 1292.

Chapter 6

Astronomy as Military Science:

The Case of Sweden, *ca.* 1800–1850

SVEN WIDMALM

> During the revolutionary era [*mansålder*], our part of the world was ravaged
> by the storms of war; the greatest Man of the age rose, because he was also
> the greatest warrior, to the most powerful position of the age, and almost
> everywhere the foremost rationale for national surveys became strategy. But
> the age itself, and each and everyone of its great men, was born and raised
> among great ideas; numerous wonderful ideas always influenced the develop-
> ment of cartographic projects during this period, even though the strategic
> purpose was foremost. It is only these manifold purposes, and perhaps above
> all the growing interest in knowledge itself, without thought for practical
> utility, which can explain the precision, far above all practical needs, that
> was now demanded from maps. Since the conclusion of peace, this [demand
> for precision] has increased rather than decreased, and in many countries
> the real purpose of national surveys is thought—since its results are dissemi-
> nated by means of engravings—to be: to enlighten the people and their own
> spheres of activity, and not only to fulfill the needs of Government.
> —CARL GUSTAF SPENS (1837).[1]

The words of Count Carl Gustaf Spens illustrate the thinking of a new class
of professionals who rose with the Napoleonic wars—military cartographers. These men were champions of the strategic thinking of total war, as
well as of a military science that was to encompass and define the nation-
state. Spens was a typical member of this group in that he was a geodetic
expert (the inventor of a popular method of cartographic projection) but
also an advocate of social reform, promoting liberal ideas in the Swedish
House of Nobility. Spens was an untypical professional in that he was a
count—most successful topographic officers were of middle-class origin—
but he was also an untypical count in that he had a Ph.D.

Early-nineteenth-century military cartography was founded on the scientific principles of physical astronomy but, as will be shown in this chapter, its scope was much broader. It may be described as a science of the nation-state, dealing with military, civilian, and scientific interests from a national point of view.[2] The development of Napoleonic military cartography provides an example of the importance of astronomy for social projects, and shows how this technically refined natural science can also be seen as a social science.

In the eighteenth century astronomy led the way in the development of new forms of scientific organization. Networks of scientific correspondence centered on metropolitan academies spread across Europe as astronomers eagerly collected observational data and information about instruments, their makers, and each other. National and international collaborations emerged around well-connected practitioners, especially the Frenchmen Jean-Nicolas Delisle and Joseph-Jérôme Lalande, the German Franz Xaver von Zach, and the Swede Pehr Wilhelm Wargentin.[3] As a result, astronomers became much more than observers and interpreters of the skies. They became information specialists, experts in collecting, communicating, and analyzing large amounts of data. This opened up a variety of fields of influence and activity—in the natural sciences, in technology, and in the social sciences. A remarkable example is the close involvement of Swedish astronomers in the production of population statistics from the 1740s onward. Pehr Wilhelm Wargentin, astronomer and perpetual secretary at the Royal Swedish Academy of Sciences, used similar statistical methods for predicting the movements of the Jovian satellites and for interpreting the vital statistics amassed at the Office of Tables (Tabellverket).[4] This is an early example of the close relationship between astronomy and the social sciences, which would be exemplified later by Auguste Comte and Adolphe Quetelet.

The case of vital statistics in Sweden is unique in eighteenth-century Europe. In other respects Sweden conformed to international trends. Astronomical practices were supported and developed in close relation with economic or military projects undertaken by the state, such as efforts to solve the longitude problem, land and coastal surveys, and the production of almanacs.[5] Astronomers watched the skies, but their main function was to map the world, and the directions taken by their mapping ventures were determined to a large extent by governmental interests.

This was particularly true of physical astronomy, or geodesy—first through the geodetic mapping of France by generations of Cassinis and many smaller geodetic measurements throughout Europe, then, in the last decades of the century, more forcefully through the establishment of the Trigonometrical, or Ordnance, Survey of Great Britain and the metric survey in France.[6] During the Napoleonic wars national geodetic surveys became more widely established in Europe, usually carried out in close association between military officers and civilian astronomers or mathematicians such as François Arago and Jean-Baptiste Biot in France, Carl Friedrich Gauss in Hanover, Friedrich Wilhelm Bessel in Prussia, and Friedrich Georg Wilhelm von Struve in Russia. By the early nineteenth century European cartography, founded on astronomical principles and coordinated with military discipline, was becoming established.[7]

The militarization of physical astronomy resulted from a perceived need to systematically collect and evaluate information thought necessary for conducting modern warfare as practiced by Napoleon I. The flexible use of conscripted mass armies was thought to rely on a detailed knowledge of geography, economic resources, and population.[8] Napoleonic total war was seen as information-based (to use an anachronistic but pertinent term), demanding a broad overview from a privileged vantage point associated with the General Staff, an institution that came into its own around 1800. Astronomy, statistics, and military history were institutionalized in the topographical bureaus established by the French in continental Europe as well as in countries such as Sweden that followed the French example. After 1815 this military information system was broadened to include the production of maps for civilian purposes, vital ingredients in the transformation of infrastructures in the nation-state. Hence the military adoption of physical astronomy illustrates a more general thesis advanced by many historians about the importance of military interests for early industrialization.[9]

This chapter deals with one episode in the development of physical astronomy as a military science during early industrialization. Focused on Sweden, it examines an area of Europe where industrialization had yet to make a mark in the first half of the nineteenth century. Swedish astronomers and military men did however make a solid attempt to keep up with continental developments in their respective trades, and the result was a military cartographic organization that was typical of what was rapidly becoming the European norm: the country's foremost astronomers oversaw geodetic work carried out by military personnel for military purposes but eventually also for civilian purposes.

In what follows, technical as well as broad social features of the military application of physical astronomy will be discussed. Special attention is paid to the growth of a new ideal of military science, which merged knowledge production for military and civilian purposes, and the penetration of military science into civilian society by means of its *social carriers*, the topographical officers. The chapter does not focus only on the astronomical aspects of this process. Rather it should be seen as an attempt to analyze the technological, social, and political contexts of astronomical expansion, which during this period were closely connected to the expansion of military geodesy and military technical expertise in general.

The French Connection

MILITARY CARTOGRAPHY

In 1798 the military engineer Gustaf Wilhelm Tibell left Sweden to join Napoleon's Italian campaign. Tibell had taken part in the Swedish-Russian war (1788-90) and later taught military science at the Military Academy of Karlsberg. An admirer of French warfare, he had become "infected with the teachings of political liberty emanating from the French Revolution."[10] In 1796 he and another radical colleague founded a society for military science that was later transformed into the Royal Academy of Military Science (1805), one of the first organizations of its kind in Europe. When Tibell joined Napoleon's forces he took a career step that was common among ambitious officers (to gain professional experience by fighting in a foreign war) but he also developed larger ambitions: to promote a new kind of military science.

After 1800 progressive (or modern) military thinkers began to view strategic thinking inspired by French warfare as an imperative to create new organizational structures for collecting and processing vast amounts of information. The hub of this information network was the General Staff. Officers that were specialists in the military information science were attached to the staff, and hence to the center of power in the new military and militaristic system: "The Napoleonic map is above all the map of the Prince."[11]

At the heart of the new military science lay cartography, not the old kind, exemplified by the sketching and reconnoitering of areas of tactical importance by officers with some education in geometry and watercolor drawing, but the geodetic kind, perceived to be of strategic rather than tactical importance. Back from the Italian campaign, Tibell became director of the Cisalpine Engineering Corps in Milan and was asked to collaborate with his French counterpart in preparing a map of the new republic. The organization created for this task was similar to the many topographical bureaus founded by the French. Its administrative center was an archive where topographic and geodetic, but also statistical and historical information was collected. This organization and others like it were modeled on the General War Depot (Dépôt Général de la Guerre) in Paris.[12]

Tibell was more than a simple channel of transmission of French military science to Italy and Sweden: he was an important actor in his own right. For example, he established an academy for military science in Milan that aimed to attract army officers who thought of modern warfare as "a broad science, founded on mathematical, physical and political knowledge."[13] In 1802 he started to pull strings to obtain a suitable commission back in Sweden. Canvassing patrons for the creation of a topographical corps on the Italian model, Tibell disowned his radical past: "here one talks of liberty, but we are more free in Sweden." In Italy, Tibell boasted, he had been in charge of "a number of astronomers, mineralogists, state economists [*Stats Oeconomister*], and other learned men," and he now wanted to see to it that Swedish military cartography was "enlivened, founded on more secure foundations, applying for this purpose the highly necessary astronomical and trigonometric operations that have formerly been neglected."[14] As other officers in Sweden had similar ideas, the scheme gained substantial support within the military leadership.

Tibell's original plan for a Swedish Topographical Corps did not include a civilian astronomer—possibly because he believed such expertise was not available. In late 1805 or 1806 it was decided, however, that an astronomer should be hired.[15] By then Jöns Svanberg had published his book about the geodetic survey carried out in 1801–3 to "correct" the value for the length of a degree of latitude reached sixty-five years earlier by the famous French expedition led by Pierre-Louis Moreau de Maupertuis. With this survey, the technology and theoretical expertise of advanced French geodesy was imported into Sweden. The scientific match with Tibell's military scheme was perfect.

GEODESY

In 1736–37 Maupertuis's expedition was a propaganda victory for Newtonian science, but most mathematical practitioners found his results shoddy. In Sweden the expedition was seen as an affront to members of the nation's scientific community because it proved that they lacked the power to sponsor such work themselves. A number of schemes to make a new measurement were put forward, and toward the end of the century plans came to fruition.[16] The result was, paradoxically, a wholesale importation of French geodetic methodology and technology as they had been developed for the metric survey.

The leader of the new Nordic survey was the mathematician Jöns Svanberg. He and Tibell were both self-made men, who had risen from humble origins to the top of their professions. They were the same age, and both were modernists in the sense that they were enthusiastic about post-revolutionary ideas of political liberty and science. For Svanberg this was expressed through a deep admiration of Kantian philosophy and Frenchified Newtonianism. By inclination a theoretician, Laplace was his hero.[17]

The Academy of Sciences in Stockholm bought a Borda circle to be used as the main expedition instrument (as it had been during the metric survey). While preparing for the expedition and later when evaluating its results, Svanberg relied on the guidance of French experts such as Lalande, Laplace, and above all Jean-Baptiste Delambre, who provided the Swedes with extensive advice concerning the use of the circle and many other details.[18] Svanberg's book (1805) presenting the results of the expedition was written in French and used the metric system throughout. This survey must be seen as one of the first important scientific projects undertaken outside of France and presented to the world in full metric guise. All in all, the expedition was an advertisement for French geodetic theory and practice, and for the metric system itself. This is probably why the institute awarded Lalande's gold medal to Svanberg even though Delambre, irritated by the Swede's love of fanciful theory and his lack of technical competence, thought the expedition had failed to fulfill its main objective—to show exactly what, if anything, was wrong with Maupertuis's measurement.[19]

Delambre's criticisms aside, Svanberg was hired the following year to lead the geodetic operations that would lay the foundation of the military survey of Sweden. The academy's repeating circle was borrowed by the military for a pilot survey between Uppsala and Stockholm. At the same time Tibell revived the Royal Academy of Military Science, which had lain dormant during his absence. An important result of this initiative was the emergence of a new group of experts who had the ambition not only of reforming military science and strategic thinking in general but of injecting the ideals of military science into civilian society on a broad scale. This laid the organizational foundation of a modern military science that included physical astronomy.

Instructions for the Topographical Corps stated that its officers should "make complete military maps of the Nation [*Riket*] founded on astronomical observations and trigonometric operations, as well as topographical, statistical and military descriptions of the land."[20] Furthermore, the corps was responsible for writing Sweden's military history. It was to gather information in a systematic and scientific manner, paying close attention to international developments.[21] In Tibell's opinion Svenberg was the perfect man to ensure a high quality education in mathematics and geodesy for the officers of the corps. Tibell painted in vivid colors the ideal for a new military science that would enjoy a social standing comparable to that of law and medicine, one that would be taught not only in military schools but also in gymnasia.[22] He summarized his views by the motto: "For a military man, it is not enough to sacrifice his life, he must also make it useful to the nation—by staying alive."[23] His close associate Carl Akrell—who would later head the Topographical Corps—likewise emphasized the encompassing nature of military science: "in past times, when armies were small and limited . . . and often active only within the borders of a province, . . . knowledge of the land was not so useful; but now, when large and extensive armies spread in an instant over wide stretches of land, when stockpiles are considered a hindrance for rapid movement, and all the resources of a country must be taken into account when calculating the options to support and move armies, then a true knowledge of the land becomes the foundation for the composition, leadership and employment of these masses, which constitute the armies of our time, wherefore the art of war, instead of being a separate branch of science, nowadays includes everything that goes by the name of political science, political economy, knowledge of countries and nations, that were earlier considered to be separate sciences."[24] This ideal was best represented by the modern military map, characterized not by any particular tactical or strategic purpose but by a scientific methodology with which maps could be produced at any scale and for any desired purpose. It was founded on an all-encompassing quantification that ensured maximum flexibility. As we shall see, qualities such as these were thought to make military cartography especially useful for civilian society—or rather, they illustrated how military and civilian needs converged.

The aggressive new self-promotion of the corps met with resistance: in 1809 attempts were made to subsume military cartography within the fortifications department. The ensuing debate in Parliament made clear, however, that the modernist ideal of military science associated with the corps had broad political support. Even though a new Engineering Corps was created into which topography and fortification were integrated, this outcome constituted a victory for the cartographers, as Tibell was chosen to head the new corps and funds appropriated for the fortifications subdivision were greatly reduced. In 1812, however, the military modernizers experienced a dramatic setback. Tibell was accused of treason (rightly, it seems) and sacked. A fortifications officer took his place as chief of the Engineering Corps, and though Tibell was restored to political favor fairly quickly, the institutional position of the topographers failed to reflect their aspirations until much later. In 1831 they were separated from the Engineering Corps and became a division of the General Staff.[25]

Their institutional position notwithstanding, military topographers as a group became influential in Swedish politics (including defense), education, communications, and statistics. From the outset topographers advocated that the trigonometric survey should be what it eventually became all over Europe, the scientific foundation for a national cartography. But they also insisted on the need for geographic and statistical information that would benefit military as well as civilian purposes. The civilian value of modern military science was one of Tibell's selling points when he canvassed support for the creation of a topographical corps, and it was on the agenda of the Academy of Military Science.

In reality, progress was slow. At the beginning short-term military concerns dictated much of the activities of the corps. Reconnaissance in the far North and in the South was carried out in response to threats from Russia and Denmark. Likewise, war in Germany and the annexation of Norway in 1814 made surveys there a military necessity. Around 1820, however, strategic planning began to guide the corps in its work. While peripheral areas of special importance had first been surveyed in line with military doctrines emphasizing fortifications and the defense of borders, after the Napoleonic wars the doctrine of "central defense" gained support in Sweden. A prominent military cartographer of the Topographical Corps, Johan Peter Lefrén, became its main spokesman. The country should not be defended at its borders, in his view, but rather made impenetrable by positioning armies in fortified strongholds strategically located in one or two central sites. The navy and fortifications witnessed a decline in importance in favor of the army and military cartography. As a consequence, military interest in the vast but thinly populated northern parts of the country decreased — and in effect, the mapping of northern Sweden ceased until the end of the century. Military cartographers meanwhile concentrated on mapping the more densely populated and economically important areas of southern Sweden, all the while striving to abandon military secrecy to make their maps useful to general political and economic concerns.[26]

Akrell, by then chief of the Topographical Corps, made the case for openness in 1841 based on his broad conception of military science:

> As the military system in later times has assumed a more comprehensive, definite character, and as widely stretched strategic operations necessitate a more general apprehension and overview of the theatre of war, topography has likewise assumed a wider field of activity, gained in military and political importance, and finally transformed from a supplementary knowledge to a complete science of war, the purpose of which is: the representation and description of countries, not only for the benefit of the commander but also for the statesman, the state economist and the scientist, who in equal measure need and benefit from the guiding intelligence it offers.
>
> The calling and power of topography should hence not be judged only from the point of view of military needs; other parts of government too make demands of its services. To fulfill the demands of both for a clear, graphic [*åskådlig*] representation of the land, and thus to prepare and promote a complete knowledge of the land, is nowadays the destiny of topography and the measure by which its worth must be gauged.[27]

Like other military cartographers, Akrell identified topography with military science in general, and military science with a universal science of the land, clearly perceived as the *nation*, including its military, political, economic, and scientific aspects.

The declassification of the military map occurred in stages and was not completed until 1857, when Parliament dramatically increased the appropriations of the Topographical Corps. Changes in infrastructure and the utilization of natural resources were important factors then: railway construction and the geological survey were also promoted at this parliamentary session. In Parliament it was pointed out that the money should be perceived not as "truly" military but "in reality" for public purposes.[28] From then and until the 1930s the General Staff Map was a fixture of Swedish society, like the Carte d'État-major in France or the maps of the Ordnance Survey in Britain.

THE TRIGONOMETRIC SURVEY OF SWEDEN

Merritt Roe Smith has written that around 1800 the military "uniformity system" that developed in the early seventeenth century became a full-fledged bureaucracy. Others have pointed to the striking similarities between the military and industrial systems of organization a century or more before the rise of Taylorism.[29] But the division of labor and mechanized, hierarchical modes of discipline and control characteristic of the uniformity system had by and large spared the tradition of military cartography. On the contrary, it relied on qualities difficult to systematize, such as tactical vision combined with artistic skill. The scientific production of topographically relevant information required this tradition to be re-engineered.

As the Swedish trigonometric survey evolved, the uniformity system was implemented in different ways, with, for instance, the creation of special detachments within the corps. Detachments for reconnoitering, instruction, geodesy, and engraving were in place by the early 1830s. By then systematic methods of topographical and statistical data gathering had been adopted.[30] A few examples will illustrate how scientific and military modes of collecting and representing topographical data were co-produced.

The graphic representation of differences in elevation and steepness of slope was a particularly thorny issue for scientific cartography. One problem with introducing quantitative methods of slope depiction was the lack of data until the 1850s. Nevertheless, in the second decade of the century Swedish military cartographers introduced Johann Georg Lehmann's "German" method, based on the use of "conventional lines" whose thickness and density varied with the steepness of slope. Widely adopted not only in Sweden but also in Germany, this method was later commended by scientific authorities including Humboldt and Arago.[31] The leaders of the Swedish survey valued the method less because it was a practical way of representing quantitative data as Lehmann had argued, but rather because, unlike older artistic methods, it could be carried out as in a drill. The method partook in the creation of a uniform cartographic language that *in principle* reflected its scientific and military foundations.[32]

The choice of scale also reflected the ambition to mechanize and discipline cartographic methods while providing more reliable knowledge. According to military modernizers, flexibility in strategic planning could be ensured by exhaustive geographic knowledge. Trigonometric surveys of the highest scientific accuracy would provide the backbone of more detailed surveys on the smallest scale — the British Ordnance Survey set the standard with its mapping of Great Britain on a scale of 1:10,560. A large country with meager resources such as Sweden could not come close to this ideal, but the principle was adopted: military cartography should produce information that could be used to give a broad overview but also to zoom into geographical areas of particular interest. At the center of military planning such information could be endlessly combined and recombined. But the choice of scale had disciplinary purposes as well.

The scale used by military cartographers in Sweden for reconnoitering purposes varied as the leaders of the corps struggled to uphold their ideals of discipline and flexibility. In 1810 a scale of 1:20,000 was adopted; in 1821 it was changed to 1:100,000; from then until mid-century it varied between 1:100,000 and 1:50,000.[33] Akrell discussed at length the virtues of large scales: not only did they provide more information, increasing flexibility, but they enabled a much stricter discipline. On small-scale maps officers could easily fill in the blanks with carelessly drawn lines indicating changes in elevation even when no ocular inspection of the area had been made. With a larger scale (Akrell would have preferred 1:20,000), carelessness in surveying was harder to conceal. A large scale would *force* officers to uphold the discipline demanded by their duties, by remote control as it were. It increased flexibility since editing the cartographic material made it possible to produce different maps for different purposes. Military cartography had been founded on the ability of skilled topographers to edit judiciously and draw convincing representations of topographic details in the field; control over this process was now removed from the field to the drawing chamber, where it could be carried out under the auspices of the military leadership.[34]

These cartographic principles led to a division of cartographic labor between relatively unskilled data gatherers and scientifically and militarily qualified management. By the late 1850s noncommissioned officers (NCOs) were hired to carry out what had by then become thoroughly mechanized data collection requiring no personal judgment or skill. Several observers noted that the NCOs would have virtually no opportunity for professional advancement, as they were never required to acquire specialist skills.[35]

Geodetic fieldwork was carried out by élite, mathematically trained officers within the corps. Following the directions from the astronomers at the Academy of Sciences, they had a professional ethos characterized by scientific ambitions. This led them to discard the navy's coastal trigonometric measurements in favor of a new triangulation following the best French practices as reflected in Svanberg's work, and used as the foundation of the first geodetic textbook published by the corps, in 1807.[36]

In reality the quality of the geodetic work performed by the corps varied. One case in point is the building and permanent marking of stations. French and, later, Russian models were purportedly followed, but resources were lacking—e.g. for the marking of station points with beautiful stone pyramids, as in France, or the building of station towers a hundred feet tall or more, as in Russia. Not until the 1820s did the Swedes care to mark their station points at all, using iron spikes or rocks. To ensure the permanence of these markers was a constant struggle—since locals often removed them. In time iron pipes sunk into the bedrock that were not easily removed or destroyed marked the observation points. This, it was hoped, would enable station points to be found years later and old measurements checked.[37]

As this example makes clear, the triangular net was *not only* a graphic representation of geometric space propped up by abstruse mathematics (e.g. spherical geometry and error analysis), and the abstract principles of physical astronomy (e.g. the notion of the geoid). It was physically anchored in the landscape itself. Much has been made by historians of cartography, and quite rightly, of the switch from a geography of place to a geography of space.[38] But the national territory defined by nineteenth-century geodesy, in Sweden and elsewhere, was in the final analysis not a mathematical abstraction. It was a technological construction, conjured into existence through the use of advanced instrumentation and physical markers that would withstand the test of time. Marking station points was in a sense like dotting the geography with innumerable observatories: potential observation points from which the authority of the physical description could be extended or reassessed. In this way the triangular net created a physical framework that helped to define the nation (figure 31).

BORDERLINES

The military cartographers' scientific ethos is illustrated by their adoption of bona fide scientific technology and theory, by the use of qualified astronomical expertise, and by their emphasis on geodetic training. But these standards were not easy to uphold in practice: resources were often insufficient and the quality of geodetic measurements compromised. On certain occasions military cartographers rose to the standards required by their scientific ideals, especially when measurements were made in collabo-

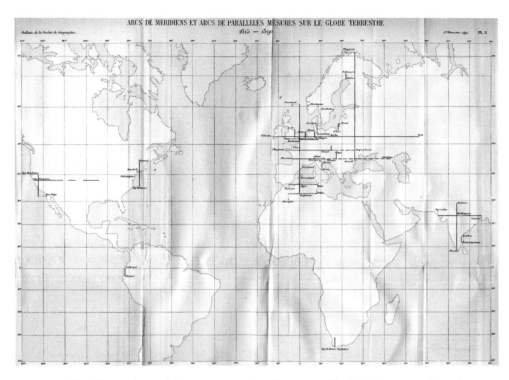

31. Meridian and parallel arcs measured on the terrestrial globe between 1615 and 1890. Source: Léon Bassot, "La géodésie française," *Bulletin de la Société de Géographie* 12 (1891), plate 2. Courtesy of the Société de Géographie.

ration with foreign colleagues, when triangulations crossed national borders, or when international comparisons were involved, such as Svanberg's correction of Maupertuis's measurements from the 1730s. From the 1830s funds for new instruments were made available for further collaborations with Russian topographers. To connect Swedish and Danish triangular nets, Bessel's new base-measuring apparatus was purchased. As one topographical officer put it, only with the help of such advanced technology could Swedish measurements "gain their full voting rights in future combinations and comparisons."[39] In the 1840s the Academy of Sciences, using both civilian and military expertise, participated in measurements of the northern part of Struve's great meridian arc, and the government financed instruments that would grant technological authority to the Swedish contribution.[40]

Collaboration with the Russians was sometimes colored by political suspicions, casting doubt on their acclaimed technological excellence.[41] Liberal cartographers like Akrell and Spens thus thought that the "oriental constitution" of Russians explained their successful geodetic work: forced labor, for example, allowed the clearing out of extensive wooden areas around observing stations. Geodetic measurements were seen as national trials of strength in which not only technical, scientific, and cultural achievements but also political systems were evaluated.[42]

In theory the Topographical Corps subscribed to very high standards of scientific excellence. In practice its ability to even approach such standards was wholly dependent on the episodes of national competition which were staged irregularly in the first half of the nineteenth century but were later institutionalized with the creation of the Mitteleuropäische Gradmessung in 1862.[43] As its founder, the Prussian military cartographer Johann Jakob Baeyer, put it, scientific progress "has always been and will always be the measure of a people's intelligence."[44] Collaboration, in this context, meant competition.

The Military Cartographers

THE SOCIAL PENETRATION OF MILITARY SCIENCE

An inventory of the professional trajectories taken by the forty-seven officers who joined the Topographical Corps before 1830 shows that they were remarkably successful as a group.[45] Tibell, for example, rose in the military hierarchy because of his professional competence, his valor (he was wounded several times), and his social acumen. Though his fortunes took a drastic downturn in 1812, he was later rehabilitated and ended his career as president of the War Office. This development exemplifies a broader trend of professionalization in the armed forces, which favored the relatively highly educated engineers.[46] In the early 1860s the two highest-ranking officers in Sweden were engineers, one of them the topographer Akrell. In successful civilian careers other officers from the Topographical Corps implemented its military scientific ideals. In leading positions in the military and in the state bureaucracy, some were influential in civilian fields such as communications, education, associations, statistics, and art—reflecting the military cartographers' professional interests. During a period characterized by social mobility as well as social unrest (fear of revolution was constant in Sweden, as in the rest of Europe, during the first half of the nineteenth century), the military used the ideal of national unity, founded on military science, as a vehicle to move upward socially.[47]

To use anachronistic but pertinent terminology again: communication and information were the cornerstones of a professional and political project merging military and civilian progress. In this sense topographical officers represented a *militaristic* view of social development. The military system provided the norm for civilian society; in particular, *science* was identified with *military* science.[48]

ROYAL SERVICE, STATISTICS, AND CARTOGRAPHY

Tibell and most other successful topographical officers served as staff, orderly, or adjutant officers. As a result they were close to the court of Gustaf IV Adolf and, after the coup of 1809, Crown Prince Jean-Baptiste Bernadotte (later Karl XIV Johan). The close association between topographical officers and the center of military power during the war years was consistent with their responsibilities. During the war the corps head served as general quartermaster, a central position in the General Staff in which other corps officers also served. In this Swedish developments mimicked those on the Continent, where from the 1790s on General Staffs were being formed around topographical units.[49]

Carl Gustaf af Forsell's career illustrates the social penetration of the new ideals of military science. Around 1800 he worked on several important cartographical ventures before the corps was founded — in the military coastal survey and in a large private project for a national map. Forsell, one of the first officers to be employed by Tibell, was asked in 1808 to draw the maps to be used in planning the greatest construction work in Sweden to date, the Göta channel, a civilian infrastructure project to be carried out with the help of military personnel. Before the coup of 1808 he was in charge of negotiating with Christian August (the Crown Prince, who died mysteriously and was replaced by Bernadotte). For the next fifteen years Forsell was close to royal power and tutored Bernadotte's son. During the continental campaign of 1813–14 he followed the Crown Prince wherever he went, which, as Forsell later pointed out, provided excellent opportunities for networking at the highest European level.[50] He was wounded at the battle of Leipzig in 1813 and ennobled three years later.

No doubt Forsell was a gifted and enterprising man, but this hardly explains his successful career. Throughout Forsell remained active as a topographical officer and a military theoretician. His career was a product of the Topographical Corps, designed to attract officers like him and to place them close to the center of military power. Had he not served as a map keeper to Bernadotte, Akrell, Lefrén, or another would have done so.[51]

In 1824 Forsell became director of the Swedish land survey, but he failed to resume its map publishing activities. Instead his time was increasingly spent on social issues connected to his expertise as military cartographer. He founded a number of "associations" (an organizational form imported from Britain in the 1820s), for example a society for disseminating useful knowledge, and a savings bank.[52] He was a leading representative of the philanthropist movement that aimed to promote national unity and prevent social disintegration. He published a popular map of Scandinavia (1815–26)—itself an expression of the vision of the Scandinavian unity he cherished—to which he added a statistical supplement which soon grew and became a successful publication in its own right (1831).

By this time Forsell's liberalism had become more pronounced, leading to a break with Bernadotte. The liberal press greeted its new ally with enthusiasm: "With his unshakable honesty and his incontrovertible *figures*, which are the despair of our conservatives, he strides along and preaches unsparingly the most determined *mouvement*."[53] The quote—with the French term *mouvement* meaning troop movement—captures the flavor of the late Forsell as a promoter of a military and scientific understanding of society. New modes of government were now needed that would promote liberal industrial interests rather than those of old or new royal dynasties.

Hence Forsell helped to forward military scientific ideals in civilian society, in particular in cartography and statistics. Akrell similarly devoted much time to civilian statistics, publishing a "statistics of the realm" (*riksstatistik*) sponsored by the Academy of Agricultural Science. In the late 1820s he was responsible for a failed project to institute a new civilian bureau of statistics in Sweden. Like Forsell and other topographical officers, Akrell was successful as a private map publisher. As chief of the Topographical Corps for a number of years, he was also an important player in Swedish cartography.[54] Furthermore, his career illustrates the interconnection between different infrastructure systems: he was deeply involved in promoting railroads and in the 1850s he headed the Swedish telegraph service.[55]

ART AND HISTORY

Most topographical officers were closely associated with the nationalistic awakening typical of the Romantic period and in Sweden with the so-called Gothicist movement. (In Sweden the term "Gothicist" has other connotations than in English, meaning primarily a nationalistic cultural movement.) This movement flourished in 1810–20 with the founding of organizations, the most important of which were the Gothicist Association (Götiska Förbundet) and the Manhem Association (Manhemsförbundet).

Almost 30 percent of topographical officers active during these years belonged to one or both. In the Gothicist Association 10 percent of members were topographical officers, one of whom was a founding member.[56]

The fact that several topographical officers also worked as artists is hardly surprising, since artistic ability was a traditional prerequisite for military cartography. The son of an engraver, Akrell was commissioned to produce several landscape engravings, e.g. for a popular voyage to Cape North. Other topographical officers also published engravings of romantic landscapes feeding on the nationalistic sensibilities of the day, such as Ulrik Thersner's mammoth-sized patriotic *Past and Present Sweden* (*Fordna och närvarande Sverige*, 1817–67). Several of these artistic officers (though not Akrell) were active in the nationalistic associations mentioned above; and a couple even contributed to the influential exhibition of nationalistic art organized by the Gothicist Association in 1818.[57]

History was also a field of interest for the Topographical Corps, and some of its officers produced historical writing on a broader scale than that demanded by their profession — again an activity in line with contemporary nationalistic sentiments. Tibell and Akrell wrote short military histories, while others compiled more voluminous historical writings, in at least one case following a direct order of Bernadotte.[58]

These cultural activities — art, history, and romantic nationalism — strengthen the impression that the topographical officers were ideologically a fairly homogeneous group united by the promotion of national unity. This should come as no surprise; they were after all military men. Of interest however is that the most successful military cartographers promoted this agenda on a variety of social levels — from the geodetic staking out of national territory to the picturesque representation of the nation's landscapes and history. They brought the same agenda to bear on a variety of social and political issues where, again, they could draw on their expertise as military scientists and their proximity to the military and political centers of power.

EDUCATION, ADMINISTRATION, AND POLITICS

The emphasis on military science in the corps corresponded to a more general interest in educational matter. Many officers in the corps were teachers or wrote textbooks; several engaged in broader issues of education and pedagogy encompassing the civilian system of education. Topographical officers wrote most of the military textbooks published in Sweden during the period. Lefrén was most successful, with his three-volume *Lectures on Military Science* (1817–18) in use for more than half a century. Lefrén's career exemplifies how concerns typical of military cartographers became entwined with their different military or civilian fields of activity.

A close associate of Tibell, Lefrén was a career-wise man who was just as successful as his mentor—working his way to the top through a close association with Bernadotte. (He tutored his son in 1819 and was named chief of staff nine years later.) When in 1824 he became chief of fortifications he was a leading figure in military science, defense policy, and pedagogy. He was a Gothicist and a major proponent of the doctrine of central defense, which he interpreted in the light of Romantic nationalism.[59] Ennobled in 1818, Lefrén was powerful in the House of the Nobility and closely allied to the conservative leadership. Unlike most of his colleagues he was directly confronted with the threat of revolution: when social unrest broke out in Stockholm in 1848, troops under his command killed eighteen people. He was also deeply preoccupied with the problem of creating national unity by other, less violent but nevertheless military means.[60]

In pedagogy Lefrén was an influential spokesman for the Lancaster Bell system, which he introduced to the Royal War Academy at Karlsberg in the early 1820s.[61] He thought that this system was designed to mold enlightened and law-abiding citizens and to counteract disruptive tendencies such as democratic thinking. As such, this was a perfect vehicle for educational reform in the civilian sector as well: "the masses contain appalling elements; their reshaping is no child's play."[62] Lefrén believed in merging military and civilian education. The main objective of general education was to "kindle and sustain the war-like national spirit"—hence military subject matters should be taught in civilian schools. Another prominent topographical officer concerned with educational matters, Janne Hazelius, believed that military education should belong to the university curriculum, so that officers could become what they were in reality—true civil servants.[63]

These ideas were put forth in a political climate in which conscription was perceived by many as a way of unifying the civilian and military components of society. The conservative Lefrén belonged to an influential group of officers who actively promoted this agenda on the political stage. A liberal colleague of his put forth the idea that a national uniform should be introduced for civilians; another supported Hazelius's scheme for introducing military education in universities, which he hoped would make education "at once *scientific, civic* and *military*."[64] Akrell developed a program for militarizing the educational system in the 1830s. Through a general education system based on the Lancaster Bell method and on military discipline, citizens would become soldiers and vice versa. Military science and ordinary science would become one. To Akrell this was part of a liberal agenda: he promoted the abolition of military secrecy in cartographic matters, hoping to lay the foundation for an open discourse on military matters by merging the military and civilian spheres of education.[65]

The original strongholds of modern military science in Sweden were Tibell's creations—the Academy of Military Science and the Topographical Corps. As the military educational system developed (as in other countries, the military led the way in setting up a system of engineering education from which civil engineering would be an offshoot) and military cartography evolved into a national mapping project, the first generations of topographical officers had a broad impact on Swedish society through activities in their various areas of expertise. They became social carriers of a specific military scientific ethos. Their impact may be described in terms of the promotion of national unity, a project that defined much of European politics and culture in the era of revolutions and romantic nationalism. At the same time it may be described as a militaristic project, in the sense that military ideals were identified with those of society as a whole. Information-based strategic considerations triggered by Napoleonic warfare were behind the new military science. Strategic information was also seen as the foundation of social development and national unity.[66] One basis of military science was physical astronomy, the theory and technology of trigonometric surveying. It defined the nation in a physical sense, both internally through the determination of numerous observation points that were in effect miniature observatories, and externally through the determination of borders by national trials of technological strength. The result was a set of data flexible enough to be used both for military and civilian needs. On the basis of these activities and related work such as statistics, an informational infrastructure for military and social control and reform was created. In the process a modern nation-state was defined.

To put a Foucauldian gloss on these developments, one might say that the military cartographers offered a panoptic view of society, determined from the perspective of the General Staff—a perspective that can be identified with that of civilian government. Published maps were but a surface expression of deeper, more penetrating information, namely the collections of data in archives. The General Staff could see all but was itself invisible; its data collections were a means of control. Recently the rise of statistical practices in the period has been discussed in terms of another Foucauldian analytical approach, that of liberal government. From this perspective governing may be seen as guiding "natural" social development—and statistics is the natural history of society, establishing its social shape just as geodesy established the nation's physical shape.[67]

The topographical officers were social carriers of a program for reform that could well be identified with the program of liberal governments, but that was also militaristic in the sense discussed above. It is noteworthy that these officers did not constitute a homogeneous group politically speaking, though they can to some extent be associated with the movement of Gothicist national revival. The most prominent were either liberal or conservative, and sometimes changed sides. In the last analysis theirs was a peculiar professional project, intent on maintaining military scientific expertise close to the center of power in the post-revolutionary nation-state.

After 1815 the universalistic ambitions of military science allowed it to become a vehicle for the coordinated planning of warfare and industrial expansion. This is especially true of infrastructures such as railroads and electrical telegraphy. The military science ideals promoted by the Topographical Corps were a product of Napoleonic warfare; they became a prerequisite for later European industrialization and for the kind of strategic thinking that found its ultimate expression in the Schlieffen plan. This plan, in the words of William McNeill, symbolizes the "central dilemma of our age—the dissonance of the whole introduced, or enormously exacerbated, by a closer harmony and superior organization of its separate parts."[68]

Notes

1. Spens, "Jemnförelse emellan de metoder at förfärdiga kartor," 305-6.
2. This chapter is based on chapters 18-21 of Widmalm, *Mellan kartan och verkligheten*.
3. Sven Widmalm, "A Commerce of Letters."
4. Karin Johannisson, *Det mätbara samhället: statistik och samhällsdröm i 1700-talets Europa* (Stockholm: Norstedts, 1988); N. V. E. Nordenmark, *Pehr Wilhelm Wargentin: Kungl. Vetenskapsakademiens sekreterare och astronom, 1749-1783* (Uppsala: Almqvist och Wiksell, 1939), 202-24, 232-69.
5. The Swedish astronomer Anders Celsius promoted his new observatory in a pamphlet that discussed all these aspects. Anders Celsius, *Nytta af et astronomiskt observatorium uti Swerige* (Upsala, 1739).

6. Much has been written about the metric survey of the 1790s and its political and economic contexts. For an overview see Alder, *The Measure of All Things*.

7. Siestrunck, "La carte militaire"; Berthaut, *La Carte de la France*; Konvitz, *Cartography in France*; Gyula Pápay, "Studien zur Herausbildung der Wissenschaftsdisziplin Kartographie"; Bialas, *Erdgestalt, Kosmologie und Weltanschauung*; Hans-Jürgen Kahlfuß, *Landesaufnahme und Flurvermessung in den Herzogtümern Schleswig Holstein und Lauenburg vor 1864* (Neumünster: Karl Wachholtz, 1969); W. Grossmann, "Gauss' geodätische Tätigkeit im Rahmen zeitgenössischer Arbeiten," *Zeitschrift für Vermessungswesen* 80 (1955): 371–84. For more detailed references see Widmalm, *Mellan kartan och verkligheten*, 129–47. On French geodesy see also Martina Schiavon's contribution to this volume; on the longitude problem see Guy Boistel's.

8. Richard A. Preston and Sydney F. Wise, *Men in Arms: A History of Warfare and its Interrelationships with Western Society*, 4th edn (New York: Holt, Rinehart and Winston, 1979); McNeill, *The Pursuit of Power*, 161–63, 170–71; Henning Eichberg, "Geometrie als barocke Verhaltensnorm: Fortifikation und Exerzitien," *Zeitschrift für historische Forschung* 4 (1977): 17–50; Vagts, *A History of Militarism, Civilian and Military*.

9. McNeill, *The Pursuit of Power*, 211; Geoffrey Parker *The Military Revolution: Military Innovation and the Rise of the West, 1500–1800* (Cambridge: Cambridge University Press), 61–75; Merritt Roe Smith, "Army Ordnance and the 'American System' of Manufacturing," *Military Enterprise and Technological Change*, ed. Smith, 39–86.

10. Carl D. Forsberg, "Minnes-Anteckningar om en under många år, Wördad och tillgifwen Förman, Presidenten och General Löjtnanten m.m. Friherre G. W. af Tibell," G. W. af Tibell's papers, vol. 2, Swedish War Archives. On Tibell in France see Lars Tingsten, *Gustaf Wilhelm af Tibell: Huvuddragen av hans liv, hans verksamhet såsom generaladjutant för armén och hans avskedand, tillika en studie av Sveriges krigföring år 1808* (Stockholm: Norstedts, 1924), 5–12.

11. "La carte napoléonienne est la carte du Prince par excellence." Siestrunck, "La carte militaire," 370.

12. Berthaut, *La Carte de la France*, 2:231, 278–85.

13. G. W. af Tibell, "Embets-berättelse om Italienska Krigsvettenskaps Akademiens stiftelse och arbeten år 1802," *Krigsmannasällskapets handlingar* (1802): 184–88, at 184.

14. All quotes above from Tibell to Bror Cederström, 16 November 1802, Cederström's papers, vol. 2, Swedish War Archives.

15. Tibell, memorandum to the King, 3 September 1805, Uppsala University Library, L 313; Tibell, memorandum to the King, 5 May 1806, Swedish National Archives, M 856.

16. Widmalm, *Mellan kartan och verkligheten*, chapters 3, 6.

17. Ibid., 235–40.

18. For details see ibid., 240–53. The correspondence, including some drafts, between on the one hand Svanberg and his mentor, the perpetual secretary of the

Royal Academy of Sciences Daniel Melanderhielm, and on the other hand diverse French astronomers is at Uppsala University Library, A 640 (Svanberg) and G 172 (Melanderhielm).

19. Svanberg, *Exposition des opérations faites en Laponie.*
20. Royal Letter, 16 April 1805, *Samling af Kongl. Maj[estä]ts nådiga bref och instruk-tioner*, ed. Tibell, 4.
21. G. W. af Tibell, "Om Kongl. Fältmätnings-corpsens inrättning," *Krigsvetenskaps-akademiens Handlingar* (1807): 3–20, at 20.
22. Tibell to Cederström, 16 June 1803, Cederström's papers, vol. 2, Swedish War Archives; Tibell, "Om militäriska undervisnings-verk i Sverige," 75.
23. "Il ne suffit point à un Militaire de sacrifier sa Vie, il doit aussi la rendre utile à sa Patrie, en vivant." Tibell, "Om militäriska undervisnings-verk i Sverige," 72.
24. Carl Akrell, *Om recognoseringar* (Stockholm, 1813), 3. Similar sentiments were expressed by Akrell in *Underrättelse till kartan öfver Sverige* (Stockholm, 1811), 1, and by one of the most experienced naval cartographers in Sweden at that time, Carl Peter Hällström, in *Tal om den tillvext fäderneslandets geographie vunnit under loppet af de sistförflutna femtio åren* (Stockholm, 1815), 62–64.
25. P. H. Enger, *Kungl. fortifikationens historia*, vol. 5, no. 1 (Stockholm: Norstedts, 1951–52), 1–17; Tingsten, *Gustaf Wilhelm af Tibell*, 114–21; Harry Donner, "Kring G. M. Armfelt och G. W. Tibell åren 1808 och 1811," *Historisk tidskrift för Finland* 15 (1930): 105–18; Tibell, "Om Kongl. Fältmätnings-corpsens inrättning," 94–95; Hazelius, "Berättelse om Fältmätnings-corpsen och dess arbeten," 280.
26. Hazelius, "Berättelse om Fältmätnings-corpsen och dess arbeten"; J. F. N. Aro-senius, "Om svenska topografiska kartverket t.o.m. år 1859," G 3367 (copy), 18–19, 25–26; Spens, "Jemnförelse emellan de metoder at förfärdiga kartor," 300; Einar Bratt, *En krönika om kartor över Sverige* (Stockholm: Generalstabens Litograf-iska Anstalt, 1958), 98–102; Akrell, "Tal hållet den 12 Nov. 1841"; Axelson, "Års-berättelse för åren 1857, 58 och 59 af föredragande för topografien," 87, 92–94; Arosenius, "Årsberättelse för åren 1846–1848 af föredraganden för topografien," 109–10.
27. Akrell, "Tal hållet den 12 Nov. 1841," 220–21.
28. *Borgarståndets protokoll* (Minutes of the Burgers' Estate), 1859–60, vol. 3 (Stock-holm, 1860), 4 April 1860, 504.
29. Smith, "Army Ordnance," *Military Enterprise and Technological Change*, ed. Smith. For an overview of the topic see Smith, "Introduction," ibid., 1–37. Cf. David F. Noble, "Command Performance: A Perspective on Military Enterprise and Tech-nological Control," ibid., 329–46; Barton C. Hacker and Sally L. Hacker, "Mili-tary Institutions and the Labor Process: Noneconomic Sources of Technological Change, Women's Subordination, and the Organization of Work," *Technology and Culture* 24, no. 4 (1987): 743–75; McNeill, *The Pursuit of Power*, 125–43; Vagts, *A History of Militarism, Civilian and Military*, 84.
30. See reports by J. G. Nisbeth (6 August 1818, 3 January 1821) and Carl Akrell (31 December 1821, 19 January 1829, 1 January 1830) in Swedish War Archives, Ar-chives of the Topographical Corps, D II–IV, documents received 1812–30. See

also Hazelius, "Berättelse om Fältmätnings-corpsen och dess arbeten," 266–67; Spens, "Jemnförelse emellan de metoder at förfärdiga kartor," 296–97; Arosenius, "Årsberättelse för åren 1846–1848," 109. The military cartographers began publishing statistical reports in 1844.

31. Bratt, *En krönika om kartor över Sverige*, 88; B. Sundquist, "Det matematiska underlaget för terrängåtergivande medelst lutningsstreck enligt Lehmanns metod," *Globen* 23, no. 3 (1944): 38–48; Pápay, "Studien zur Herausbildung," 21–22, 43–52, 83–107.

32. Spens, "Jemnförelse emellan de metoder at förfärdiga kartor," 294; C. Stål, *Lärobok i topografien för linie-officerare* (Stockholm, 1835), 70–73.

33. Carl Akrell, "Ödmjukt memorial," memorandum dated 22 November 1827, Swedish War Archives, Archives of the Topographical Corps, D II–IV, documents received 1812–30.

34. Ibid.; Stål, *Lärobok i topografien för linie-officerare*, 68; Carl Magnus Thulstrup, *De första grunderna af mathematisk geografi och topografi* (Stockholm, 1855), 216; Spens, "Jemnförelse emellan de metoder at förfärdiga kartor," 292; Arosenius, "Om svenska topografiska kartverket t.o.m. år 1859," 18.

35. Axelson, "Årsberättelse för åren 1857, 58 och 59 af föredragande för topografien," 84. This issue gave rise to a lively parliamentary debate. For details see Widmalm, *Mellan kartan och verkligheten*, 380–83.

36. Tibell, "Om Kongl. Fältmätnings-corpsens inrättning och arbeten," 69–71; C. G. Tawaststjerna, *Föreläsningar i topografien* (Stockholm, 1807), 17, 45–46, 101, 103. Several engravings in this textbook are copied directly from Svanberg, *Exposition des opérations faites en Laponie*. The scientific ethos of the early corps is expressed for example in Ulrik Tersner, "Rapport," 1 July 1815, Swedish War Archives, Archives of the Topographical Corps, D II–IV, documents received 1812–30.

37. Hazelius, "Berättelse om Fältmätnings-corpsen och dess arbeten," 288–89; Per Gustaf Rosén, "Om de geodetiska och astronomiska ortbestämmelserna i Sverige," *Svenska sällskapet för antropologi och geografi: Geografiska sektionens tidskrift* 1, no. 9 (1879): 1–4, 14; Carl Magnus Thulstrup, "Anförande i Kungl. Krigs-Vetenskaps-Akademien, den 28 april 1837," *Krigsvetenskapsakademiens Handlingar*, 1837, 187–268, at 210–11; Arosenius, "Om svenska topografiska kartverket t.o.m. år 1859," 8. On the technology of marking station points see also an anonymous history on the topic from ca. 1930, G 3367.

38. E.g., J. B. Harley, "Silences and Secrecy: The Hidden Agenda of Cartography in Early Modern Europe," *Imago Mundi* 40 (1988): 57–76.

39. Carl Magnus Thulstrup, "Årsberättelse 1838 och 1839," *Krigsvetenskapsakademiens Handlingar*, 1840, 213–38, at 216. See also Abraham Häggbladh, "Om barometriska höjdmätningar," *Krigsvetenskapsakademiens Handlingar* (1837): 325–35, at 326–27; Struve, *Expédition chronométrique exécutée par ordre de Sa Majesté l'Empereur Nicolas 1er*, 2–3.

40. See F. G. W. Struve, *Arc du méridien de 25°20′ entre le Danube et la Mer Glaciale*, xv–xxxv.

41. Political issues came to the fore during collaboration on chronometric testing in

1833, when the seemingly imposing display of precision technology and discipline by the Russians was described by Akrell as a bluff indicative of the low level of Russian civilization. Carl Akrell, *Minnen från Carl XIV's, Oscar I's och Carl XV's dagar*, 131–32. On Russian geodesy see also Simon Werrett's contribution to this volume.

42. Spens, "Jemnförelse emellan de metoder at förfärdiga kartor," 276. See Alder, *The Measure of All Things*; Alder, "A Revolution to Measure: The Political Economy of the Metric System in France," *The Quantifying Spirit in the Eighteenth Century*, ed. Frängsmyr et al., 39–71; Sven Widmalm, "Accuracy, Rhetoric and Technology: The Paris-Greenwich Triangulation, 1784–88," ibid., 179–206; and John L. Heilbron, "The Measure of Enlightenment," ibid, 207–42.

43. Bialas, *Erdgestalt, Kosmologie und Weltanschauung*, 241–46. On this organization see also Schiavon's article in this volume.

44. Baeyer, *Über die Grösse und Figur der Erde*, 109–10.

45. For details see Widmalm, *Mellan kartan och verkligheten*, 317–43.

46. Bengt Abrahamsson, *Militärer, makt och politik* (Stockholm: Prisma, 1971), 7–15, 23–49; Klas Borell, *Disciplinära strategier: En historiesociologisk studie av det professionella militärdisciplinära tänkesättet, 1901–1978* (Stockholm: Almqvist och Wiksell, 1989), 15–18.

47. The most successful topographical officers were middle class at least and most were ennobled. Tibell was the son of a lieutenant, Akrell of an engraver, Lefrén of a professor of theology at Turku, Forsell of a captain, and Janne Hazelius of a tradesman. Akrell's uncle was an influential politician; otherwise this group seems to have lacked strong family ties to the higher strata of society. Akrell, *Minnen från Carl XIV's, Oscar I's och Carl XV's dagar*, 42–43.

48. This is the notion of militarism used in Vagts, *A History of Militarism, Civilian and Military*, 13–17.

49. Tibell, *Samling af Kongl. Maj[estä]ts nådiga bref och instruktioner*, 14–15; J. R. Western, "Armed Forces and the Art of War 2: Armies," *The New Cambridge Modern History* (Cambridge: Cambridge University Press, 1965), 8:190–217, at 201–2.

50. On Forsell's career see Carl Gustaf af Forsell, "Carl Gustaf af Forsell," *Biographiskt lexicon öfver namnkunnige svenska män* (Upsala, 1835), 1:13–26; Sörbom, *Läsning för folket*, 57–75.

51. Jörgen Weibull, *Karl Johan och Norge 1810–1814: Unionsplanerna och deras förverkligande* (Lund: Gleerup, 1957), 66–167.

52. Torkel Jansson, *Adertonhundratalets associationer: Forskningar och problem kring ett sprängfyllt tomrum eller sammanslutningsprinciper och föreningsformer mellan två samhällsformationer c:a 1800–1870* (Stockholm: Almqvist och Wiksell, 1985), 112–15.

53. *Aftonbladet*, 1 September 1835. Quoted from Sörbom, *Läsning för folket*, 74.

54. On Akrell's biography see Akrell, *Minnen från Carl XIV's, Oscar I's och Carl XV's dagar*; Henning Hamilton, "Carl Fredrik Akrell," *Lefnadsteckningar öfver Kongl Svenska Vetenskapsakademiens efter år 1858 aflidna ledamöter* (Stockholm, 1878–85), 2:43–56; L. W:son Munthe, "Carl Akrell," *Svenskt biografiskt lexikon* (Stockholm:

Svenskt Biografiskt Lexikon, 1918), 1:347–56. On his statistical work see Höjer, *Svenska siffror*, 75–79, 114.

55. Hans Heimbürger, *Det elektriska telegrafväsendet, 1853 1902* (Göteborg: Televerket, 1938), 22–164; Axelson, "Årsberättelse för åren 1857, 58 och 59 af föredragande för topografien," 100–101; Sverker Oredsson, *Järnvägarna och det allmänna: Svensk järnvägspolitik fram till 1890* (Lund, 1969), 55, 73–75.

56. Rudolf Hjärne, *Götiska förbundet och dess hufvudmän* (Stockholm, 1878), 1:107–21; Greta Hedin, *Manhemsförbundet: Ett bidrag till göticismens och den yngre romantikens historia* (Göteborg: Göteborgs Högskola, 1928), 346–52.

57. On artists in the corps see *Svenskt konstnärslexikon* (Malmö: Allhem, 1952–67), 1:236, 1:289, 3:32, 4:421–22, 5:376–78.

58. Akrell, *Minnen från Carl XIV's, Oscar I's och Carl XV's dagar*, 105–7.

59. Lefrén's activities in the field of defense policy are described in Alan Jansson, *Försvarsfrågan i svensk politik från 1809 till Krimkriget* (Uppsala, 1935).

60. On Lefrén's career see Carl Magnus Thulstrup, "Johan Peter Lefrén," *Lefnadsteckningar öfver Kongl. Svenska Vetenskapsakademiens efter år 1854 aflidna ledamöter* (1869–73), 1:479–504; Alf Åberg, "Johan Peter Lefrén," *Svenskt biografiskt lexikon* (Stockholm, 1977–79), 22:442–4.

61. Thor Nordin, *Växelundervisningens allmänna utveckling och dess utformning i Sverige till omkring 1830* (Stockholm, 1973).

62. Johan Peter Lefrén, "Tal hållet i Sällskapet för vexelundervisningens befrämjande inom fäderneslandet," *Redovisning och berättelser aflemnade vid allmänna årssammankomsten i Sällskapet för vexel-undervisningens befrämjande* (1825), 15.

63. J. A. Hazelius, "Svar på Kgl. KrigsVetenskaps-Akademiens fråga: Huru kan det ändamål sävinnas, at i framtiden alla blifvande officerare må . . . inhemta erforderliga kunskaper i krigsvetenskapens olika delar," *Krigsvetenskapsakademiens Handlingar*, 1824, 1–83, at 20, 24, 60. The Swedish word for civil servant, *ämbetsman*, has a military connotation in Sweden, unlike its English counterpart. Cf. Wilhelm Sjöstrand, *Pedagogikens historia*, vol. 3, no. 2 (Lund: Gleerup, 1965), 73–76, 84–98; Johan Peter Lefrén, "Tal om nyttan och nödvändigheten af en krigisk national-anda, samt medlen att väcka och vidmakthålla den," *Krigsvetenskapsakademiens Handlingar*, 1818–21, i–xxiii, at viii, xi–xxii.

64. Gustaf Montgomery, *Anmärkningar och idéer i anledning af friherre Anckarsvärds Tankar om Sveriges försvarsanstalter* (Stockholm, 1829), 47–48.

65. Carl Akrell, "Om allmän krigsbildning såsom villkor för allmänt national-försvar," *Krigsvetenskapsakademiens Handlingar*, 1835, 164–80.

66. Höjer emphasizes that statistics in general had, or was thought to have had, this function in the first half of the nineteenth century. Höjer, *Svenska siffror*, esp. 87, 249–59.

67. Frans Lundgren, *Den isolerade medborgaren: Liberalt styre och uppkomsten av det sociala vid 1800-talets mitt* (Hedemora: Gidlunds, 2003), 34–39, 85–134. Lundgren focuses on criminal statistics and their use, together with photography, in the process of creating new social categories.

68. McNeill, *The Pursuit of Power*, 306.

Chapter 7

Geodesy and Mapmaking in France and Algeria:

Between Army Officers and Observatory Scientists

MARTINA SCHIAVON

TRANSLATED BY CHARLOTTE BIGG

AND DAVID AUBIN

"*Ceterum censeo, gradum esse dimentiendum.*"[1] With these words Giovanni
Virginio Schiaparelli, the head of the Brera Observatory near Milan, instructed the astronomer Francesco Porro, of Turin, who was then em-
barking on a mission to establish an astronomical observatory in La Plata,
Argentina. As a century of work in geodesy had shown, one profitable first
step in the foundation of a permanent astronomical observatory was the
measurement of a meridian arc of the earth. During the nineteenth century
a great number of western states came to rely upon geodesy to establish
and exert territorial power, and geodesy had become a central component
of the work done at major European observatories. Geodesy, this chapter
aims to show, is a prime example of how techniques developed by, or in
collaboration with, observatory scientists were mobilized in the service of
nineteenth-century states.

Geodesy was principally concerned with determining the figure of the
earth with respect to stellar positions, but it also supplied astronomy with
a unit of length for measuring the distance of celestial bodies and the dimensions of the universe. Geodesy was further fundamental to the earth
sciences (cartography, geology, and geophysics) and was often crucial for
military and colonial undertakings. Whether for measuring meridians or
parallels, or determining the astronomical position of geodetic stations,
military officers used on earth the techniques created to observe the stars,
adapting skills and knowledge characteristic of the observatory to the measurement of terrestrial objects.[2]

Yet historians have seldom paid attention to the interaction between the observatory and geodesy. Perhaps this is because geodesy was above all a field science and accordingly seems remote from astronomy's traditional concerns. A more subtle explanation might be that geodesy's main practitioners were often not scientists but military officers. The historiography has accordingly tended to treat them as executants of other people's orders rather than innovators in their own right.[3] It would be equally misleading, though, to assume that geodesy was a contested terrain between astronomers and military officers. Indeed, it may have fostered contact and collaboration between the two.

This chapter highlights the significance of the observatory's material and conceptual resources for military practice, for the administration and control of territories, and ultimately for the constitution of the modern western technoscientific state. Building on more than fifty years of expertise in geodetic measurement, French savants launched ambitious expeditions at the time of the French Revolution to measure the meridian of Paris running from Dunkerque to Barcelona.[4] Here I focus on the significance of these expeditions for military cartography and the General Staff map of France (Carte d'État-Major). The Napoleonic wars demonstrated to the whole of Europe the value of precise maps. But while in the following decades German astronomers and geodesists were commissioned by their governments to carry out precise surveying *à la française* and developed new methods, the French lagged. The crushing defeat of 1871 opened a new era in French geodesy, which saw the introduction of foreign geodetic techniques and new forms of organization.

After reviewing these developments, I focus on one particularly illustrative undertaking of the late nineteenth century: Captain François Perrier's measurement of an arc of the earth's meridian in France and Algeria between 1870 and 1895. This case demonstrates how highly precise techniques—some of which were closely related to the observatory—were integrated into the apparatus of the state. In this period French geodesy almost completely turned into a military science, and the values of precision that pervaded the observatory became standards for officers. An investigation of the processes that led Perrier to adopt new instruments, such as the reiteration circle discussed below, shows that observatory scientists were directly involved in the scientific, military, and political issues raised by late-nineteenth-century geodesy.

The Measurement of Meridian Arcs in France before 1870

In the eighteenth century the measurement of a meridian arc, a segment of the imaginary circle running through both poles of the earth, had been supported by the Paris Academy of Sciences as a means of solving the controversial question of the shape of the earth. In the early nineteenth century, as old and new European states set out to establish their shifting borders more precisely, geodetic studies became crucial for detailed as well as for global cartography. The measurement of an arc is useful not so much for drawing local land property maps as for juxtaposing the different sheets produced by different state organizations, and for drawing maps of a whole country, of Europe, or of colonies. If the surface to be represented exceeds about one hundred by one hundred kilometers, the curvature of the earth must be taken into account. For the elaboration of large-scale maps, geodesy is therefore indispensable.

The first French map based on meridian arc measurements was the General Staff Map (GSM). The drawing of this map of France on a scale of 1:80,000 was supervised by a scientific committee headed by Pierre Simon Laplace. In 1792–98 Jean-Baptiste Delambre and Pierre Méchain measured an arc for this purpose, which also served as the theoretical basis for the French meter. Under Napoleon, Colonel Louis Puissant of the French Academy of Sciences oversaw the production of the map, for which operations were carried out in the field by officers from a technical corps created in 1808. These *ingénieurs géographes*, as they were called, were recruited from among the best students at the École Polytechnique and trained at the École Militaire de Géographie (founded 1802) before they joined the Dépôt Général de la Guerre, the body in charge of producing and safekeeping military maps.[5]

Drawing the GSM involved several steps. First, the officers reconstituted the triangles originally used by Delambre and Méchain, whose sides were on average thirty to fifty kilometers long. In the field the length of a section of the meridian was measured indirectly by triangulation (that is, a chain of triangles defined between the two ends of the arc to be measured). This in turn required three crucial measurements: the latitude of both ends determined by astronomical observation, the length of one or several "bases" measured on the ground by using chains or other surveying tools, and finally all the angles involved in the triangulation. Trigonometric identities were then used to determine the length of all triangle sides, as well as the

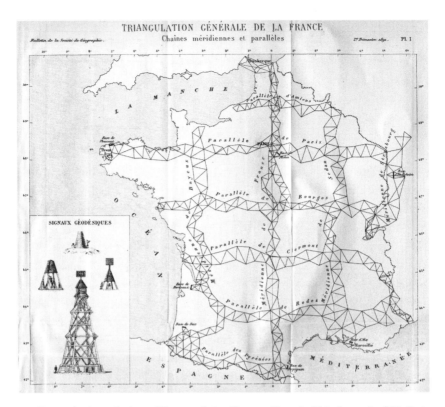

32. General triangulation of France. Léon Bassot, "La géodésie française," *Bulletin de la Société de Géographie* 12 (1891), plate. Courtesy of the Société de Géographie.

length of their projection onto the meridian. The sum of these projections yielded the length of the arc. The ingénieurs géographes referred to this chain of triangles as the "first-order triangulation" because of its primary role in cartographic drawing, but also because its points were determined in the field following especially precise astronomical procedures. The first-order chain was then overlaid with less precise "second-order" triangulation networks covering the whole territory to be mapped (figure 32).

The observation of angles required great care, because small errors could have significant repercussions. While insignificant for a single angle, a small inaccuracy in positioning the instrument at the center (eccentricity) or along either side of the angle to be measured could considerably affect the precision of a triangulation involving more than a hundred triangles. The precision of an angular measurement depended to a large extent on the instrument used. For the GSM the ingénieurs géographes used an instrument made by Henri Prudence Gambey following the instructions of the mathematician and navigator Jean-Charles de Borda.[6] The instrument consisted of a repetition circle placed exactly at the origin of the angle to be measured. By rotating the circle around its axis, one could measure vertical as well as horizontal angles, allowing both astronomical and terrestrial observations to be made. However, the common center tended to wear out because of friction, thereby increasing the eccentricity.

In the nineteenth century astronomers showed that the best approximation of the figure of the earth was an ellipsoid with an equatorial radius of approximately 6,366 kilometers and a polar axis about twenty-one kilometers smaller.[7] Yet each nation adopted its own values for the ellipsoid and each chose to refer all astronomical observations back to a national system of primary stations. In addition, to make maps each nation used different projections of the ellipsoid onto the plane. Because an ellipsoid cannot be projected exactly onto a plane, each projection addressed one particular type of error at the expense of others. Navy maps, for instance, facilitated tracing a ship's route by projecting meridians and parallels onto straight lines. Artillery maps, on the other hand, used a different projection that accurately represented the physical appearance of the territory, particularly with regard to altitude differences (leveling).[8]

For the GSM Delambre's ellipsoid was adopted, and the system of projection was worked out by the hydrographic engineer Rigobert Bonne. In this system parallels were concentric and equidistant while meridian lines were "transcendent" (not straight but curved). An important consequence of the Bonne projection for the GSM was an increase in the angular deformation as one moved away from the central meridian line. Close to the German border, for example, as the points are most distant from the Paris meridian, angles are most distorted in the Bonne projection. At the end of the nineteenth century this drawback would become a major problem for firing artillery which was progressively augmenting the distance of its shooting points.

Despite its shortcomings the GSM project was routinely pursued until after the middle of the nineteenth century. Problems were not addressed for several reasons. First, some members of the Paris Academy were afraid that acknowledging the errors would negatively affect the theoretical foundation of the French unit of length.[9] Second, François Arago (who together with Jean Baptiste Biot had extended Delambre's and Méchain's meridian arc to the Spanish island of Formentera in 1806) firmly opposed introducing new instruments in geodetic practice.[10] In 1830 the corps of ingénieurs géographes was integrated into the General Staff, revealing a shift of priorities away from the GSM project. The number of officers assigned to the task decreased steadily from then on, and the final sheet of the general staff map was not printed until 1880. This reorganization also meant that the Military School of Geography was closed down and the *Mémorial du Dépôt de la Guerre* (a publication on the physical appearance of territories for the purposes of waging war and supplying troops) was restricted to an internal distribution. With the incorporation of the ingénieurs géographes into the General Staff, an institution that had fostered collaboration between military officers and scientists was abolished. Resuming relations would prove difficult.

A New Kind of Geodesy: German
Instrumental and Computing Methods

Meanwhile other European nations had begun undertaking arc measurements and developed geodetic techniques. In the first half of the nineteenth century the most important innovations came from the German astronomers Carl Friedrich Gauss, Friedrich Wilhelm Bessel, and Wilhelm Struve. Gauss, director of the Göttingen observatory, redefined the ellipsoid, improved the precision of his instruments and invented new methods for reducing and organizing data. His most original work in geodesy was his method for the treatment of error: the method of least squares.[11]

Bessel's method of error analysis proved particularly useful for linking the triangulation systems used in different countries but also for angular observations. The Germans developed new methods for obtaining quick and accurate angle measurement that relied on several observations carried out at different times of the day, under different atmospheric conditions, using different instruments and observers. Bessel's method was useful in attributing different weights to each measurement. Such a procedure was contrary to French practice.[12] Around 1830 the Königsberg observatory director Bessel began collaborating with Johann Jakob Baeyer of the Trigonometric Bureau of the Prussian General Staff (Trigonometrisches Büro des grossen Generalstabes). Between 1821 and 1830 Baeyer had carried out the first-order triangulation of the Berlin region and of the provinces of Silesia, Poznan, and Western Prussia. Together Baeyer and Bessel improved the reliability of angular measurement instruments and made systematic studies of the errors caused by inclination, eccentricity, the division and reading of circles, and the pointing and collimation of telescope oculars.

Bessel introduced the notion of "instrument constant" that defined its precision, and he recommended that each instrument be calibrated accordingly. He believed that scientific instruments were intrinsically imperfect, but that their degree of precision could be estimated and improved: errors caused by each component (screws, for example) could be determined and reduced by technical and mathematical means. The German scientific community received Bessel's work favorably, adopting in particular the astronomer's view that giving an estimate of the measurement error reflected the experimentalist's moral integrity and that the existence of error revealed not failure but the honesty of the practitioner.[13] Bessel's work helped to make precision measurement more objective. In this way geodetic measurement became more independent of the observer's personal judgment.

Bessel's main objective was the geodetic connection between Prussia and Russia, subsequently to be extended to the other German states as well as France, England, Austria, and Denmark. He strove to establish a first-order network for the whole continent as a precondition for producing the first accurate map of Europe. Faced with very heterogeneous geodetic data, Bessel however had to renounce his original project of linking the individual countries' existing networks. Instead he advocated a new triangulation using German methods and instruments. He defined an arc crossing several countries, to be measured by national astronomical bodies. In 1862 Baeyer was put in charge of the project and the Prussian government authorized him to launch an international collaborative program. The other German states and several European nations joined in this project, known as the Central European Arc Measurement (Mittel-Europäische Gradmessung).[14] France initially declined the invitation, probably out of reluctance to adopt German methods and techniques. Nevertheless, the creation of the confederation testifies to the importance gained by geodesy by the mid-nineteenth century.[15]

Besides Gauss and Bessel, the director of the Dorpat Observatory, Struve, also significantly contributed to the renewal of geodetic techniques. Between 1816 and 1831 he carried out the triangulation of the Baltic arc. Adopting Gauss's least-squares method, Struve paid special attention to angle measurement instruments. A critic of repeating circles (that is, French techniques), he introduced a new method, "the method of directions," later known as reiteration. Mentioned for the first time by William Mudge in 1799, the reiteration method was used by George Everest in 1823 in his triangulation of western India.[16]

French Debates about German Methods:
A Case for Geological Geodesy?

In 1857 Struve visited several European countries, hoping to convince their government that they should participate in the measurement of a meridian arc of 69 degrees and a parallel arc of 55 degrees of longitude. The French government sought advice from the Academy of Sciences, whose scientists expressed a well-founded fear that Struve's ambitions went beyond mere arc measurements. Marshal Jean-Baptiste Vaillant warned his colleagues that this work foreboded the drawing of a map of Europe. Other scientists attacked Struve's methods, with his method of error analysis described by Biot as a "kind of speculative polish, an artificial compensating formula concealing the irregularities of the earth's figure that studies should aim at identifying."[17] On the other side the director of the Paris Observatory, Urbain Le Verrier, insisted on the value of arc measurements for determining the ellipsoid approaching the true figure of the earth and as an opportunity for extending his telegraphic data-gathering network for meteorological purposes.[18]

But the most ardent supporter of Struve's project was the astronomer Hervé Faye, who saw it as an occasion for reviving an old project of his own, long opposed by Arago: to revise Delambre's and Méchain's meridian arc measurement by using new geodetic instruments, including the reiteration circle. In 1857 Faye put forward his geodetic program, this time connecting it to geological investigations, which gained him the support of the perpetual secretary of the academy, Léonce Elie de Beaumont, a mining engineer by training and the author of a theory of mountain range formation.[19] With the help of Beaumont's theory Faye thought that he could explain the significant deviations from the vertical measured in several geographical sites and study anomalies in the figure of the earth. The definition of latitude—that is, the angular distance between a geographical site and the equator—involved several definitions of the vertical: the *normal vertical* was perpendicular to the theoretical surface of the earth (ellipsoid), whereas the *actual vertical* was perpendicular to the real and irregular surface of the earth (geoid), which is given at each point by the direction of a plumb line. Consequently the length of a meridian arc varies significantly depending on whether the latitude measured at each end refers to one or the other definition of the vertical.

Astronomers called "local attraction," or deviation of the plumb line, the angle between the actual and the normal vertical, which indicated the presence of an "anomaly" or bump in the geoid. Conventionally, and following Isaac Newton's interpretation, a deviation of this sort was thought to result from the attraction of the great accumulation of matter that made up mountains. The Kheops pyramid was thought to deviate a plumb line by 0.7 seconds. A pyramid of similar constitution but ten times larger was expected to increase this effect tenfold, to seven seconds. Nineteenth-century astronomers accordingly believed that most mountains would cause much greater deviations. Yet doubt was cast over the Newtonian theory when in 1855 the Reverend John Henry Pratt measured a deviation of four seconds close to the Himalayas instead of the expected twenty-eight seconds.[20]

Reviewing these results, Faye argued for the necessity of studying the relationship between geodesy and geology. He believed that new arc measurements were needed to improve instrumentation but also to provide data supporting Élie de Beaumont's theory, according to which mountains had a specific structure and internal layers. Before any conclusions could be reached, Faye suggested, the density of mountainous masses should be estimated, by using local geology at every geographical point, and compared to the average density of the earth.[21]

Instead of advocating the study of forces acting on the plumb line, as the British did, Faye believed that mountains were not solid masses and that motions in fluid layers could sometimes cause small deviations of the plumb line. Starting from geological makeup, Faye therefore was led to pay attention to the terrain's local physiognomy. In his view German error theory failed to take these specific characteristics into account. French geodesists should instead rely on direct field experience and carry out geological studies and thorough terrain surveys, without which a "truly scientific triangulation" was impossible. Faye saw geodesists as field scientists investigating local anomalies rather than cabinet observers relying on computations. Thus while German astronomers endeavored to make geodetic measurement more independent of the observer's personal judgment, Faye argued that in making calculations the French geodesists should combine experience in the field with their personal sensibility.

A New Impetus for French Geodesy: François Perrier and the War Depot

In 1854 the French government severed the link between the Bureau of Longitudes and the Paris Observatory, and the bureau was now solely responsible for "all developments in the vast field of [astronomy's] applications to navigation, geodesy, geography, and geophysics."[22] Looking for new fields of action and under pressure to justify its existence, the bureau opened its doors to military officers of the army and navy. To French scientists, military officers were precious collaborators whose training at the École Polytechnique they often shared and who made important contributions, together with instrument makers, to the study and development of scientific instrumentation.

Captain François Perrier was one such scientifically trained officer. The son of a baker, Perrier was admitted to the École Polytechnique in 1853. After additional training at the General Staff School (École d'État-Major) and in Algeria, Perrier returned to France in 1861 to work on the geodetic junction of France and Britain through the Pas-de-Calais. This operation aimed to establish precise borders and to adjust the cartographic sheets prepared separately and simultaneously by both nations, based on a new triangulation. The Anglo-French operation was directed by geodesists of the Ordnance Survey and by a former ingénieur géographe, Hippolyte Louis Levret. This operation provided an opportunity for comparing both nations' practices, instruments, and methods. While the French used Borda circles, the British used Ramsden theodolites and a heliotrope for filtering light through the misty skies of Calais. Introduced into geodetic practice by Bessel, this instrument was, according to the ingénieur géographe Jean-Jacques Levallois, of no use in the repetition method.[23] This would suggest that the British, critical of Delambre's and Méchain's old measurement, were experimenting with the reiteration method introduced by Struve. It is likely that Perrier first came into contact with the method on this occasion.

After 1861 Perrier ordered from the Brünner brothers an azimuthal reiteration circle.[24] The instrument was used solely for azimuth angle measurements, a practice symptomatic of the use of a different instrument for each geodetic operation. The instrument's name refers to its main feature: a reiteration circle on which each angle is measured several times, equidistant lines around the limb (the graduated edge of a circular measurement instrument divided in grads) serving as the origin of each measurement.[25] The starting and ending points were spread across the circumference to eliminate any systematic errors in the limb's division (figure 33). For each angle eight values were thus obtained (four in one position of the telescope, four in the position obtained with the optical axis reversed). In this way several additional points of the limb came into play and the whole circle was used. The reiteration method involved more readings than the repetition

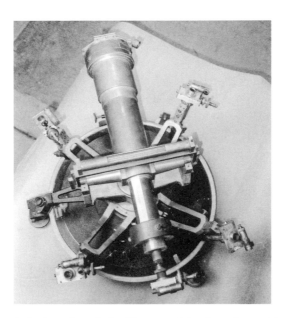

33. Azimuthal reiteration circle. Photograph by the author, with permission of the Institut Géographique National.

method, but it lent itself more easily to the study of errors using the least-squares method. Moreover, measurements were quicker than with the repetition method, since the circle and the telescope rested on a single support, enhancing the stability of the instrument and facilitating its manipulation in the field.

While French Academicians debated the practical validity of error analysis, Perrier had come into contact with it in different settings. At the École Polytechnique Faye taught the method in his courses on geodesy and astronomy, but it did not figure in the examinations. In his military school training Perrier had been exposed to the problem of efficiency in connection with aiming accuracy. At the War Depot a few officers had become aware of the latest innovations in German geodesy through the work of the Belgian geodesist Jean-Baptiste Liagre.[26]

Sent again to Algeria in 1867 in order to measure a parallel arc, Perrier further familiarized himself with the reiteration circle and error analysis. The bloody conflict of 1830, the French conquest of the Algerian inland, ruled by Emir Abd el-Kader, had cost the French army 106,000 men, excluding those who died of sickness. Not until the 1860s could France, to use official terminology, "pacify" and settle the Algerian territory. Perrier's work along the Mediterranean coastline between the Tunisian and

Moroccan borders was to help the settlement of French colonists. At first the French high command was able to implement a controlled exploitation of the Algerian land that did not allow settlers to act freely. The land was administered by appointed or confirmed native chiefs in charge of collecting tax. But the growing number of immigrants, especially after the French agricultural crisis of 1868, complicated matters. In the first ten years of colonization more than 400,000 hectares were taken from native inhabitants, creating a pressing need for a detailed map of Algeria for the first census.[27]

In 1868, as he was returning one evening to the Tlemcen station in Algeria, Perrier sighted the Spanish coast on the horizon: "seduced by the unexpected and incomparable beauty of the spectacle taking place before my eyes, and despite my extreme tiredness after a day of strenuous observations, I hurried to set up my circle in the observing station and to calculate the azimuth with respect to the still visible Tessala summit." Perrier's surprise stemmed less from his ability to sight the Spanish coast 270 kilometers away than from his recollection of the project, about which he had been briefed by Levret, to extend the French meridian arc to Algeria without going through the Gibraltar straits. He recalled that when Arago and Biot extended the French meridian to the island of Formentera in 1806–8, they had also considered this possibility, claiming that "if ever European civilization succeeds in settling on the African coasts, nothing will be easier than to bridge the Mediterranean with a few triangles, by extending our [geodetic] chain westwards to the Gata cape."[28]

Perrier was a man of deeds. Back in Paris in January 1869, he hurried to get hold of a large-scale map of the Algerian and Spanish coasts. With Faye's support he asked the Geodetic Commission of the Bureau of Longitudes to back his attempt at verifying Delambre's and Méchain's astronomical chain and extending it to Algeria. Perrier shrewdly tied the junction of France to Algeria with the improvement of the French arc, pointing out that the nation's prestige was at stake, since the meridian had been used to define the meter. A collaborator of Le Verrier, the astronomer Antoine Yvon-Villarceau, had moreover found in 1864–65 that geodetic and astronomical angle measurements did not coincide, an error that Perrier attributed to repeating instruments.[29]

Perrier's project squared well with contemporary plans for mapping Algeria. The new French meridian was an immense arc whose amplitude exceeded twenty-seven grads, running from the Shetland Islands to the North to Laghouat in the Sahara Desert. One only need consider its extension to realize that this junction had a practical as well as a symbolic importance for establishing colonies in Africa. The French-Algerian junction would bring metropolitan time to the colony and become the basis for three additional meridian chains around which railways and roads could be established. Thus together with demographic control, cartography was a means of increasing metropolitan control over agricultural capitalism as well as the exploitation of mineral resources.[30] The Algerian maps representing areas of agricultural exploitation and transport lines established between 1860 and 1900 demonstrate that geodesy, as a preliminary to cartography, facilitated the appropriation of territories, their administration, and their exploitation.[31]

Although the Bureau of Longitudes was largely favorable to the project, its president Vaillant believed that Perrier lacked the aptitude required for carrying out the triangulation. Without the *ingénieurs géographes*, the War Depot no longer possessed qualified personnel nor resources. Vaillant's initial reluctance notwithstanding, the bureau submitted a favorable report to the minister of public instruction Victor Duruy and to the war minister Marshal Adolphe Niel. Perrier had explained to Niel that this operation would be an ideal occasion for reorganizing the War Depot, winning the support of the astronomers and instrument makers in the Bureau of Longitudes to improve instruments and computation methods. The operation would supply an exact triangulation of France and the basis for a more detailed map of the country. In the end Perrier successfully argued that the War Depot had the material and human resources to carry out this operation.[32]

The Lessons from the War with Prussia

A budget was allocated for the New French Meridian shortly before war with Prussia broke out in 1870. Even if it brought all geodetic operations to a halt, the conflict was nonetheless important in that it revealed the lack of military cartographic supplies and the War Depot's shortcomings. Soon after mobilization the depot was emptied of its personnel: all officers, including Perrier, were assigned to different army corps, expected to take commanding positions for which they were ill prepared, while the troops in the field received incomplete maps. The Rhine army was sent bundles of maps of the right bank of the Rhine, while the Metz army only received maps of the immediate surroundings of their position, most of them outdated, and had to buy maps from the city's libraries. In 1872 the minister of public instruction Jules Simon launched a survey to find out the cause of "French geographic inferiority."

After the defeat Perrier insisted that "geodetic science, which was created in France" had been neglected for too long, while Germany, by cultivating it, had achieved victory. During the conflict cannons that loaded from the rear were introduced, allowing armies to move more rapidly. Rifled cannon tubes that shot rotating projectiles with more stable trajectories and greater perforation power were introduced in 1858. Fitted with such tubes, cannon ranges now reached between three and eleven kilometers.[33] These new technologies increased the precision of artillery. In this context maps were not only useful for preparing battle plans but essential for accurate shooting. The strategic importance of geodesy was thus reinforced by the war.

Although Perrier's first concern was geodesy, his project to measure a meridian was also helped by the government's renewed interest in geography. Between 1870 and 1880 a group of Republican politicians led by Léon Gambetta and Jules Ferry became enthusiastic advocates of colonial expansion.[34] While geodetic officers merely accompanied the navy and artillery in their conquest of new colonial territory, they played a decisive role in organizing settlement and trade. Through his contacts with Gambetta and Ferry, Perrier was able to attract support from the new regime for his arc measurement.

Perrier transformed the War Depot into the Service Géographique de l'Armée. Putting forward geodesy as the precise description of the terrain on which military operations took place, Perrier promoted it as a profession for officers, requiring in-depth scientific training, intimate acquaintance with the field, and practical experience in manipulating instruments. Albeit conventional in many respects, Perrier's discourse after the defeat of 1870 helped raise the status of geodetic studies in the army.

Fieldwork for the New French Meridian

Perrier's geodetic project involved revising the French meridian arc and its junction with Algeria. The first operation was carried out between 1872 and 1888 and the second, dependent upon the Spanish government's approval, in 1879. The last great operation, the extension of the Algerian coastline meridian to the Sahara, was completed in 1895. The work took much longer than the three or four years initially planned. The geodesists were few and were simultaneously carrying out the triangulation of Algeria. At the same time, Perrier also surveyed the territory, tried out new instruments, and trained observers. Measurement of the arc was a complicated operation which took years to carry out in the field and required long-term financial commitment from the state.

Perrier began with a study of the territory to identify appropriate locations for setting up observing stations. He also determined a reference peak from which all other altitudes were determined in order to compute the coordinates of artillery shootings. Called a "leveling of the territory" (*nivellement du territoire*), this operation had not been performed for the GSM, in which relief was only approximately represented using different kinds of shading. In tune with the kind of geological study imagined by Faye, this type of leveling was also useful for civil engineering. In 1878 a committee set up by the Ministry of Public Works (on which Perrier sat) decided that the War Department would produce "high-precision" leveling alongside the main railways, roads, and waterways. In 1884, to facilitate high-precision leveling, the government set up a Service du Nivellement Général de la France, made up of military officers led by Charles Lallemand, a former collaborator of Perrier.[35]

After completing his study of the territory Perrier's next step was to check the existing first-order triangulation network, which was derived from three bases: Cassel in the North, Melun close to Paris, and Perpignan in the South. He measured all the bases using modern apparatus made by the Brünner brothers. Though Perrier could rely on a well-trained team of forty men and four observers who could measure up to one hundred meters an hour, work was very slow and the data to be analyzed filled three thick volumes. The results were encouraging. The measurement of the Perpignan base revealed an error made by Delambre and confirmed the increased precision of Perrier's instruments.

The new triangulation network was to overlap roughly with earlier ones, but Perrier added a few sites chosen for their strategic importance. Angles were no longer measured from the top of belfries or towers, since in such observatories angles could not be measured at the center, which was usually occupied by the bell. Instead Perrier built concrete pillars or solid woodwork to support observing instruments. The instrument was placed inside the scaffolding and centered on the reference position.[36]

Perrier also experimented with observing conditions, introducing nighttime observations, carried out two or three hours after sunset. This helped to increase the number of measurements.[37] Together with his student Léon Bassot, Perrier showed that these observations were particularly valuable for azimuth measurements when powerful projectors were used. Through

Yvon-Villarceau and the maker Louis Breguet, Perrier found out about Colonel Charles Mangin's collimators, initially built for optical telegraphy.[38] The device was easy to set up and calibrate, and a box protected the collimator against the accidents that might occur during transfers. The lamp burned petrol, a cheap combustible easily found in villages. Once lit, the lamp was well protected against the wind, and it could be seen with the naked eye from as far as seventy kilometers in good weather. Officers valued these technical characteristics, and the projectors were later in common use for nighttime exchanges between the stations. The collaboration of astronomers, military officers, and instrument makers here proved decisive.

For angle measurements Perrier set the reiteration circle in the vertical position to carry out meridian observations. The meridian circle, long used by astronomers, was thereby turned into a portable instrument.[39] Perrier used it to verify the triangulation network and to connect it with the French fundamental station (the Paris Observatory). To extend the chain to Algeria he made two improvements on the reiteration circle: he used prisms reflecting zenithal light towards the microscopes to light its divisions and, following Yvon-Villarceau's advice, adapted to the eyepiece of the telescope a "mobile thread" similar to those used by astronomers.

To extend the triangulation chain to Algeria, Perrier had to find a light source of such power that it could be seen from a distance of 270 kilometers. After experimenting with different chemical mixtures (oxygen, hydrogen, and magnesium), which proved satisfactory but had the disadvantage of requiring the "preliminary education" of the soldier in charge of overseeing the flame, Perrier settled for electric light (even though it was more expensive and required Mangin to modify his collimators accordingly). In 1873 Perrier, now a member of the Bureau of Longitudes, met Louis Breguet's son Antoine, who owned the manufacturing license for Zénobe Gramme's magnetoelectric machine. The Breguet workshops began the industrial production of the machine.[40] New roads then had to be built in steep terrain to transport the machine and the steam engine (disassembled into three parts, the heaviest weighing five hundred kilograms), as well as to supply water and coal to the stations (see figure 7, page 17).

Perrier's ideal observer was to combine "the skill and *coup d'œil* that will make of him a good, honest observer observing only in favorable conditions."[41] Good instruments alone did not produce good observations; training and experience were also necessary to calibrate the eye for geodetic observation: "when the eye points the same object many times in a relatively short time span, it becomes used to always point in the same way."[42]

Arc measurements therefore took time not only because bad weather hampered the visibility of signals but also because of the need to train soldiers for fieldwork.

In 1874 Perrier set up a permanent pavilion for this purpose near Algiers. Fitted with up-to-date geodetic instruments, the pavilion was used to train depot officers sent to the colonies. Algeria was, in the words of the geodetic officer Robert Bourgeois, turned into a "great school for colonial geodesy and topography." The pavilion was not originally intended for astronomical observations. But when it was made the origin of the Algerian map—the initial station of the whole geodetic network—its importance grew and justified its conversion into a permanent astronomical observatory.[43]

Perrier also established a training site in Paris. In 1875 the Bureau des Longitudes was authorized by the City of Paris to use a space in Montsouris Park for the astronomical training of navy officers and travelers. With the support of the bureau, the War Ministry reserved a space for Perrier to train depot officers in the theory and practice of geodetic measurement. These officers were expected to attend the courses given by astronomers including Yvon-Villarceau, Le Verrier, and Maurice Lœwy. In 1874 Bassot was adjoined to Le Verrier on a mission to determine astronomical azimuths at the Bastide du Haut-Mont (Cantal). From 1876 to 1892 Bassot performed different "astronomical observations," such as the telegraphic determination of twenty-one longitude differences, some of which were important international collaborations: Geneva-Lyon (1877), Leyden-Paris (1884), Paris-Madrid (1886), and Paris-Greenwich (1888 and 1892). Only in 1879, after a long period of training, was Bassot judged fit enough by Perrier to take over a station in the France-Algeria junction.[44]

The Geodetical Crossing of the Mediterranean

The extension of the European meridian to Africa was at the core of Perrier's project. A giant four-sided figure spanned the Mediterranean Sea, its summits occupied by Tética and Mulhacén in Andalusia, Filhaussen and M'Sabiha in the Oran province of Algeria. While Bassot and Perrier were posted on the two Algerian summits, the measurements were carried out in the first two locations respectively by Miguel Melchor Merino, astronomer at the Madrid Observatory, and Colonel Joaquín Barraquer, geodesist at the Instituto Geográfico in Madrid (figure 34).[45] It had been necessary to instruct the Spanish personnel in Perrier's observation methods. Though the Spanish officers were under the command of General Carlos Ibáñez de

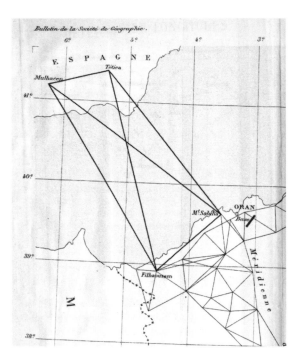

34. Geodetic junction of France and Algeria. Léon Bassot, "La géodésie française," *Bulletin de la Société de Géographie* 12 (1891), plate. Courtesy of the Société de Géographie.

Ibero, it was Perrier who directed the fieldwork. Having selected instruments made in France for carrying out the operations, Perrier invited the three Spanish scientists to come beforehand to Paris to pick up the instruments and learn to manipulate them.[46] A professor of geodesy at the Instituto Geográfico, Merino was an expert in the astronomical measurement of latitude and longitude differences. At the time he was busy establishing the first Spanish triangulation network.[47]

While training at the Montsouris Observatory (figure 30, page 165), Merino stayed for two months in Paris and acquainted himself with the reiteration circle. Perrier and Merino also took advantage of their common stay in Paris to measure their respective personal equations by exchanging signals between the Montlhéry Tower and the Paris Observatory. During one month they observed "several thousand signals," exchanging their telescopes every forty observations. The personal equation between the officer and the astronomer amounted to 0.124 seconds, an acceptable value when considering that transit observation could vary from one to ten tenths of a second. From this Perrier concluded that both men were good observers "by nature." Meanwhile he had shown that the personal equation was not negligible in the observation of intermittent light signals.[48]

Despite powerful electrical projectors, it took twenty days to make the junction. On 9 September 1879, after an abundant rainfall had "purified" the atmosphere, light signals were finally sighted. From the next day until 1 October geodetic observations were carried out every day. Once the geodetic junction was made, the astronomical junction was undertaken, that is, the difference of longitude between Tetica and M'Sabiha was computed on the basis of light signals exchanged over fifteen evenings, following the procedure developed in Paris.

The operation was a great success and brought high honors to Perrier and the Brünner brothers.[49] From then on the reiteration circle became a standard tool for geodetic work in France. The reiteration circle was not entirely Perrier's invention, but a combination of his own astronomical knowledge and skills and his collaborative technical work with the Brünner brothers.[50] What made the instrument remarkable was less its makeup than in its adaptation to geodetic uses; it was generally recognized that by correcting technical defects Perrier had been the first to adapt the instrument to the difficult conditions of fieldwork. Building on his geodetic field practice, he had moreover studied it using German error theory. When in December 1879 François Perrier was admitted to the Academy of Sciences in the section of Geography and Navigation, Antoine d'Abbadie insisted that the academy had the "moral duty" of welcoming geodesists within its ranks (figure 35): "The mathematician works alone following his own rhythm and inspiration. The chemist and the physicist can almost always create an experiment that materializes the idea that obsesses them. The astronomer foresees the moments he will be able to work and has no other adversary than the clouds. He too can observe alone. Hydrographers and geodesists however need to rely, in addition to expensive instrumentation, on a staff that has to be instructed and disciplined until the weather and the state of the atmosphere will allow them to work. The public rarely justly values their efforts, the detail of their most beautiful results, and the enormous difficulties in achieving them."[51] Their election to the academy took on a symbolic meaning. It signified acknowledgment of the value of an observatory science, geodesy, and of an institution, the Bureau of Longitudes, which had both proved their utility to the state. In encouraging geodetic work the academicians promoted the diffusion of astronomical techniques. Indeed, geodetic work confronted theoretical values with experience in

35. General François Perrier, member of the Académie des Sciences, geography and navigation section. Courtesy of the Académie des Sciences–Institut de France.

the field; it juxtaposed an ideal of precision in scientific measurement with reality. Geodetic practice also testified to the importance of error in scientific measurement, of the instruments used, and of computation methods.

In conclusion, the case of geodesy shows that science in general, and astronomy in particular, are not simply constituted of statements describing the world or of knowledge independent of human intervention. Instead one might describe science as a system of knowledge and practices designed to master and intervene in the natural and human world: to understand it but also to act upon and modify it. Instrumentation and technical objects played a decisive role in that they enabled more efficient action in these pursuits. This is why they never ceased being a concern for all powers, whether scientific, economic, or military.

Perrier became head of the Service Géographique de l'Armée in 1887. The service comprised three sections corresponding to the three stages of mapmaking. But foremost was the geodesy section in charge of verifying calculations, preparing the *Mémorial*, and administrating instruments and archives. Unlike the depot the Geographical Service was in direct contact with the war ministry and the scientific bodies (especially the Academy of Sciences and the Bureau of Longitudes). And it was not dismantled during the First World War. Indeed the service was considerably expanded during the war, which constituted a decisive moment for spreading observatory skills and techniques on the battlefield, as well as the moment when geodesy became a highly specialized, autonomous field. In the end geodesy loosened its ties with astronomy, becoming essentially military in nature.

Notes

The author wishes to thank Charlotte Bigg and David Aubin for their comments, as well as the director of the archives at the Académie des Sciences and the staff of the Institut Géographique National for permission to use their resources and reproduce images in their possession.

1. "Yet I believe that the degree [the length of the meridian arc] should be measured." Giovanni Virginio Schiaparelli to Francesco Porro, 1 December 1905, AOB Schiaparelli Papers, folder 445, file 2.

2. On the history of geodesy in France see Léon Bassot, "La Géodésie Française," *Bulletin de la Société de Géographie* 12 (1891): 162–89; Berthaut, *La carte de la France*; Georges Perrier, *Petite Histoire de la géodésie: comment l'homme a mesuré et pesé la Terre* (Paris: Presses Universitaires de France, 1939); Michael Rand Hoare, *The Quest for the true Figure of the Earth: Ideas and Expeditions in Four Centuries of Geodesy* (Aldershot: Ashgate, 2005). For a social history of geodesy in Sweden see Sven Widmalm's chapter in this volume.

3. On field science see Kuklick and Kohler, eds., "Science in the Field"; on travel and field science see Bourguet, Licoppe, and Sibum, eds., *Instruments, Travel and Science*; Raj, *Relocating Modern Science*; Pang, *Empire and the Sun*; and Pyenson, *Civilizing Mission*; on military expeditions specifically see Bourguet, Lepetit, Nordmann, et al., eds., *L'Invention scientifique de la Méditerranée*; and Lewis Pyenson, "On the Military and Exact Sciences in France," *National Military Establishments and the Advancement of Science and Technology: Studies in Twentieth-Century History*, ed. Paul Forman and Jose M. Sánchez-Ron (Dordrecht: Kluwer, 1996), 135–52.

4. See Alder, *The Measure of All Things*.

5. See Henri Berthaut, *Les ingénieurs géographes militaires, 1624–1831* (Paris: Service Géographique de l'Armée, 1902); Patrice Bret, "Le Dépôt général de la Guerre et la formation des ingénieurs-géographes militaires en France, 1789-1830," *Annals of Science* 48 (1991): 113–57; and Valeria Pansini, "L'œil du topographe et la science de la guerre: travail scientifique et perception militaire (1760–1820)" (Ph.D. thesis, École des Hautes Études en Sciences Sociales, 2002).

6. On this instrument see Danielle Fauque, "Un instrument essentiel de l'expédition pour la mesure de la Terre: le quart de cercle mobile," *La figure de la terre du XVIIIe siècle à l'ère spatiale*, ed. Lacombe and Costabel, 209-21; and Rondeau-Jozeau, "Géodésie au XIXème siècle."

7. This is the value given by Schiaparelli in one of his regular lectures on geodesy, 13 November 1865, Osservatorio astronomico di Brera, Milan, Schiaparelli papers, folder 437, file 3.

8. See Bernard Lepetit, "Missions scientifiques et expéditions militaires," *L'invention scientifique de la Méditerranée*, ed. Bourguet, Lepetit, Nordmann, et al., 97–116.

9. On the French unit of length see Alder *The Measure of All Things*.

10. Arago was convinced that the introduction of new instruments might "depreciate the work on the map of France." See Hervé Faye, "Note sur les propositions de M. de Struve, et sur la question académique qu'elles ont soulevée," *Comptes rendus des séances hebdomadaires de l'Académie des Sciences* 45 (1857): 670.

11. On Gauss see Walter Bühler, *Gauss: A Personal Study* (Berlin: Springer, 1981); on his geodetic work see Oscar B. Sheynin, "Carl Friedrich Gauss and Geodetic Observation," *Archives for the History of Exact Sciences* 46 (1994): 253–83; S. M. Stigler, "Gauss and the Invention of Least Squares," *Annals of Statistics* 9, no. 3 (1981): 465–74; George Rassias, ed., *The Mathematical Heritage of C. F. Gauss: A Collection of Papers in Memory of C. F. Gauss* (Singapore: World Scientific, 1991).

12. Rondeau-Jozeau, "Géodésie au XIXème siècle," 144.

13. On Bessel see Kathryn M. Olesko, "The Meaning of Precision: The Exact Sensibility in Early Nineteenth-Century Germany," *The Values of Precision*, ed. Wise, 103–34; Kathryn M. Olesko, ed., "Science in Germany," *Osiris* 5 (1989) [special issue]. On Baeyer see Baeyer, *Über die Grösse und Figur der Erde*; Ernst Buschmann, *Aus Leben und Werk von Johann Jacob Baeyer* (Frankfurt: Inst. für Angewandte Geodäsie, 1994).

14. In 1867 the confederation comprised all European states and took the name of European Arc Measurement (Europäische Gradmessung). In 1886 Mexico, Chile, Argentina, the United States, and Japan joined in, and it became the International Earth Measurement (Internationale Erdmessung). See J.-J. Levallois, "L'Association Internationale de Géodésie: notice historique," *Bulletin géodésique* 54 (1980): 248–70; Wolfgang Torge, "The International Association of Geodesy, 1862 to 1922: From a Regional Project to an International Organization," *Journal of Geodesy* 78 (2005): 558–68.

15. Baeyer's project was presented to the Paris Academy of Sciences by Hervé Faye and Urbain Le Verrier: Faye, "Rapport verbal sur le protocole de la conférence géodésique tenue à Berlin en avril 1862," *Comptes rendus des séances hebdomadaires de l'Académie des Sciences* 55 (1863): 28–34, and Le Verrier, "Quelques remarques à compléter le rapport de Faye," ibid., 34–37. For further discussions see also pages 66, 116, 154, 158, 193, 249, and 372 in this volume.

16. William Mudge, *An Account of the Operations Carried on for Accomplishing a Trigonometrical Survey of England and Wales* (London, 1799). For more on Struve's geodetic work and his adoption of the reiteration method see Simon Werrett's contribution to this volume.

17. These words are reported by Urbain Le Verrier in "Remarques relatives à la communication de M. Biot, insérée au *Compte rendu* de la dernière séance," *Comptes rendus des séances hebdomadaires de l'Académie des Sciences* 45 (1857): 674–78, at 677.

18. Ibid., 675. On meteorology at the Paris Observatory and its connections with the telegraph see John L. Davis, "Weather Forecasting and the Development of Meteorological Theory at the Paris Observatory, 1853–1878," *Annals of Science* 41 (1984): 359–82; Locher, "Le nombre et le temps"; and Aubin "The Fading Star of the Paris Observatory in the Nineteenth Century."

19. See A. Birembaut, "Élie de Beaumont, Jean-Baptiste-Armand-Louis-Léonce, 1798–1874," *Dictionary of Scientific Biography*, ed. Charles C. Gillespie (New York: Scribner, 1981), 3:347–50.

20. See John Henry Pratt, "On the Attraction of the Himalaya Mountains and of the Elevated Regions beyond Them, upon the Plumb-line in India," *Philosophical Transactions of the Royal Society of London* 145 (1855): 53–100. The Kheops pyramid was used as an example by Schiaparelli, "Sulle anomalie della gravità: discorso letto alla Società italiana di scienza naturali in Milano, 1 marzo 1896," AOB Library.

21. Faye, "Sur les instruments géodésiques et sur la densité moyenne de la Terre," *Comptes rendus des séances hebdomadaires de l'Académie des Sciences* 56 (1863): 557–66.

22. Faye, "Sur l'état actuel de la géodésie et sur les travaux à entreprendre par le Bureau des Longitudes"; on the Bureau des Longitudes see J.-M. Feurtet, "Le Bureau des Longitudes (1795–1854): de Lalande à Le Verrier" (thesis, École Nationale des Chartes, 2005).

23. Levallois, *Mesurer la terre: 300 ans de géodésie française, de la toise du Châtelet au satellite* (Paris: École Nationale de Ponts et Chaussées, 1988), 132.

24. On the Brünner dynasty see Paolo Brenni, "The Brünners and Paul Gautier," *Bulletin of the Scientific Instruments Society* 49 (1996): 3–8.

25. It is worth noting that French military geodesists, alone in this custom, adopted the grad, the metric angular measurement unit. A full circle comprises four hundred grads. The prototype of the azimuthal circle used for measuring the Algerian parallel had a limb thirty-two centimers wide divided into decigrads (0.09°). The angular reading was the average of the readings of two microscopes two hundred grads apart. The binocular's focal distance was fifty-four centimeters, with an objective of 5.3 centimeters. The instrument used for measuring the French meridian had a diameter of forty-two centimeters and four opposed microscopes, one hundred grads apart from each other.

26. On probability theory at the École Polytechnique and at military schools see Pierre Crépel, "Le calcul des probabilités: de l'arithmétique sociale à l'art militaire," *La formation polytechnicienne, 1794–1994*, ed. B. Belhoste, A. Dahan-Dalmedico, and A. Picon (Paris: Dunod, 1994), 197–215, and Bernard Bru, "Le problème de l'efficacité du tir à l'École de Metz," *L'École d'Application de l'Artillerie et du Génie de Metz (1802–1870)*, ed. Bruno Belhoste and Antoine Picon (Paris: Musée des Plans-Reliefs, 1996), 61–69; on the familiarity of the depot's officers with German geodesy see Peytier, "Nouvelle description géométrique de la France"; on Liagre see Rondeau-Jozeau, "Géodésie au XIXème siècle."

27. Daniel R. Headrick, *The Tools of Empire: Technology and European Imperialism in the*

Nineteenth Century (New York: Oxford University Press, 1981); C. A. Julien, *Histoire de l'Algérie contemporaine* (Paris: Presses Universitaire de France, 1964); and Robert Schnerb, "Le monde musulman de l'Asie centrale russe au Maroc," *Histoire générale des civilisations: le XIXe siècle: l'apogée de l'expansion européenne (1815–1914)*, ed. M. Crouzet (Paris: Presses Universitaires de France, 1968), 6:347-79.

28. Perrier, "Nouvelle méridienne de France," x, viii.

29. Antoine Yvon-Villarceau, "Comparaison des déterminations astronomiques des longitudes, latitudes et azimuts terrestres, faites par l'Observatoire impérial de Paris, avec les positions et azimuts géodésiques publiés par le Dépôt de la Guerre," *Comptes rendus des séances hebdomadaires de l'Académie des Sciences* 62 (1866): 804-9.

30. To oversee the exploitation of mineral resources (mainly iron and phosphates), a Service de la Carte Géologique d'Algérie was created.

31. These maps are reproduced in C. Canestrari, L. Colajanni, and M. Pazienti, *La colonizzazione del territorio: il modello francese: l'Algeria tra il 1830 e il 1962* (Milan: Franco Angeli / Storia Urbana, 1983). On the scientific settlement in Algeria see Pyenson, *Civilizing Mission*, 87-127.

32. See François Perrier's personal file, Archives of the Academy of Sciences, Paris.

33. See "Artillerie," *Dictionnaire des inventeurs et inventions*, ed. T. de Galiana and M. Rival (Paris: Larousse, 1996), 86-87; J. Carpentier, "Les progrès techniques et le sort des batailles terrestres, des origines à 1918," *Revue scientifique et technique de la défense* 55 (2002): 3-22.

34. On the development of geography in the colonial context after 1870 see Olivier Soubeyran, "Imperialism and Colonialism versus Disciplinarity in French Geography," *Geography and Empire*, ed. A. Godlewska and N. Smith (Oxford: Blackwell, 1994), 245-64; and Michael J. Heffernan, "The Science of Empire: The French Geographical Movement and the Forms of French Imperialism, 1870–1920," ibid., 92-114. On Gambetta's and Ferry's role in rekindling colonization see C.-R. Ageron, "Gambetta et la reprise de l'expansion coloniale," *Revue française d'histoire d'outre-mer* 59 (1972): 165-204; and Ageron, "Jules Ferry et la colonisation," *Jules Ferry: fondateur de la République*, ed. François Furet (Paris: Presses Universitaires de France, 1986), 191-206.

35. Charles Lallemand, "La réfection du cadastre et la carte de France," *Géographie* 3 (1900): 33-45; Charles Lallemand, *Nivellement général de la France* (Paris: Marchadier, 1885); Charles Lallemand and Eugène Prévot, *Le nivellement général de la France de 1878 à 1926* (Paris: Imprimerie Nationale, 1927). See also Jean Vignal, "Charles Lallemand, 1857-1938," *Annales des Mines* 14 (1938); Jean-Jacques Levallois, *Mesurer la terre: 300 ans de géodésie française* (Paris: Ecole Nationale des Ponts et Chaussées, 1988).

36. Perrier, "Travaux," François Perrier's personal file, Archives of the Academy of Sciences, Paris.

37. François Perrier and Louis Bassot, "Étude comparative des observations de jour et de nuit," *Comptes rendus des séances hebdomadaires de l'Académie des Sciences* 84 (1877): 1312-15, 1383-86.

38. See Paolo Brenni, "Louis Clément François Breguet and Antoine Louis Breguet," *Bulletin of the Scientific Instrument Society* 50 (1996): 19-24.

39. The azimuthal circle fit in a single box, and because it weighed less than sixty kilograms it could be carried by one man or on a donkey's back. One could only carry a meridian circle by splitting it into two parts (circle and support).

40. In the 1880s workshop prospered thanks to navy orders for projectors; see Brenni, "Louis Clément François Breguet and Antoine Louis Breguet," 22.

41. Louis Bassot's personal file, Archives of the Academy of Sciences, Paris.

42. François Perrier, "Description géométrique de l'Algérie," *Mémorial du Dépôt de la Guerre* 10 (1874): 18.

43. The pavilion should not be confused with the astronomical observatory directed by Charles Bulard, set up in Algiers in 1858 by the French navy to distribute time. See Pyenson, *Civilizing Mission*, 93–103; Françoise Le Guet Tully and Hamid Sadsaoud, "La création de l'observatoire d'Alger," *Revue du Musée des Arts et Métiers* 38 (2003): 26–35. On Algeria as a school of geodesy see R. Bourgeois, "Discours du président de l'Académie: séance publique annuelle du lundi 12 décembre 1932," *Comptes rendus des séances hebdomadaires de l'Académie des Sciences* 195 (1932): 1113–23.

44. Because of their reputation as good observers, officers were commissioned to carry out several other scientific missions on behalf of the Academy of Sciences: in December 1882, for example, Bassot and Defforges took part in the Florida mission to observe the transit of Venus. See Martina Schiavon, "Astronomie de terrain, entre monde académique et armée," *Cahiers François Viète* 11–12 (2007): 129–45. For the services he rendered during the observation of this astronomical phenomenon, the academy rewarded Bassot with the Lalande Prize. G. Perrier, "Notice sur les travaux de Bassot" (1935), Léon Bassot's personal file, Archives of the Academy of Sciences, Paris. On Montsouris see Guy Boistel's chapter in this volume.

45. A member of Madrid's Royal Academy of Exact, Physical and Natural Sciences, Colonel Barraquer was a reputed geodesist in Spain and the author of numerous scientific volumes. Together with Ibáñez he carried out the Spanish triangulation as well as many gravimetric measurements. See "Barraquer (Joaquín)," *Enciclopedia universal ilustrada europeo-americana* (Madrid: Rois Rosas, 1930), 904.

46. Perrier, "Travaux," François Perrier's personal file, Archives of the Academy of Sciences, Paris.

47. At the age of twenty-one Miguel Melchor Merino was hired as a computer at the Escuela Preparatoria de Ingenieros de Caminos y de Minas y Arquitectos in Madrid. Later he was sent to the School of Telegraphy, also in Madrid, until 1859, when he was named second astronomer at the Madrid Observatory. Promoted to first astronomer in 1864, he was nominated to the Statistics Commission and to the directorship of the geodetic brigade at the Geographic Institute. As secretary of the Royal Academy of Exact and Physical and Natural Sciences in 1882, Merino was his country's delegate to the International Geodetic Association. See "Merino y Melchor (Miguel)," *Enciclopedia universal ilustrada europeo-americana* (Madrid: Rois Rosas, 1930), 938.

48. Perrier, "Travaux," François Perrier's personal file, Archives of the Academy of Sciences, Paris.

49. The Brünner brothers considered each instrument "individually," as the achievement of a single, highly qualified workman and a true "artist." According to the instrument maker Gustave Yvon, mechanization was hindered in nineteenth-century France by this "artistic conception" of the scientific instrument. Yvon, "Les industries françaises de l'optique," *Revue d'optique théorique et instrumentale* 25 (1946): 177–94. On the French scientific community's particular attitude vis-à-vis the manufacture and manipulation of scientific instruments see in particular Paolo Brenni, "L'industria degli strumenti scientifici in Francia nel XVIII e XIX secolo," *Gli strumenti*, ed. G.L'E. Turner (Milan: Banca Popolare di Milano / Mondadori Arte, 1990), 450–63; Dörries, "Balances, Spectroscopes, and the Reflexive Nature of Experiment"; Blondel, "Les physiciens français de l'électricité industrielle à la fin du XIXe siècle"; and Bigg, "Behind the Lines."

50. François Perrier, "Réponse à une réclamation de priorité à propos de ma candidature" (1879), François Perrier's personal file, Archives of the Academy of Sciences, Paris.

51. Antoine d'Abbadie, "Rapport sur les travaux de F. Perrier" (December 1879), François Perrier's personal file, Archives of the Academy of Sciences, Paris.

Chapter 8

Michelson and the Observatory:

Physics and the Astronomical Community

in Late Nineteenth-Century America

RICHARD STALEY

American scientists in the nineteenth century sometimes regarded the youth and geography of their country as having especially favored the development of both the field sciences and astronomy. This is how the mathematical physicist Willard Gibbs explained the work of his professor of mathematics Hubert Newton in the 1850s: "In the attention which has been paid to astronomy in this country we may recognize the history of the world repeating itself in a new country with respect to the order of the sciences, or it may be enough to say that the questions which nature forces on us are likely to get more attraction in a new country and a bustling age, than those which a reflective mind puts itself. . . . At all events, when Professor Newton was entering upon his professorship, the study of the higher geometry was less consonant with the spirit of the age in this country than the pursuit of astronomical knowledge."[1] Indeed American astronomers and astronomical observatories often participated in a series of survey networks developed by government agencies, whose diverse tasks indicate developing federal interest (and scientific involvement) in different forms of conquest. The Coast Survey was established in 1807, followed by the U.S. Naval Observatory in 1830 (as the Depot for Charts and Instruments), the Army Corps of Topographical Engineers in 1838, and, much later, the Geological Survey in 1879. In addition to their work in observational astronomy and providing time services, astronomers contributed to geodetic surveys and developed geophysical and meteorological measurements, and their observatories were significant loci for coordinating a wide range of information. If the vigor of astronomy in the United States was due in part to the success of astronomers in meeting diverse local needs,

Stephen Brush has described the period until 1876 as one when Americans approached the level of European knowledge in astronomy and established their interest in new technologies. Two particular features soon stood out as distinctively American strengths: the construction of large telescopes (many for new observatories) and the commitment of a significant group of American scientists to the research goals and institutional foundations of a new specialty or subdiscipline variously described as the "new" astronomy and "astrophysics."[2]

Americans built observatories enthusiastically and on an emphatic scale. A first wave of observatories in the 1830s was followed from the 1870s by a boom in college observatories, and the construction of a few extremely large and expensive instruments. The Naval Observatory in Washington (with a twenty-six-inch objective lens, 1873), the Lick on Mount Hamilton (thirty-six-inch, 1887) and the forty-inch Yerkes instrument in southern Wisconsin (1897) all held records for being the largest working refractor telescopes, while Mount Wilson in Pasadena marked the move to giant reflector telescopes in the early twentieth century. Usually drawing on private patronage, the big telescopes were worked by hierarchically organized ensembles in which an entrepreneur director and several astronomers relied on unskilled or semiskilled workers to carry out extensive calculations or to interpret and catalogue mass-produced photographs. Several historians have argued that as an early instance of applying factory methods of mechanized data production, American astronomy also provides an important model for the later emergence of big science. By 1886 the United States had 40 observatories with 128 astronomers and computers, while the comparable figures for Germany were 26 and 81, and for Britain 32 and 70.[3]

The flowering of a new specialty in astronomical practice in the late nineteenth century was clearly dependent on the increasingly widespread use of new tools and methods to explore the physical and chemical constitution of celestial objects. From the 1880s the fine new grating spectroscopes produced by the physicist Henry Rowland of Johns Hopkins found important uses in solar research and joined photographic studies of the brightness of stars as central tools for astronomers who analyzed celestial light in a search for clues as to the distances of stars from the earth, their motion in the line of sight, and their magnitude, constitution, and evolutionary history. Indeed when George Ellery Hale, the most significant observatory builder and a tireless advocate of the new astronomy, began the *Astrophysical Journal* to support the endeavor in 1895, he chose a subtitle that described it as an international review of spectroscopy and astronomical physics. Given the scepticism with which some findings of the new astronomy were greeted, and the occasionally bitter exchanges between proponents of the new methods and adherents of the so-called old astronomy (which was based on measuring the positions and motions of the stars), historians have often described this merger of physics and astronomy as raising important questions about the identity of astronomers themselves.[4]

This chapter explores the importance of observatory techniques in mediating between the heavens and the earth—and between astronomy and other sectors of late-nineteenth-century America—by taking a sideways glance at leading representatives of that neighboring discipline of physics. Without directly taking up the complex questions that astrophysics raised for astronomers, this perspective will help me outline significant features of the meeting grounds between two disciplines. As the premier experimental physicists of the period, Henry Rowland and Albert Michelson are well known both for their research and their role in leading new laboratories at Johns Hopkins University and the University of Chicago. Physics laboratories too were burgeoning as part of the general rise of research schools initiated by Rowland's university in 1875; but both men have long been regarded as isolated and exceptional figures who overcame the institutional weakness then plaguing physics in the United States.[5] Here I will argue that thinking of them largely in terms of the professional identity they chose to represent, and accepting too readily the rhetoric of purity that they sometimes deployed to describe their endeavours, has obscured important features of the disciplinary environment within which they worked. Evans and Warner have usefully described both figures—together with a much lesser-known astronomer, William Rogers—as academic mechanicians, thus linking them (however indirectly) to the industrial manufacturing that played a leading role in America's technological progress.[6] Recently this point has become recognized more generally, especially in discussions of Rowland and his research school. George Sweetnam has delineated the engineering basis to laboratory work on spectra at Johns Hopkins, and Norton Wise has pointed to engineering training as a key feature shared by Rowland, Michelson, and Willard Gibbs, the third major American physicist of the period.[7]

But a second factor is perhaps as significant for the disciplinary profile that Rowland and Michelson cultivated. Their research and instrument making also linked both very closely to the astronomical community, then certainly one of the most highly visible and institutionally powerful embodiments of science in the United States (dwarfing their own discipline).[8]

Again, the use of Rowland's gratings for astrophysical studies has been the most evident mark of these important connections. This chapter will enrich our understanding of Rowland's work and demonstrate the subtle reach and multidirectional nature of the bonds between astronomy and other disciplines by developing an account of Michelson's debts to the observatory. The network of collaborators and the culture of circulation on which Michelson drew early in his career were dominated by astronomers, and his experimental program developed within a framework set primarily by astronomical concerns. Indeed his first major experiments were devoted to providing terrestrial measurements of a key astronomical constant (the velocity of light) and to turning the earth in its orbit into an ether observatory; and they also generated new instruments for astronomical measurements and spectroscopy. Like Rowland, Michelson emphasized light as a tool of measurement and thereby addressed a complex of issues that bridged mechanics and astronomy. Taking up Brush's suggestion that American research in other sciences was stimulated by the success of astronomy, my aim will be to show how fully the visions of these leading physicists were shaped by astronomy, and to use this observation to explore the cultural position of both disciplines in the late nineteenth century.[9] My analysis will show firstly that the grounds shared by physics and astronomy were broader than would be suggested by focusing narrowly on the particular techniques central to astrophysics, and secondly that *physicists* benefited far more deeply from those common grounds than is usually supposed.

The Velocity of Light

As a student and then an instructor in the U.S. Naval Academy in Annapolis from 1869 to 1879, Albert Michelson had an early career that took shape at a time when teaching in the Department of Physics and Chemistry was being reformed, in the belief that one could enhance the application of science to military aims by exposing naval recruits to more intense study. But the physical sciences were overshadowed by the more powerful disciplines of mathematics and astronomy. Michelson became an instructor of physics and chemistry (under Captain William T. Sampson) in 1876. In the following year he decided to demonstrate Léon Foucault's experiment on the velocity of light for a lecture in optics.[10] Soon moving beyond this pedagogical aim, Michelson began researches that proved spectacularly successful, in part as a result of developing a new instrumental arrangement that improved on experiments previously carried out in France by Hyppolite

Fizeau, Foucault, and Alfred Cornu. Michelson combined the advantage of Fizeau's and Cornu's experiments—sending light over a long pathway—with the key component that Foucault had used—a revolving mirror—and cobbled together apparatus at hand to undertake trials on Naval Academy grounds in March 1878. Revealingly, Michelson used a lens that had first been used on an expedition of the Naval Observatory to observe the transit of Venus.[11] The requisition is symbolic of the nature of his project, employing a terrestrial arrangement—an experiment—to deliver a new value for the velocity of light, a constant that had long been a subject of astronomical concern, most often approached through transit observations. Looking back on Fizeau's work, Cornu later wrote of the reversal that this experiment represented: for the first time physics had provided astronomy with the constant that determined the absolute unit of length for measuring celestial space.[12] Michelson had taken his first steps into a program of research that bridged heaven and earth by experimental manipulation of light (and similarly, many of his later studies were first suggested or carried out by members of the French community of optical physicists).[13]

In addition to his instrumental ingenuity, Michelson was fortunate in choosing a problem that had already attracted significant attention within the naval astronomical community. Simon Newcomb was director of the Nautical Almanac Office in nearby Washington, a faculty member at Johns Hopkins, and one of the most respected, influential, and well-known scientists in the United States. As well as organizing the high-profile campaign in the United States to observe the transit of Venus in 1874, Newcomb since 1867 had advocated new experiments to measure the speed of light and had begun making his own plans for such experiments in 1876.[14] Captain Sampson sent Newcomb an enthusiastic report on Michelson's progress in March 1878, and a month later Michelson wrote to ask for Newcomb's advice.[15] The intense working relationship that they formed strongly shaped Michelson's development as a physicist, and the problems that he pursued in the next decades. Newcomb himself had drawn on a wide web of support to get his project under way. His early proposals were developed in collaboration with George Barker, a physicist at the University of Pennsylvania, and Edison had offered to help carry out the experiment at Menlo Park.[16] Michelson's early trials in Annapolis showed the viability of the venture, encouraging Newcomb to seek government funds from the Naval Observatory budget *for his own experiment.*

Newcomb's primary aim was to redetermine the solar parallax in order to rewrite the astronomical tables. He had been engaged in the central tasks of traditional, positional astronomy for over twenty years of research and observational work in the Nautical Almanac Office and Naval Observatory, analyzing the theoretical models and observational data on the relative positions of the earth, moon, and planets. The accuracy of the data depended crucially on an exact knowledge of the earth's position; he now turned his attention to reevaluating the reference system of the stars and the many, interrelated astronomical constants (known to varying degrees of certainty, and difficult to reconcile).[17] His funding applications highlighted the poor state of current astronomical tables, which Newcomb described as "so far from perfect that scarcely any result in the national Ephemerides can be accepted as definitive when the greatest precision is required. Even in operations so simple as running a boundary line, it generally happens that the correction of astronomical data is a considerable part of the labor of calculation."[18] A terrestrial determination of the velocity of light would be less costly and time-consuming than expeditions to measure the transit of Venus; and Newcomb went as far as promising that if his measurements could be made in time, an expedition abroad to measure the transit of Venus in 1882 would not be necessary. Yet Foucault and Cornu had delivered values for the velocity that differed by nearly 1 percent, and there was no certainty as to which to trust. This was the weak link in the chain of reasoning behind the astronomical tables.

There were broader objectives also. Newcomb told the National Academy of Science that the experiment might deliver knowledge respecting "the qualities of the luminiferous ether and the relation between light, electricity and other visible forces," in addition to its "more generally recognized" importance for determining the distance of the sun.[19] Michelson's less expensive project stepped into this network of interests and was supported by his father-in-law after unsuccessful attempts to gain funding from Congress. He also motivated the endeavor by describing the velocity of light as "one of the simplest and most accurate means of ascertaining the distance of the sun from the earth."[20]

This astronomical agenda provided the primary rationale for carrying out experimental work on the velocity of light. When Michelson began turning trials into a full-fledged determination, in addition to benefiting from the guidance provided by Sampson's physics curriculum (especially because of its emphasis on understanding experimental error), astronomers and their instrument makers provided two critical instrumental resources: his lens and his measurement scale.[21] Michelson replaced the lens from the transit observations with a larger one that had a greater focal length, purpose-built by the noted telescope makers Alvan Clark & Sons in Cambridgeport, Massachusetts.[22] His scale came from William Rogers of the Harvard College Observatory (who recalibrated it after the experiment). So Michelson went to Cambridge to obtain the best lenses and scales available. There, as we shall see, he met people whose careers exemplified the institutional traffic between the physical sciences and astronomy in the United States, and whose research underlined the strategic role played by optics in general and the mechanics of measurement in particular in linking physics to astronomy—and linking the American scientific community to industry.

Rogers, for example, had been professor of mathematics and astronomy (and taught industrial mechanics) at Alfred University in western New York from 1859 to 1870. In 1870 he became an assistant at the Harvard Observatory, taking charge of the newly erected eight-inch meridional circle. Work on the problems of microscopy and ruling scales led Rogers to develop considerable expertise in precision distance measurements and in the construction of standard plates and diffraction gratings. The Coast and Geodetic Survey controlled standard weights and measures in the United States, but uncertainties in the relations between their measures and practical standards in use elsewhere prompted widespread concern. Finding that there were no well-defined relations between the yard and meter standards then in use in the United States, Rogers undertook intercomparisons of the standard length measures of the United States, Britain, and the International Bureau of Weights and Measures in Sèvres. By the late 1870s he had become the foremost American authority on length standards, and his concerns were matched in industrial circles.[23]

Histories of scientific involvement in late-nineteenth-century modernization have drawn attention to standardized time telling and described astronomical observatories as a force for rationalization in response to the firm-bound railway times that had proliferated throughout the western world. Selling accurate time, observatory scientists helped push for time zones based on geographical position, rather than local sun time on the one hand or network times that followed the reach of the different railway company lines on the other.[24] In the same period that the United States moved toward standard railway time zones (adopted in 1883), the railways faced a similarly important need for standards in another realm: that of distance measurements. When railway manufacturers found there were no well-defined relations between nuts and bolts nominally of the same dimensions but supplied by different firms, the machine tool firm Pratt &

Whitney was engaged in a project to reform the standards available to industry. They enrolled William Rogers, asking him to join a representative of their company, George Bond, in designing a comparator. Two exemplars were built, with one held in Pratt & Whitney's gauge room while the other went to the Harvard Observatory; the Rogers-Bond comparator was also offered for sale in the Pratt & Whitney catalogue.[25]

Rogers was a key figure in the bridging of astronomical measurement and instrument making with mechanics, but he was not the only scientist concerned with this complex of issues. His director at the observatory, Edward Pickering, had a similarly broad background. Pickering began his career teaching experimental physics at MIT, and in 1873 he published a manual based on the approach he had developed. *Elements of Physical Manipulation* exemplified the increasing role of the student laboratory in the discipline, as careers using the skills of the trained physical scientist began to open up through the telegraphy and the power and lighting industries. It soon became the most important laboratory training text in the United States and helped to define the emerging discipline of physics. In 1877 Pickering took over leadership of the Harvard Observatory. There he became a leading representative of the "new astronomy," developing a program that photographically surveyed the brightness of stars (and later their spectra), and giving increasing responsibility to the women who carried out much of the routine computational work of the observatory and classified thousands of stars.[26]

Michelson picked up his new lens in Cambridge, but also enquired whether Pickering could host Newcomb's velocity of light measurements at the Harvard observatory (in the event, convenience led Newcomb to carry out the experiment in Washington).[27] Newcomb is usually regarded as a representative of the old astronomy, but unlike some of his colleagues at the Naval Observatory he was by no means antagonistic to the emerging astrophysics.[28] He was also clearly open to the traffic between physics and astronomy. Like Rogers, Newcomb devoted his attention to consolidating standards, describing the possibility of testing standards of length as a subsidiary goal of providing accurate determinations of the velocity of light.[29] The complex of concerns that Rogers and Newcomb articulated and addressed in the 1870s were taken up in different ways by both Michelson and Rowland in the following decade.

In the early 1880s Henry Rowland moved from experiments aimed at redetermining the mechanical equivalent of heat to building a ruling engine and producing diffraction gratings. His sudden concern with optics and spectroscopy seems to have been sparked by his recognition that gratings ruled on concave surfaces would render unnecessary the focusing lenses of normal grating spectroscopes. On the level of research goals this marks a distinct and barely explicable change of emphasis, but Otto Sibum has shown significant continuities at the level of experimental techniques. Rowland's thermal experiments had exploited inscribed lines to register performance and control errors in the apparatus (one facet of Rowland's work to improve physics by introducing engineering rigor into experimental arrangements).[30] Those continuities were also present in Rowland's community more generally: in his work building a new ruling engine, Rowland drew liberally on Rogers's design innovations, diamond ruling tools, and expertise in standards.[31]

Rowland distributed the gratings at cost, and used his best exemplars to make standard photographic maps of the solar spectrum, offered in twenty three-foot sections at a scale of about five angstroms to the inch. One of many students and assistants whose research took up related topics, his colleague Louis Bell, used diffraction gratings to provide a standard of length in terms of the wavelength of light. These were important developments in the emergence of the physics department at Johns Hopkins as a standards-based research entity, providing and distributing both the key instrument and spectral maps and wavelength tables that could set definitive standards for the pursuit of spectroscopy in other laboratories and observatories. Rowland's concentration on the sun was strategically important. Assuming the congruence of solar and terrestrial elements that Bunsen and Kirchhoff had first outlined in 1859, his maps of the sun's spectral lines could equally set reference lines for observatories studying stars and laboratories analyzing terrestrial elements.[32] As Rowland's new instruments began to be publicly known, in 1883 he took the stage at the AAAS to make a strident "Plea for Pure Science," which he thought was in danger of being neglected in favor of applications and swamped by mediocrity. Edison was a specific target in Rowland's campaign to promote the values of scientific research in the new country; and to improve its hold in educational institutions Rowland advocated the concentration of resources in a few strong universities worthy of that exalted name.[33] His inspiration surely came in part from his experience of the old world, but even more strongly from a source closer to home, the observatory: "That simplest of the departments of physics, namely, astronomy, has now reached such perfection that nobody can expect to do much more in it without a perfectly equipped observatory; and even this would be useless without an income sufficient to employ a corps of assistants to make the observations and computations."[34]

The endowments of the observatories had rendered astronomy the most perfect branch of physics (befitting its simplicity), and Rowland had no doubt of the success of similar institutions for other branches of the discipline, "or, better, to include the whole of physics." His ideal was the endowment of a grand laboratory that would, following the pattern of the observatory, be large, with perfect equipment, large revenues, a corps of professors and assistants, and a machine shop. As yet such an institution did not exist anywhere in the world.[35]

For American physicists like Rowland and Michelson the observatory was a model institution, and astronomical patronage and research imperatives provided a path to developing a strong research profile. In turn the experimental values, standard maps, and instruments that their physics laboratories could provide complemented the new directions being opened up in astronomy. With significant common ground and strong links to be made at the different levels of institutions, research imperatives, and practices, physics and astrophysics in the United States could develop symbiotically, as Michelson's career began to show.

An Ether Observatory?

Astronomy in general and naval astronomy in particular provided the most important context for Michelson's first experimental work, but his present position could not offer a future in research. Michelson first sought a navy professorship of mathematics and astronomy (a post that Newcomb held). On learning that this position required a test of proficiency in both subjects, Michelson instead advertised his interest in an academic professorship, declaring himself neither mathematician nor astronomer but a physicist looking for a fair position "with a respectable physical laboratory, and if possible also a respectable salary."[36] Surely the reforming concern with physics and chemistry teaching in the Naval Academy helped Michelson to carve out his chosen specialty, but his vision of physics continued to be shaped by the strength of the astronomical community in the United States. When his search for a new position failed to bear immediate fruit, Michelson took leave of absence from the Navy, setting out on a two-year tour of Europe that had him working in the laboratories of Helmholtz and Quincke in Berlin and Heidelberg, and later Cornu, Mascart, and Lippman in Paris (a group whose optical interests were particularly important for Michelson). He arrived in Berlin with an idea probably stimulated by a letter that James Clerk Maxwell had sent to the Nautical Almanac Office in March 1879, subsequently published in *Nature*. Maxwell stated that the velocity of the earth with respect to the stationary ether alters the velocity of light by a quantity depending on the square of the ratio of the earth's velocity to that of light (a second-order effect), which was "quite too small to be observed."[37] Newcomb thought similarly, hoping that a "crucial experiment" might still detect a first-order effect, perhaps by measuring wavelengths through a transparent ruled plate.[38] Thus Michelson set out to solve a problem that was well recognized in the office of the Nautical Almanac; but interference phenomena gave him a rare confidence that a second-order effect would be "easily measurable."[39]

Several French physicists had used measures of the speed of light in different media to explore the implications of the wave theory of light and to investigate the relations between light, the ether (as the medium in which light moved), and matter. Their arrangements exploited the displacement of the "interference" fringes of light and darkness that are created when light from a common source is split up, sent over pathways that differ in length or in the media that they traverse, and then recombined. Mascart's work on the refraction of gases provides an example, which Pickering had pressed on Michelson in August 1879.[40] Following Jules Jamin's design, Mascart used parallel mirrors to split a beam of light into two components that traveled different pathways parallel to one another and in close proximity (each passing through a different gas). The device made it possible to detect tiny relative differences in the velocity of the light, by observing fringe displacements that were fractions of the wavelength of light. French refractometry offered Michelson a measuring technique precise enough to address Maxwell's query; and he devised a geometrical arrangement capable of revealing any differences in velocity due to orientation to the ether. Michelson would send two rays on different round-journeys, perpendicular to each other. Orienting the instrument so that one light beam traveled in the direction of the earth's motion in its orbit (East-West) while the other travelled perpendicular to this (North-South) would maximize the difference in velocity due to any ether wind resulting from the earth's motion. Changing the orientation of the instrument by turning it would then indicate whether the velocity of the light in the different arms changed, by shifting the interference patterns created by the recombination of the light rays.

Realizing his plan proved difficult. Michelson only found a location remote and stable enough to prevent the footfall and commerce of the street from disturbing the play of light in his apparatus by leaving Berlin's premier physical laboratory for the new Astrophysical Observatory in nearby Potsdam. Michelson did not look to the sky but paced around a stone pier in the basement underneath the telescope, looking for shifts in the position of the interference fringes that his device created. But his plan would turn the apparatus itself into an ether observatory: discerning the magnitude of the effect of the ether wind in different directions could be expected to give a new measure of the motion of the earth in its orbit, the motion of the solar system through the fixed stars, and perhaps the relative exposure of different laboratory locations.

Finding a null result meant that Michelson never became the ether observer he originally envisaged, but despite its nature as a practical and short-lived expedient, his use of the observatory is richly symbolic of how Michelson subsequently came to configure his career. Taking light as his principal research object, and as his principal research tool the apparatus at the heart of his ether-drift experiment (the refractometer or, later, "interferometer"), Michelson would work with light as a physicist. Among physicists this would render him something of an aristocrat, independent of the more clearly practical concerns with electricity and thermodynamics that were prompting the expansion of the discipline. But it would continue the experimental, instrument-based, and personal associations with the astronomical community that we have already seen so clearly in Michelson's early work; allow his physics to preserve many of the cultural associations of astronomy as an élite discipline in the United States; and also bring Michelson himself back to the observatory every now and then.

The identification of Michelson's career with light was only strengthened when he returned to take up his long-desired leadership of a physics laboratory, at the Case School of Applied Science in Cleveland. First Newcomb demanded that Michelson repeat his velocity-of-light measurements to resolve differences between their two values; and then Michelson expanded his study of the relations between light, the ether, and different media by repeating first Fizeau's ether-drag experiment, and then his own ether-drift experiment (with the urging of William Thomson and Lord Rayleigh). The instrument makers he used for these projects followed the Cambridgeport recipe, linking astronomy and mechanics to combine popular profile with profitable business. Worcester Warner and Ambrose Swasey had both worked in the Pratt & Whitney workshops in the 1870s before leaving the company in 1880 to set up independently in the machine tool trade, initially in Chicago and then in Cleveland to be nearer their most important customers.[41] From the outset they involved themselves in the astronomical instrument business and particularly telescope manufacture, which quickly became a high-profile adjunct to their more routine work. In 1881 Warner & Swasey began negotiations for and eventually won the contract to build the mounting and dome for a thirty-six-inch refractor for the new Lick Observatory. Shortly before he died, the real estate entrepreneur James Lick had been persuaded that building the world's largest telescope would be a more suitable way of adding luster to his (mixed) reputation than erecting a pyramid in California. Simon Newcomb led early consultations with the Lick trustees, Edward Holden (also from the Naval Observatory) was named director, and Warner & Swasey completed the telescope in 1887. In 1891 they were awarded the reconstruction of the twenty-six-inch refractor at the Naval Observatory (which needed an improved mechanical mounting when the observatory relocated), after an acrimonious tussle with Alvan Clark & Sons, who had designed the original mounting. Thereby the firm won a key role in the observatory-building boom of the period, with college observatories mushrooming across the United States at the same time that a few key institutions played patrons off each other to build the largest astronomical instruments. In the meantime Warner & Swasey filled many orders for Michelson and for Edward Morley, the chemist at the nearby Western Reserve University, who soon joined Michelson in his ether-drag and drift experiments (figures 36–37).[42]

Confirming the original results of these two experiments made for a difficult conundrum. Fizeau's experiment implied that the ether is only partially dragged by matter, while the null result of the ether-drift experiment was most easily explained by assuming that the ether was completely dragged by the earth. But establishing this theoretical problematic through more accurate repetitions was only one of Michelson's goals. His new arrangements improved on the original procedures in several important respects. After attempting to rework Fizeau's arrangement with the same basic optical components (using a masked lens to split the light into two parallel beams that were sent through pipes of flowing water), Michelson decided to employ once more the technique he had pioneered in his ether-drift experiment, in which a half-silvered mirror sent the light in two directions.[43] This allowed the two pencils of light to be radically separated from

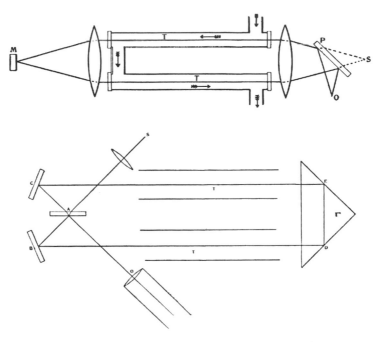

36–37. Michelson's figures illustrate a critical difference between the ether-drag instruments conducted by Fizeau in 1851 and Michelson in 1886. Fizeau used two portions of a lens to split light into two parallel beams, while Michelson used a part-silvered mirror and arrangement of mirrors. A. A. Michelson, *Studies in Optics* (Chicago: University of Chicago Press, 1927), 145, 147.

each other, a feature that Michelson had trumpeted as the most important instrumental advance of his new form of interferential refractometer. But his brief use of the masked lens might have helped underline the relations between refractometers and traditional optical instruments. At the same time that he set out to show just how valuable the half-silvered mirror could be in a number of different arrangements (focusing on using the wavelength of light as a natural standard of length), Michelson also worked on out-lining a theoretical understanding of refractometry.[44] That understanding would integrate refractometry with the more familiar optical instruments of the microscope and telescope—precisely by showing how a lens could be turned into a refractometer—while incorporating under a common framework the various refractometer arrangements that other scientists had used.

Michelson began to outline the broader grasp of his research in an address to the AAAS in 1889. Echoing Rowland, it was entitled "A Plea for Light Waves" and contrasted the practical achievements and enthusiasm surrounding the study of electricity with the relative neglect of the "sister department" of light, which might already be thought "fairly completed." With no fear that his siren call would drain the lifeblood from the current development of electricity, Michelson hoped to persuade a few brother physicists and investigators to follow him into research in optics.[45] Rowland's plea was a generalized call for "pure science" emphasizing scientific institutions. The inducement for Michelson's more specific call was his new apparatus and the limitations of its existing forms. In contrast to the near practical perfection of the microscope, telescope, spectroscope, and camera, Michelson regarded the interferential refractometer as offering many possibilities for further development, and a wide range of applications. These included the manufacture of plane surfaces and accurate measures of coefficients of expansion and elasticity, and of the density and optical properties of solids, liquids, and gases (this was the field in which French and German researchers had developed refractometry most fully). But Michelson's own primary concerns lay in his attempt to establish light as a standard of length, and the spectroscopic information that his instrument had begun to reveal in the search for a practical standard.

Refractometer measures of wavelength promised to exceed by far the accuracy of results obtained from the most perfect diffraction gratings; and Michelson explicitly contrasted his technique with the recent efforts of the Johns Hopkins laboratory. Using his new diffraction gratings, Rowland's measures offered an accuracy of one part in half a million for relative determinations, and Bell's elaborate research on absolute wavelengths claimed but one part in two hundred thousand. Michelson's refractometer, however, would be accurate to one part in two million even for absolute comparisons between the wavelength of light and the meter.[46] Such an absolute comparison became Michelson's primary goal as he moved from Case in Cleveland to Clark University in Worcester, Massachusetts, and searched for a light source homogeneous enough to serve as a standard while refining the design of an apparatus that would allow rigorous comparisons over the length of a meter. This project had the potential to establish light as the "ultimate" standard of length, a natural unit that would depend only on the properties of the vibrating atoms and the universal ether, independent of any particular conventional, material standard.[47]

Michelson's search for a reliable light source was already yielding new information. While sodium light was a suitable standard for diffraction gratings, it showed too complex a structure to be used for this purpose within the refractometer. Indeed Michelson's observations of the varying visibility of the interference fringes that he tracked for several spectral lines allowed him to infer a fine structure not visible with diffraction gratings: the red line of hydrogen and the green line of thallium, for example, were in fact close doubles. The remainder of his address to the AAAS explored how the temperature, thickness, and density of the source might affect the behavior of spectral lines (in particular causing symmetrical or asymmetrical broadening), thereby perhaps offering insight into the molecular motions that caused spectral lines.

Thus in 1889 Michelson situated himself as being somewhat isolated by his primary concern with light. His physics would be distinguished by its claim for unprecedented accuracy of measurement. Like astronomy and in contrast to the electrical solutions that were so often pursued as the remedy for modern needs, this was science on a hilltop rather than science in the service of practical applications. Rather than fill practical needs directly Michelson would provide scientific techniques that might in the future be applied in precision industry, and scientific standards that would link his work to research in many realms. He would provide an ultimate measure for evaluating the international meter itself, Rowland's wavelength tables, and—one step further along—the industrial standards realized through the Rogers-Bond comparator.[48]

As he widened his study of refractometry, Michelson developed a comparison that helped demonstrate the formal properties of his field of research, and also led him into explicitly astronomical measurements. Speaking on "Measurement by Light-Waves" before the Jefferson Physical Laboratory at Harvard, he outlined an analogy in the fundamental properties of optical instruments by treating the refractometer as a masked lens.[49] The resolution and definition prized in the microscope and telescope depend upon the accuracy with which all parts of the lens contribute to make the elementary waves reach the focus in exactly the same phase. Consider a microscope with an objective consisting of a single perfect lens. Magnification of an image increases its indistinctness (since the image of a luminous point is actually composed of small concentric rings of light), but by increasing the size of the interference fringes in relation to the crosshairs of the eyepiece, such magnification allows a great increase in the accuracy with which the position of the point can be established. In the microscope this magnification decreases the illumination rapidly enough to render the technique fruitless. But taking only the external annular ring of the lens, or better still two small portions at opposite ends of the diameter, will allow the size of the interference fringes to be increased up to any point without affecting the intensity of the light. The portions need not be curved and may be either plane mirrors or prisms: the telescope or microscope has been converted into a refractometer, sacrificing the resolution and definition of an image for the sake of accuracy of measurement.[50] Further, a similar analogy holds with the spectroscope: using only the extreme portions of the surface increases accuracy at the expense of resolution and definition. Michelson had outlined a kind of instrumental unity, showing how the refractometer could encompass a whole spectrum of instrumental forms. In each case, and quite literally, his new instrument would allow the realization of accuracy in the extreme.[51]

Astronomical Measurements and Astrophysical Contributions

In the early 1890s, while preparing apparatus to render light as a standard of length, Michelson carried out laboratory investigations at Clark University to establish the feasibility of using interference methods to measure the diameter of astronomical objects. Michelson fitted a fine four-inch refractor telescope with adjustable slits and experimented with a light source of variable diameter at a distance of ten meters, publishing a theoretical study in the *Philosophical Magazine*.[52] Then he received Pickering's approval to try out the technique on the fifteen-inch refractor at Harvard. When those tests failed, Michelson turned to Edward Holden at the Lick, seeking better atmospheric conditions. Working with the twelve-inch refractor on Mount Hamilton the following summer, Michelson measured the diameter of the satellites of Jupiter (figure 38). Then he looked to scale up, constructing a mask for the thirty-six-inch refractor that he sent to Mount Hamilton in December 1891. But there the project halted. Michelson of course had another iron in the fire, traveling to France to determine the length of the international meter in cadmium wavelengths in 1892; and the Lick had such a heavy schedule that William Campbell had no time to work with the apparatus. Interferometric astronomical observations would await Michelson's personal involvement before they were developed in 1920, by which time there was considerable interest in the possibility of di-

38. The Lick Observatory on Mount Hamilton, photographed during construction of the main dome by Warner & Swasey (completed in 1887). In 1891 Michelson undertook interferometric measures of the diameter of Jupiter's moons using adjustable slits placed over the objective of the twelve-inch equatorial telescope in the dome to the right. Lick Observatory Archives, courtesy AIP Emilio Segre Visual Archives.

rectly measuring the diameter of giant stars like Betelgeuse, stimulated by Eddington's large estimates of the apparent diameter.[53] In some ways this was characteristic of Michelson's relation to the astronomical community (and spectroscopic practice) throughout his career. It was mediated principally through the provision of a value (the velocity of light, precise measurements of the absolute wavelengths of three spectral lines of cadmium) or the demonstration of a method (interferometric measurements), rather than through the integration of Michelson's own techniques into working practices. Rowland provides a contrast. Rowland himself mapped a great number of spectral lines, and the distribution of his instruments, together with the publication of his maps and wavelength tables, made the task of assigning wavelengths to particular lines more direct, and the task of correlating the work of different laboratories and observatories significantly easier. Thereby Rowland's work became part of the fabric of research in spectroscopy, including astronomical spectroscopy. But even in the 1890s Michelson's failure to enlist others in the project of interferometric methods of observation characterizes only one facet of his relationship with the astronomical community.

Michelson's papers on his early attempts to measure astronomical diameters thanked Pickering, Holden, and Campbell at the two observatories, but also referred to the aid of Frank L. O. Wadsworth. A fellow at Clark University, Wadsworth assisted Michelson's astronomical and spectroscopical research, as well as supervising the construction of Michelson's apparatus for determining the meter. His association with Michelson helped Wadsworth make the transition from a training in mechanics and engineering to a career in astronomy. In 1893 Michelson left Clark to become the founding director of the new physics laboratory at the University of Chicago, bringing Wadsworth with him as an assistant professor. Once again Michelson found astronomy strongly represented in the young but ambitious advocate of astrophysics George Ellery Hale, who planned an observatory for the university to be funded by the transportation financier Charles Yerkes. In the competitive stakes of patronage Yerkes wanted to put his name to a telescope "to beat everything of its kind." His money paid for a forty-inch lens (which had originally been ordered by the University of Southern California), the two disks of glass made by Mantois of Paris and ground by Alvan Clark & Sons (figure 39). Warner & Swasey built the mounting and dome to house the telescope in Williams Bay, Wisconsin. Hale took the continuum between physics and astronomy (for which he often sought common ground) and between the physics laboratory and the astronomical observatory in a direction opposite from that pursued by Rowland, arguing that the astrophysical research institution needed to become more like a laboratory than an observatory.[54] A related aim was to render the Yerkes observatory largely independent of instrument makers by building up its own workshop facilities, so Hale hired Wadsworth.[55] The two began constructing a ruling engine, but Wadsworth left Yerkes with the engine still incomplete to become the director of the Allegheny Observatory in Pittsburgh in 1899.

In 1895, two years before Yerkes observatory was completed in 1897, Hale founded a journal for the new astronomy, long poorly served by his compatriot Benjamin Gould's *Astronomical Journal* (which refused to take space away from mathematical and positional astronomy). William W. Payne's *Sidereal Messenger* had provided a second American forum for research contributions. In it papers in astronomical spectroscopy and the like

39. Astronomers on the steps of Yerkes Observatory on the occasion of its opening and the First Conference of Astronomers and Astrophysicists, on 18 October 1897. F. L. O. Wadsworth, John A. Brashear, Edward Pickering, and George Ellery Hale are among them (and Newcomb and others arrived later in the day). Yerkes Observatory, University of Chicago, courtesy of AIP Emilio Segre Visual Archives.

were juxtaposed with articles intended for amateur astronomers and general readers, but the magazine began losing subscriptions when it formalized the practice with a new title in 1892, *Astronomy and Astro-Physics*. Within a year Payne had founded a second journal to better proselytize the field to amateurs and school teachers, *Popular Astronomy*; and Hale entered negotiations to buy out *Astronomy and Astro-Physics* and dedicate the publication to astrophysics alone.[56] Given the cultural importance of astronomy in the United States it is as important to recognize the continued links between research and popular astronomy as the growing separation of the two—in just the period when astrophysics emerged as a distinct field (as Charlotte Bigg points out in her contribution). That process was accompanied by vigorous and sometimes vicious debates on the identity of the astronomer, mostly centering on the different characteristics of the new and old astronomy. What for the budding astrophysicist was a field dominated by new instruments and photographic methods, and by research conquests studded with discoveries, could be regarded as largely speculative and faddish by some purists who emphasized the careful refinement of methodological approaches to traditional concerns in celestial mechanics.[57] Maintaining robust links to the public was certainly important to astrophysicists, but in some distinction to more traditional astronomy, the relationship would be cultivated by endowments and popularization rather than amateur participation in observational work itself. Some purists saw a strong connection between the nature of the new astrophysics and its patronage. Eastman, for example, argued that the new astronomy was endowed so heavily because such (spurious) novelty conformed more fully to patrons' need for immediate results in enhancing their public image than did the methodical, careful refinements that characterized traditional astronomy. Noting this debate, it is important to recognize also that a significant middle ground of astronomers saw value in both old and new methods; and that the endeavor of many promoting astrophysics in particular was to forge links with a second discipline, as much as it was to argue for any strong distinction from traditional astronomy.[58]

Intended as an "international review of spectroscopy, and astronomical physics," the *Astrophysical Journal* was turned over to the University of Chicago Press. Hale enrolled Keeler as a joint editor and assembled a board of associate editors that included among its American contingent Michelson, Rowland, and the physicist and optical expert Charles S. Hastings from Yale, as well as Pickering and the astronomer C. A. Young from Princeton. The international complement included the French physicist Alfred Cornu together with several astronomers affiliated in one way or another with the new astronomy: William Huggins (England), Pietro Tacchini (Italy), Hermann Vogel (Germany), and Nils P. Dunér (Sweden). The first article in the first issue was Michelson's, "On the Conditions Which Affect the Spectro-Photography of the Sun," and he published five times by 1900.[59] Rowland's tables of twenty thousand solar absorption-line wavelengths began appearing in the first volume, and in the third the journal's editors announced that they favored the use of these standards in the *Astrophysical Journal*—and all similar publications. Both Michelson and Rowland commonly published in the *American Journal of Science* and the *Philosophical Magazine* in Britain (and Michelson also offered versions of his work to the *Journal de physique théorique et appliquée*). In contrast to their support for the astrophysical journal, neither scientist published in the *Physical Review*, the journal founded in 1893 by physicists at Cornell to provide a forum for the publication of American physics. That journal did not gain significant international prestige until the 1930s.[60] Finally, in the early 1900s the *Astrophysical Journal* hosted a debate on the relative merits of Rowland's wavelength standards and Michelson's cadmium determinations, a debate which provided a significant impetus for Hale's development of a new International Union for Cooperation in Solar Research, and which in 1907 was resolved in Michelson's favor—helping him gain the Nobel Prize in that year.

The last quarter of the nineteenth century saw rapid change in the institutional foundations of both astronomy and physics. Rowland and Michelson were inaugural professors for their subject in new universities, and Michelson broke ground three times. Both were deeply concerned with purity, carefully managing relations with industry through standards and upholding research ideals that sometimes linked them more closely to astronomers than to colleagues in their own discipline. Given the rise of astrophysics, its search for common ground between physics and astronomy, and the explicit battles sometimes joined between different forms of astronomy, it would be easy to assimilate the two physicists to the astrophysical community in particular. But while both Michelson and Rowland clearly contributed to and supported that specific endeavor, their debts to astronomy were far greater—in ways that trouble any easy distinction between old and new astronomy. After all, those debts began at the still more fundamental point at which mechanics and optics met, in research goals and practical instrumental achievements that fashioned important links between astronomy and industry. Michelson's and Rowland's joint concerns with light as a tool of measurement sustained active research programs through decades and across communities. Further, whether or not this was because as a physicist he raised no specific challenge to astronomers' identity, Michelson's work also won support across the lines between old and new, throughout his career. Newcomb was his first patron, while Gould pressed his case as the American delegate to the International Bureau of Weights and Measures (and in both cases it was regarded as a point of national pride that an American had measured the velocity of light and provided an absolute measure of the international meter). Similarly, Michelson's concentration on standards was one cultivated in the measurement practices of old-school astronomers, just as it was a pressing need in the new fields of observation opened up by astrophysicists. In the service of extreme accuracy, Michelson's contributions were to be definitive experiments, definitive (re)determinations, or new standards. Rather than offer practical remedies, or even map fields of research as Rowland had done with the spectral lines of the sun, Michelson sought to engage his community by setting standards that could be practical, absolute, and natural. While Michelson's search for ultimate standards has all the austerity of astronomy as an élite activity, Rowland's attention to incorporating them in all levels of practice—and in the work of a research school—drew on astronomy as an institution, in the model of the observatory. Whether or not they did so because it was the simplest science, two of America's premier experimental physicists in the late nineteenth century looked first to astronomy in their search to forge new scientific ideals for the growing country.

Notes

I wish to thank the editors of this volume, Karl Hufbauer, Peter Susalla, and my colleagues at the University of Wisconsin, Madison, for their careful reading of earlier versions of this paper. My archival studies of Michelson were supported by the Pollock Award of the Dudley Observatory.

1. J. Willard Gibbs, "Hubert Anson Newton" (1897), cited in Karen Hunger Parshall and David E. Rowe, *The Emergence of the American Mathematical Research Community, 1876–1900: J. J. Sylvester, Felix Klein, and F. H. Moore* (Providence, R.I.: American and London Mathematical Societies, 1994), 22. When American mathematics began to gain strength in the late nineteenth century pure mathematics was dominant, and the separation of pure and applied mathematics was described as astonishing. John W. Servos, "Mathematics and the Physical Sciences in America, 1880–1930," *The Scientific Enterprise in America: Readings from Isis*, ed. Ronald L. Numbers and Charles E. Rosenberg (Chicago: University of Chicago Press, 1996), 141–59, 148.

2. Brush, "Looking Up."

3. For overviews see Herrmann, *The History of Astronomy from Herschel to Hertzsprung*, esp. chapter 4, and contributions to *Astrophysics and Twentieth-Century Astronomy to 1950*, ed. Owen Gingerich; Hoskin, *The General History of Astronomy*, 4B; for astronomy as big science see for example Lankford and Slavings, "The Industrialization of American Astronomy"; the statistics are from Lankford with Slavings, *American Astronomy*, 384.

4. On astrophysics see Hufbauer, *Exploring the Sun*, chapter 2; A. J. Meadows, "The New Astronomy," *Astrophysics and Twentieth-Century Astronomy to 1950*, ed. Gingerich, 59–72; Lankford with Slavings, *American Astronomy*, chapter 3; David H. DeVorkin, "The Pickering Years," *The American Astronomical Society's First Century*, 20–36; Joann Eisberg, "Solar Science and Astrophysics," *The Cambridge History of Science*, vol. 5, ed. Mary Jo Nye (Cambridge: Cambridge University Press, 2003), 505–21; but note Susalla's argument that a considerable middle ground existed between the extremes, Susalla, "The Old School in a Progressive Science."

5. This is the interpretation presented in Kevles, *The Physicists*. See in particular his chapter "The Flaws of American Physics," but note the more nuanced picture of the formal and informal structures supporting physicists' work in Albert E. Moyer, "American Physics in 1887," *The Michelson Era in American Science*, ed. Goldberg and Stuewer, 102–10. For studies of Michelson that focus on his instruments see Staley, "The Interferometer and the Spectroscope"; Staley, "Travelling Light," *Instruments, Travel and Science*, ed. Bourget, Liccoppe, and Sibum, 243–72. For his contributions to relativity physics see Richard Staley, *Einstein's Generation: The Origins of the Relativity Revolution* (Chicago: University of Chicago Press, 2008).

6. See Chris J. Evans and Deborah Jean Warner, "Precision Engineering and Experimental Physics: William A. Rogers, the First Academic Mechanician in the U.S.," *The Michelson Era in American Science*, ed. Goldberg and Stuewer, 2–12.

7. Sweetnam, *The Command of Light*, esp. introd. by M. Norton Wise, "Henry Rowland and American Physics," xv–xxv.

8. Here I differ with Lankford, who describes the founding generation of astrophysics as regarding physics more highly than astronomy, Lankford with Slavings, *American Astronomy*, 39.

9. Brush, "Looking Up," 58. A key ambiguity in Lankford's study of the astronomical community stems from his inclusive definition of astronomers. For example, he includes both Rowland and Michelson because of their positions on the editorial board of the *Astrophysical Journal*, and computers because they were on observatory payrolls, thereby obscuring the dynamic between astronomers and other sectors of the population in his quantitative studies. Interestingly, despite briefly comparing chemistry and astronomy he does not analyze the much closer relations between physics and astronomy. For a critical review see Karl Hufbauer, "Counting Astronomers (Essay Review)," *Minerva* 38 (2000): 352–62.

10. For an excellent study arguing that his involvement in pedagogy at Annapolis was the most important factor shaping Michelson's investigative style see Kathryn M. Olesko, "Michelson and the Reform of Physics Instruction at the Naval Academy in the 1870s," *The Michelson Era in American Science*, ed. Goldberg and Stuewer, 111–32.

11. The lens was five inches in diameter with a focal length of thirty-nine feet. Michelson to Newcomb, 18 December 1878, in *Science in Nineteenth-Century America*, ed. Reingold, 280–81. See also Michelson, "Experimental Determination of the Velocity of Light," 116.

12. A. Cornu, "Notice sur l'œuvre scientifique de H. Fizeau," *Annuaire du Bureau des Longitudes*, 1898, C.1–C.40, at C.16. For a discussion of the merger of astronomical and physical practices in Fizeau's work see Frercks, "Creativity and Technology in Experimentation."

13. See Bigg, "Behind the Lines."

14. On Newcomb see Moyer, *A Scientist's Voice in American Culture*; and Norberg, "Simon Newcomb's Early Astronomical Career." For a discussion of the two pathways to solar parallax, Newcomb's role, and the significance of the transits

for disciplinary relations see Richard Staley, "Conspiracies of Proof and Diversity of Judgment in Astronomy and Physics: On Physicists' Attempts to Time Light's Wings and Solve Astronomy's Noblest Problem," *Cahiers François Viète* 11–12 (2007): 83–97.

15. Michelson wrote to Newcomb on 26 April 1878. *Science in Nineteenth-Century America*, ed. Reingold, 278–79.

16. See the correspondence from Joseph Wharton to Simon Newcomb, 17, 24, and 26 July and 4 September 1876, and George Barker to Simon Newcomb, 26 February 1876, 30 April 1877, and 28 April 1878, all in the Library of Congress, Simon Newcomb Papers 57, subject file: "Eclipses, Electricity and Light," folder: "Velocity of Light"; and George Barker to Simon Newcomb, 22 August 1879, Library of Congress, Naval Historical Foundation Collection 22.

17. See Norberg, "Simon Newcomb's Early Astronomical Career," 224.

18. Newcomb to Com. R. W. Thompson, secretary of the navy, 12 December 1878, Library of Congress, Newcomb Papers 57.

19. Newcomb, "On a Proposed Modification of Foucault's Method of Measuring the Velocity of Light," communicated to the National Academy of Sciences, April 1878, Library of Congress, Newcomb Papers 57.

20. Michelson, "Experimental Determination of the Velocity of Light," 71.

21. On the importance of his teaching tasks for the evolution of Michelson's approach to research see Olesko, "Michelson and the Reform of Physics Instruction at the Naval Academy in the 1870s," 122–25.

22. His mirror was made by Fauth & Co. in Washington, but he later recommended Alvan Clark & Sons in preference. Michelson to Newcomb, 15 July 1882, *Science in America*, ed. Reingold, 303–4.

23. On Rogers see Evans and Warner, "Precision Engineering and Experimental Physics"; Morley, "Biographical Memoir of William Augustus Rogers."

24. Bartky, *Selling the True Time*.

25. Pratt & Whitney published a collection of papers and reports advertising their success in obtaining a reliable standard and practical methods of making comparisons suitable for precision industrial use: George M. Bond, ed., *Standards of Length and Their Practical Application* (Hartford, Conn.: Pratt & Whitney, 1887).

26. On Pickering's manual see Olesko, "Michelson and the Reform of Physics Instruction at the Naval Academy in the 1870s"; on his contributions to astronomy see *The American Astronomical Society's First Century*, ed. DeVorkin, 20–36; Howard Plotkin, "Edward C. Pickering and the Endowment of Scientific Research in America, 1877–1918," *Isis* 69 (1978): 44–57. On women in American astronomy see Lankford with Slavings, *American Astronomy*, chapter 9.

27. See Pickering to Newcomb, 6 January 1879, Harvard University Archives / UAV 630.14 Harvard College Observatory/E.C. Pickering, Director/Letterbooks (outgoing) / B1, February 1877–July 1880.

28. For an extended discussion of the fields of conflict between the old and new astronomy see Lankford with Slavings, *American Astronomy*, chapter 3, and sources cited in note 4, above.

29. Simon Newcomb, "Measures of the Velocity of Light Made under Direction of the Secretary to the Navy during the Years 1880–'82," *U.S. Nautical Almanac Office Astronomical Papers* 2, no. 3 (1891): 107–230, 109.

30. H. Otto Sibum, "Exploring the Margins of Precision," *Instruments, Travel and Science*, ed. Bourget, Licoppe, and Sibum, 216–42.

31. See Randall Brooks, "Towards the Perfect Screw Thread: The Making of Precision Screws in the 17th–19th Centuries," *Transactions of the Newcomen Society for the Study of the History of Engineering and Technology* 64 (1993): 101–20; Evans and Warner, "Precision Engineering," 5; D. J. Warner, "Rowland's Gratings: Contemporary Technology," *Vistas in Astronomy* 29 (1986): 125–30.

32. On the solar-terrestrial comparison see Sweetnam, *The Command of Light*, chapter 4. On the importance of wavelengths and Rowland's relation to existing practices in spectroscopy see Bigg, "Spectroscopic Metrologies."

33. Rowland, "A Plea for Pure Science." On Edison see David A. Hounshell, "Edison and the Pure Science Ideal in 19th-Century America," *Science* 207, no. 4431 (1980): 612–17; Dennis argues persuasively that Rowland's vision of the moral economy of pure science must be seen in relation to the combined disciplinary and pedagogical role of the research university that with sufficient endowment could be insulated from the marketplace mentality. Michael Aaron Dennis, "Accounting for Research: New Histories of Corporate Laboratories and the Social History of American Science," *Social Studies of Science* 17 (1987): 479–518.

34. Rowland, "A Plea for Pure Science," 246. Clerke depicted a similar continuum between physics and astronomy, writing of the science of heavenly bodies as "a branch of terrestrial physics, or rather a higher kind of integration of all their results." Clerke, *A Popular History of Astronomy during the Nineteenth Century*, 463.

35. Rowland, "A Plea for Pure Science," 248. Rowland recognized limitations in the present endowment of observatories, advocating a stronger concentration of resources there also.

36. Michelson to A. M. Mayer, 26 June 1880, *Science in Nineteenth-Century America*, ed. Reingold, 286–87. See also Michelson to Pickering, 7 July 1880, Harvard University Archives HUG 1690.15 E.C. Pickering Private Letters 1850–1883.

37. James Clerk Maxwell to David Peck Todd, 19 March 1879, *Nature* 21 (1880): 314–15.

38. See the letters from Newcomb to Michelson of 2 May and 2 June 1881 in *Science in Nineteenth-Century America*, ed. Reingold, 290–91, 292, and 296–97.

39. See A. A. Michelson, "The Relative Motion of the Earth and the Luminiferous Ether," *American Journal of Science* 22 (1881): 120–29, 121.

40. See Michelson to Newcomb, 29 December 1878 and 16 January 1879, *Science in Nineteenth-Century America*, ed. Reingold, 283–85. Also Pickering to Michelson, n.d. [*ca.* 21 August 1879], Harvard University Archives / UAV 630.14 Harvard College Observatory / E. C. Pickering, Director/Letterbooks (outgoing) / B1, February 1877–July 1880. Pickering noted experiments that he had discussed with Michelson earlier, and mailed Michelson some of his own articles. The references were to Eleuthère Mascart, "Sur la réfraction des gaz," *Comptes rendus des séances*

hebdomadaires de l'Académie des Sciences 78 (1874): 617–21; Mascart, "Sur la dispersion des gaz," ibid., 679–82; Mascart, "Sur la réfraction de l'eau comprimée," ibid., 801–5. Pickering provided letters of introduction for Michelson's European tour and in 1903 recommended Michelson for a Nobel Prize. Michelson to Pickering, 14 November 1903, Harvard University Archives UAV 630.17.7.

41. On Warner & Swasey see Edward J. Pershey, "The Early Telescopes of Warner & Swasey," *Sky and Telescope* 67 (1984): 309–11; and Pershey, "Warner and Swasey at the Naval Observatory: A View of the Science-Technology Relationship," *Beyond History of Science: Essays in Honor of Robert E. Schofield*, ed. Elizabeth Garber (Bethlehem: Lehigh University Press, 1990), 220–31.

42. On Morley's and Michelson's relations with Warner & Swasey see Darwin H. Stapleton, "The Context of Science: The Community of Industry and Higher Education in Cleveland in the 1880s," *The Michelson Era in American Science*, ed. Goldberg and Stuewer, 13–22. John Brashear of Pittsburg was responsible for the optical components in Michelson's arrangements.

43. A. A. Michelson and E. W. Morley, "Influence of Motion of the Medium on the Velocity of Light," *American Journal of Science* 31 (1886): 377–86.

44. A. A. Michelson and E. W. Morley, "On a Method of Making the Wave Length of Sodium Light the Actual and Practical Standard of Length," *American Journal of Science* 34 (1887): 427–30.

45. A. A. Michelson, "A Plea for Light Waves," *Proceedings of the American Association for the Advancement of Science* 37 (1889): 67–78, at 67–69.

46. Michelson, "A Plea for Light Waves," 74.

47. For these descriptions see A. A. Michelson and E. W. Morley, "On the Feasibility of Establishing a Light-Wave as the Ultimate Standard of Length," *American Journal of Science* 38 (1889): 181–86; A. A. Michelson, "Comparison of the International Meter with the Wave Length of the Light of Cadmium," *Astronomy and Astrophysics* 12 (1893): 556–60, at 560.

48. When his student Frank Baldwin Jewett chose to join what later became the Bell Telephone Laboratories, Michelson described such industrial research as "prostituted physics." See Nathan Reingold, "Physics and Engineering in the United States, 1945–1965: A Study of Pride and Prejudice," *The Michelson Era in American Science*, ed. Goldberg and Stuewer, 288–98, at 288–89; S. Weart, "The Rise of 'Prostituted' Physics," *Nature* 262 (1976): 13–17.

49. A. A. Michelson, "Measurement by Light-Waves," *American Journal of Science* 39 (1890): 115–21.

50. Ibid., 116–18.

51. Michelson was primarily concerned with the pragmatic unity of method that he could advance by promoting the interferometer as an instrument, but Maxwell's electromagnetic theory of light had invested particular importance in understanding the medium that transmits electromagnetic disturbances. Later Michelson described the grand speculative framework within which much research found another form of unity, as explorations of the various manifestations of the ether. See A. A. Michelson, *Light Waves and Their Uses* (Chicago: University of Chicago Press, 1902), 162–63.

52. A. A. Michelson, "On the Application of Interference Methods to Astronomical Measurements," *Philosophical Magazine* 30 (1890): 1–21.

53. See David DeVorkin, "Michelson and the Problem of Stellar Diameters," *Journal for the History of Astronomy* 6 (1975): 1–18; A. A. Michelson, "Measurement of Jupiter's Satellites by Interference," *Nature* 45 (1891): 160–65.

54. G. E. Hale, "The Aim of the Yerkes Observatory," *Astrophysical Journal* 6 (1897): 310–21, 311; G. E. Hale, "The Solar Observatory of the Carnegie Institution of Washington," *Astrophysical Journal* 21 (1905): 151–72, discussed in Lankford with Slavings, *American Astronomy*, 70.

55. Livingston, *The Master of Light*, 188–90.

56. Donald E. Osterbrock, "Founded in 1895 by George E. Hale and James E. Keeler: The *Astrophysical Journal* Centennial," *Astrophysical Journal* 438 (1995): 1–7.

57. Gould, Seth Chandler, John Ritchie, and Eastman offered this kind of attack; Pickering and the Harvard Observatory were particular targets; and Langley, Keeler, and Hale were prominent advocates of astrophysics. See Lankford with Slavings, *American Astronomy*, chapter 3.

58. J. R. Eastman, "The Neglected Field of Fundamental Astronomy," *Proceedings of the American Association for the Advancement of Science* 43 (1892): 17–32. For the middle ground see Susalla, "The Old School in a Progressive Science."

59. A. A. Michelson, "On the Conditions Which Affect the Spectro-Photography of the Sun," *Astrophysical Journal* 1 (1895): 1–9.

60. Paul Hartman, *A Memoir on the Physical Review: A History of the First Hundred Years* (New York: American Institute of Physics, 1994).

Chapter 9

Even the Tools will be Free:

Humboldt's Romantic Technologies

JOHN TRESCH

Humboldt's Aesthetic Anxiety

Alexander von Humboldt's *Cosmos: A Sketch of a Physical Description of the Universe* was an overview of all past and present knowledge of the earth and heavens. In taking on such a staggeringly vast topic, Humboldt confessed in his first paragraph to an anxiety with a "two-fold cause": "The subject before me is so inexhaustible and so varied, that I fear either to fall into the superficiality of the encyclopedist, or to weary the mind of my reader by aphorisms consisting of mere generalities clothed in dry and dogmatical forms." The solution he found was an aesthetic one: "Nature is a free domain," he wrote at the end of the introductory paragraph, "and the profound conceptions and enjoyments she awakens within us can only be delineated by thought clothed in exalted forms of speech, worthy of bearing witness to the majesty and greatness of the creation."

Many have emphasized that Humboldt's "exalted forms of speech" in *Cosmos*, like the beautiful colored prints that accompanied his books, appeal to the senses and attempt to convey, indeed to produce, the "different degrees of enjoyment presented to us by the aspect of nature and the study of her laws" that *Cosmos* announces as one of its topics. Yet as we know from many recent studies, the concept of "the aesthetic" has philosophical and political implications that go well beyond sense pleasure and formal harmony.[1] The tension that Humboldt expresses between two opposed tendencies—one leaning toward the limitless play of particular empirical phenomena, the other to dogmatic, formal generalities—lines up precisely with the oppositions that troubled Friedrich Schiller in *Letters on the Aes-*

40. Telescopes observe the eclipse of 1842. J. J. Grandville, *Un autre monde: trans-formations, visions, incarnations, ascentions, locomotions, explorations, pérégrinations, excursions* (Paris: H. Fournier, 1844), 94. Reproduction by permission of Herzog August Bibliothek Wolfenbüttel: Ars Libr. 4: Grandville 16. 4° 185.

thetic Education of Man (1794). Schiller, the poet, philosopher, and playwright who was closely associated with Goethe and the Humboldt brothers, balanced these oppositions in an intermediary "aesthetic state." Between the barbaric formal emptiness of excessive refinement and the slavish attachment to savagery lies a middle ground where the two fundamental urges of humanity—the formal drive and the sense drive—are reconciled. This third drive, the "play drive," roves freely between form and sense, and expresses itself in the beautiful appearances of fine art. Schiller's view of the aesthetic was inspired by Kant's third critique, in which aesthetic judgment is examined as a subjective universal intimately connected with the goal of culture and *Bildung*.

Alexander von Humboldt had a sustained exposure to this set of concerns. Schooled in French mathematical physics and German romanticism, Humboldt brought about a revolution in the organization and direction of the sciences throughout Europe and the Americas in the first decades of the nineteenth century. Like his brother Wilhelm, who in his study of languages strove for "a conception of the world in its individuality and totality," Alexander described the different spaces of the globe in their singularity and their interactions, studying the reciprocal combinations of forces on a cosmic scale.[2] He used a vast range of instruments to measure geophysical phenomena; in addition to findings taken on his own voyages, he helped to establish international networks of observers to collect similar data from around the world. New techniques of combining data into striking visual images allowed him to chart the changes in average temperature across the globe and changes in vegetation in different milieus. With the beautiful images accompanying his texts and the evocative language that filled them, he sought to format and diffuse the sciences to reach the widest audience possible. His hope was that exposing readers to the pleasures of nature and the methods of the sciences would contribute to moral and political reform.[3] While his nomadic geophysical wanderings set a heroic example, he always remained tethered to the observatory. The very idea of an assembly of techniques that can be identified as "observatory science" owes much to the heterogeneous regime of concerns and approaches that Humboldt tied together in this period.

This chapter explores the relation between the "aesthetic" dimension of Humboldt's work and his instrumental practices, in which observatories played such a central role. In particular, I will concentrate on the ways in which Humboldt adopted the language and assumptions of Kant's and Schiller's moral and political philosophy in his new mode of conducting the sciences. Their conception of the aesthetic had a significant impact on Humboldt's scientific work, not only on his evocative writing style and his visual representations but on the very content of his "physical description of the universe" and the theory that it provided for his practice of observation and his efforts to organize the sciences. Describing how precision instrumentation and its numerical data helped to create a common language for an extended community of researchers, Marie-Noëlle Bourget has written of Humboldt's "republic of instruments."[4] I suggest that some of the key "citizens" of this republic were the instruments themselves. I begin by looking at themes and approaches in Kant and Schiller that Humboldt drew upon to construct his polity of science. In keeping with Humboldt's political concerns, a guiding question of this genealogy of the morals of the observatory science of the first half of the nineteenth century is the relation between objectivity, as both theory and practice, and the views of *freedom* that were associated with it.

This examination of the moral, aesthetic, and political underpinnings of Humboldtian science is a contribution to the history of scientific objectivity, yet it raises questions for certain accepted claims. The early nineteenth century plays a crucial transitional role in Lorraine Daston's and Peter Galison's influential explorations of the multifarious historical and conceptual strata of the ideal of objectivity. In this period, they claim, the notion of objectivity as an escape from an individual perspective was transferred from the domain of moral judgments to epistemology; at the same time, an idealist meaning of objectivity, as the internal agreement in thought with a formal truth, was externalized and materialized. The outcome of this phase, "mechanical objectivity," exemplified by the daguerreotype, aimed at removing the influence of individual prejudice and human error by means of images produced by machines.[5]

In both of these transitions Humboldt played a central role. Yet neither "aperspectival" nor "mechanical" objectivity fully grasps Humboldt's practice and the philosophical discourses which nourished it. Like many late Enlightenment thinkers, Kant in his reflections on instrumentality acknowledged a boundary between the mechanical and the organic or human, yet he would eventually blur and undo this distinction; Schiller developed this dialectic between the organic and the mechanical into a view of human artifice as part of a vital cosmic circuit integrating humans and nature. Influenced by both Kant and Schiller, Humboldt did not perceive the machines that he used to make readings of natural phenomena as the antithesis of the human, organic, or natural, as they would be for many.[6] Furthermore, the version of objectivity which appears in the writings of Schiller and Humboldt was not a theory of escape, negation, or disconnection. For Humboldt the best way to know the world was to multiply mediations and observers, not eliminate them. The kind of "freedom" pursued in the Humboldtian regime of observatory science, and the version of "objectivity" that it realized, were associated with interdependence, mediation, and community.

Humboldt's phrase "nature is a free domain" nonchalantly restates the central problem of Kant's *Critique of Judgment* and Schiller's *Aesthetic Education*.[7] Despite the occasionally dry and "anaesthetic" nature of any discussion of the central themes in Kant and Schiller, such a discussion is essential if we want to understand the particular tensions that Humboldt's observatory science confronted and sought to overcome.[8]

Kant's Objectivities

Kant's influence on late-eighteenth- and nineteenth-century circles of philosophers, literary authors, and practitioners of *Naturphilosophie* in German-speaking lands was unrivaled: the cultural upsurge of romanticism owes much to the problems that Kant posed, as well as to dissatisfaction with the answers he provided. In a famous distinction, for Kant all objects of the world have both a phenomenal (sensible) and a noumenal (supersensible) dimension. Accordingly, humanity also has two distinct natures: man "belongs to the world of sense" and is "subject to laws of nature" as an empirically observable, determined phenomenon within the system of physical, mechanical causes. But "insofar as he belongs to the intelligible world subject to laws which, independent of nature, are not empirical but are founded only on reason" man also possess a supersensible nature, a free will, which gives him the capacity to legislate laws himself.[9]

This split in humanity corresponds to a divide among the "autonomous" faculties responsible for making sense of different kinds of experience.[10] Judgments about nature come from the faculty of the *understanding*; they are "objective" to the extent that they are brought under an a priori concept. Because of their conformity to the system of categories that are the condition of the possibility of experience, they are universally valid. Thus in the realm of the theoretical—explored in the *Critique of Pure Reason*— "objectivity" arises not because of the activity of a robust object upon a passive receiver but from the logical processes of the human mind that "constitute" the object. The "Copernican revolution" of Kant's epistemology lies in the *active spontaneity* with which the understanding attributes a concept to these sensory givens, shaping them as "objects of experience."[11] The "objectivity" of the judgments made by the understanding is internal, a function of the intellect, having no essential dependence on any sensory ("aesthetic") givens.

The "objectivity" that pertains to laws regulating desire takes a different form: instead of laws of nature, we have here to do with laws of freedom.[12] As analyzed in *The Critique of Practical Reason*, these are legislated by the faculty of reason, not understanding. Ordinarily reason leaps ahead of us, inquiring about things of which certain knowledge is impossible: the origins of the universe, the existence of God, the life of the soul. Although the freedom of the will is just such an Idea of Reason, Kant claims that here reason is justified in its certainty that we can freely choose our actions.

But for Kant "freedom" does not mean doing whatever one wants. Morality demands a purely "good will" acting only from duty, without any reference to benefits or outcomes of the act. A will that chooses to follow its duty by behaving in a given circumstance as if under the compulsion of a universal law possesses autonomy; this obedience to the law that one gives oneself is for Kant synonymous with freedom.[13] How can we know this universal law? We formulate it by abstracting from our circumstances, expressing the act we are considering as a general maxim; only if no logical contradiction follows from the idea of a world in which *everyone* followed such a maxim can it be considered an acceptable action. Our freedom as rational beings means we may choose *not* to behave in such a way, and choose instead to follow our desires, our "hypothetical" motivations for doing what we want. Yet this very freedom—an essential attribute of the supersensible side of our nature—demands that we hold ourselves to a law that is *categorical*, that is, one that holds with the same kind of determining, *objective* compulsion as the a priori categories (concepts) of the understanding, as mapped out in the *Critique of Pure Reason*. The *categorical imperative* thus demands that "I should never act except in such a way that I can also will that my maxim should become a universal law." To act morally, one must already act as if a "kingdom of ends" exists: this means treating all other rational beings with the dignity they deserve, *not as mere instruments* or means but as ends—as rational beings also endowed with freedom. Since any necessary and universally valid "law of freedom" is one that we could choose not to obey, such a law legislates over desire in a way different from the way in which the determining, a priori concepts of the understanding rule over knowledge. Although both are "objective"—both assume the form of a universally valid law prescribed by the intellect—their modes of objectivity are completely different: the one concerns objects determined by mechanical causality, the other rational subjects free to obey or not.

Because we are both empirical beings in the system of causes and free subjects of reason, we are subject to both kinds of law; yet the vast difference between these two kinds of causality begs the question: How can the empirically perceivable, determined system of causes that is nature be brought into line with the ideal law of the kingdom of ends? In other words, how can freedom act within nature? *The Critique of Judgment* (1790) seeks to bridge this divide with a third faculty, one lying between the understanding, which legislates over *knowledge*, and the reason, which legislates over *desire*. This is the faculty of judgment. It legislates over *feeling*—for instance, the pleasure that comes from the "free play" of our faculties when confronted with a beautiful scene of nature, or the pleasure that we derive from the harmony between our faculties and the objects of the world. In neither case can we pass from subjective pleasure to a universally valid, objective intellectual concept—a judgment either that "this is beautiful" or that "there must be a creator responsible for the harmony between human faculties and the world." The system of a priori concepts involved *determinant* judgments; these were the necessary forms that any judgment about natural phenomena must take to be considered knowledge. But judgments of taste and judgments that involve the postulation of final ends make use of the *reflective* judgment, in which a priori categories are not involved. Our evaluation that some object is beautiful is based not on a concept but on feeling; we feel that everyone would assent to our view in matters of taste, but this feeling is merely a "subjective universal," not authorized by an objective a priori concept. Similarly, in teleological judgments we attribute a final end, and thus a purpose, to natural entities and to nature as a whole. Such finality cannot be observed directly in the entities being considered; nor does it involve the kind of mechanical causality of causes and effects that is cognized as one of the a priori categories of the understanding. Nevertheless, teleological explanations of the reflective judgment are necessary as *regulative principles* for the understanding. They allow us to grasp the intrinsic interrelations of parts within living organisms (and eventually nature as a whole), giving us a principle for conceiving of *generation*, a phenomenon that purely mechanical explanations cannot grasp.

Teleological explanations bring with them a version of instrumentality that highlights the difference, running throughout Kant's system, between a machine and a living being. In both a watch and an organism the motion of each part causes and is caused by every other, the parts working in tandem as an organized whole.[14] Yet it is only in an organism that the different parts actually engender each other by self-organizing development, growth, and repair: "the part must be an organ *producing* the other parts—each, consequently, reciprocally producing the others."[15] The efficient cause of the watch is the person who made it, while the efficient cause of the organism is identical with its final cause: *itself*. Each part of the watch is at most an instrument of *art*, potentially designed and built by an external agent and set mechanically to perform some task; in the organism, however, each organ is an instrument of *nature*, formed by the organism itself. In an organism the part is both means and end.

Further, the *extrinsic* relations among organisms, such as the complementarity of the sexes, or the interactions between an organism and its habitat, encourage us to seek out the interconnections among the various domains and empirical laws of nature. Humboldt's *Cosmos* is guided by the idea—which in Kant's terminology is just that, an *Idea of Reason*—that diverse empirical laws form a single system.[16] We have an obligation, Kant says, to pursue interconnections and push our explanations of mechanical principles of cause and effect as far as we can; but at the limit of these mechanical explanations, we reach beyond the realm of appearances, to the supersensible support and goal of the system, and ultimately encounter ourselves.[17] "What is the end and purpose of these and all the preceding natural kingdoms? For man, we say, and the multifarious uses to which his intelligence teaches him to put all these forms of life. [Man] is the ultimate end of creation here upon earth, because he is the one and only being upon it that is able to form a conception of ends, and from an aggregate of things purposively fashioned to construct by the aid of his reason a system of ends."[18] Man's *understanding* allows him to clear forests, build houses, dam rivers, and bring the empirical givens of nature into conceptual order as sciences. Yet such tasks address only man's habitat and material needs; they do not draw upon his unique possession, the free use of *reason*.

This aspect of his nature calls man to a higher goal than "individual happiness," one that surpasses "the mere fact that he is the chief instrument for instituting order and harmony in irrational external nature." Humans are the "lords of nature" because they can conceive and create a "kingdom of ends"; this highest human calling takes the form of *culture*, "the production in a rational being of an aptitude for any ends whatever of his own choosing." Hence "it is only culture that can be the ultimate end which we have cause to attribute to nature in respect of the human race."[19] In the nation culture takes the form of "a constitution so regulating the mutual relations of men that the abuse of freedom by individuals striving one against another is opposed by a lawful authority centered in a whole, called a *civil community*." At the next level up we find a *cosmopolitan whole*, a law-bound "system of all states that are in danger of acting injuriously toward each other."[20] Although such a perpetually peaceful system may only be possible as an ideal, much like the kingdom of ends in *The Critique of Practical Reason*, the activities undertaken toward this goal have immediate benefits. Kant's privileged examples are "fine art and the sciences," which, "if they do not make man morally better, yet, by conveying a pleasure that admits of universal communication and by introducing polish and refinement into society, make him civilized." Art and science, activities which are undertaken in the world of sensuous particularity but are universally communicable, prepare us for the rule of free reason.[21]

In "What Is Enlightenment" (1799) this progress appears as a metamorphosis from mechanical determinism to the rule of freedom. In current society the Enlightenment's audacious quest for knowledge is possible only in the "public" side of man's life; only there can the scholar and cosmopolitan "world citizen" question old dogmas and advance new doctrines. On the "private" side, however, duties to the state must be fulfilled; "some sort of mechanism" is needed to ensure obedience to communal laws. Eventually, however, the pursuit of freedom will allow the supersensible to act upon the mechanical: "at last," he foresees, "free thought acts even on the fundamentals of government and the state finds it agreeable to treat man, who is now *more than a machine*, in accord with his dignity."[22] The causality of freedom acts upon mechanical causality, creating a polity in which all may freely discern and obey the law that determines their duty, investigate the empirical relations of the phenomenal world, and create and enjoy works of beauty. In their function as instruments of nature, such citizens are liberated, and are now both means and ends.[23]

The Critique of Judgment united the two sides of man's being by the discovery of a supersensible goal that must nevertheless be realized within nature. *Culture* (or its cognate, *Bildung*)—the creation of institutions and practices for developing the aptitudes of mankind in keeping with reason—fuses the laws of both freedom and nature and their autonomous modes of objectivity. Despite these optimistic sketches of human destiny, many found in Kant's system only austerity, self-denial, and a grim refusal of pleasure; his works seemed to demand a superhuman refusal of natural inclinations while expressing a disdain for temporal change, human variability, and sensory enjoyment.[24] It is this impression that Schiller sought to combat by refashioning Kant's philosophy in *The Aesthetic Education of Mankind*.

Schiller's Aesthetic State and Its Citizens

Schiller acknowledged the influence of "Kantian principles" on his argument but paid little heed to Kant's neat divisions and rigorously patrolled boundaries.[25] *The Aesthetic Education* melded the two types of objectivity in Kant—moral and theoretical, laws of nature and laws of freedom—in Schiller's notion of the aesthetic state. His discussion's central term was *Selbständigkeit*, "self-standingness," translated and reappropriated as "autonomy" in the critical tradition that has followed him.[26] This does not, however, mean a pure, godlike state of disengagement and self-sufficiency. Schiller took up and developed Kant's reflections on instrumentality and means and ends in *practical* terms; his view of autonomy implied a connection with other humans and linked pure form with sense and inclination. In focusing on the movement, exchange, and activity of this intermediate realm, he also brought readers' attention to concrete practices and materials. As much as it was a theory of an intermediate state between universality and particularity, Schiller's *Aesthetic Education* was a theory of the action of material *mediation*.

In the *Aesthetic Education* the tension in the architecture of Kant's theory between nature and freedom became the engine for an unfolding argument whose harmonic elaboration mirrored the dialectical development of humanity from savagery through barbarism to culture and morality. Schiller set out binary pairs that were sublimated, overcome, reconciled in a higher unity or dialectical reconciliation (to speak like Hegel, who learned much from him). The form numerologically restated the content: the dyads that became triads, in the classic symphonic form of A, B, A′, are echoed throughout the twenty-seven (3^3) letters. Where Kant laid out a static map of the regions of the mind, Schiller conceived the faculties as drives and forces in active conflict; the development of the individual, like the development of society, involved a dramatic movement and struggle among faculties. He asserted a fundamental tension between the passive "sense drive," "life," the empirical world with its desires and impressions, and on the other hand the active "form drive," which concerned abstraction and active, eternal principles in science or art. Between the two lay the "play drive" which relishes the mere appearance of things—the autonomous (self-standing) "*schöne Schein*," detached from the desire either to possess or to freeze as timeless knowledge. Writing in 1794, after the storming of the Bastille and the Terror, Schiller identified the root of the political imbalance — the viciousness of the mob as much as the arrogance and indifference of the rulers — as a fundamental imbalance within mankind. Cure the latter and you cure the former. Thus his solution demanded a transformation at once artistic, emotional, political, and moral.

Kant mocked those who would seek morality in a principle outside of reason, in their own subjective inclinations or desires, or for the rewards it would bring; he consigned them to "a dream of sweet illusions (in which not Juno but a cloud is embraced)." Early in *The Aesthetic Education* Schiller rejects the self-abnegation implied by Kant's claim that the pure will was the only possible basis of morality. "If then, man is to retain his power of choice and yet, at the same time, be a reliable link in the chain of causality, this can only be brought about through both these motive forces, inclination and duty, producing completely identical results in the world of phenomena." Duty and inclination, which for Kant were divided by the same gulf that separated man's spiritual nature from his physical nature, must fuse; law and sense must harmonize. Like Kant, Schiller aims to unite the two systems of causality laid out in the first two critiques (and their respective modes of objectivity) by means of culture, art, and science. Yet to Schiller, for freedom to play a role in nature, reason must agree with desire.[27]

An instrument (*Werkzeug*) is required to bring about this transformation. Schiller proposes fine art, the analysis of which recapitulates the fusion of *duty* and *inclination* in the realm of morality: form and matter must balance each other in a dynamic equilibrium. His key example is the famous and quite large Roman head, the "Juno Ludovisi" made into a recurrent topos of classicist and romantic criticism by Winckelmann. In what can be read as a reply to Kant's dismissal of the delusions of consequentalist ethics by reference to Juno, Schiller describes this concrete goddess as the embodiment of the ideal physiognomy: the statue balances physical and spiritual perfection, Grace and Dignity, woman and God, the sensible and the supersensible. Desire pushes us toward formal purity, and formal perfection guides our desire. The statue leads those who contemplate it into the *Aesthetic State*, one of "both utter repose and supreme agitation, and there results that wondrous stirring of the heart for which mind has no concept nor speech any name." Here the aesthetic is a specific psychological state balanced between opposed forces, combining "melting beauty" with "energizing beauty." In contradiction with Kant, moral freedom no longer requires humanity to depart from the world of the senses. "We need, then, no longer feel at a loss for a way which might lead us from our dependence upon sense towards moral freedom, since beauty offers us an instance of the latter being perfectly compatible with the former, an instance of man not needing to flee matter in order to manifest himself as spirit." In works of beauty, man is "already free while still in association with sense"; the "objective" moral law is joined with the "stuff" of sense and matter.[28]

This psychological state paves the way for a new political state, in which the desires of the individual will harmonize with the demands of civil society: "Once man is inwardly at one with himself, he will be able to preserve his individuality however much he may universalize his conduct, and the State will be merely the interpreter of his own finest instinct." Schiller again recounts a history of humanity, this time in terms of the development of an individual's consciousness: the appearance of the play-drive takes the subject from a first state of immersion in mere sense to an awareness of the distinction between self and object and a delight in everything that offers "material for possible shaping." This new relation with things accompanies a new relation with people. Relations between men and women go from mere satisfaction of physical desire to an intellectual and existential exchange of recognition: "from being a force impinging upon feeling, [man] must become a form confronting the mind; he must be willing to concede freedom because it is freedom he wishes to please." This reciprocity spreads throughout "the complex whole of society, endeavouring to reconcile the gentle with the violent in the moral world," opening up a new, autonomous realm between "the fearful kingdom of forces" and "the sacred kingdom of laws." Schiller calls this autonomous zone "a third joyous kingdom of play and of appearance, in which man is relieved of the shackles of circumstance." This aesthetic realm is characterized by a specific notion of freedom.[29]

For Schiller freedom is not merely a state but an activity, one that can only be brought into being by the presence and reciprocal involvement of others: "To bestow freedom by means of freedom is the fundamental law of this kingdom."[30] This conception of freedom—and its vicissitudes—was a central theme in Schiller's work, from *Wilhelm Tell* and *Don Carlos* to the poem that he wrote for the chorale of Beethoven's Ninth Symphony. While the poem is known as "Ode to Joy," the original title (as those who selected it as the jingle of the European Union surely knew), was "Ode to Freedom":

Thy magic reunites those
Whom stern custom has parted;
All men will become brothers
Under thy gentle wing.

May he who has had the fortune
To gain a true friend
And he who has won a noble wife
Join in our jubilation!

Yes, even if he calls but one soul
His own in all the world.
But he who has failed in this
Must steal away alone and in tears.

Even the last lines' stinging *Schadenfreude* captures the essential point of Schiller's notion of freedom. In Kant's *Critique of Practical Reason* freedom was the property of an individual in his innermost isolation, abstracted from all social ties and fellow-feeling. For Schiller, freedom (and the morality that is founded upon it) cannot exist outside of active relations with others. In this exchange of recognizing gazes, this phenomenology of mutual and reciprocal self-possession, freedom is given to others and received back from them, implying ownership and being owned—as in Rousseau's social contract, under which each gives himself to all and receives the others back.

Schiller's conception of the state that incorporates this freedom involves a change in the definition of and relation between the "objective" and the "subjective." In Kant the "subjective" pertained to the sensuous and changing, both empirical sense givens as well as judgments not regulated by a universally valid concept; "objective" judgments involved a priori concepts, thereby constituting an object of knowledge. In Schiller the terms begin to take on more familiar contours. Again, years before Hegel, Schiller calls the state "the objective and, as it were, canonical form in which all the diversity of individual subjects strive to unite." But as if in anticipation of those who have seen in the line of political thought that grew out of this statement (from Hegel to Marx and their followers) a quasi-totalitarian theory of uniformity imposed from above, Schiller takes pains to note that "the State should not only respect the objective and generic character in its individual subjects; it should also honour their subjective and specific character." Again he seeks an autonomous middle ground: the good society will preserve what he calls here the "subjective" character of a citizen who is "inwardly one with himself" and is therefore able "to preserve his individuality however much he may universalize his conduct." Thus the "State will be merely the interpreter of his own finest instinct." As noted with regard to morality, Schiller mixed the "objectivity" of the good will that acts in conformity to reason alone with the "subjective" desires, inclinations, and experiences of the individual. The political sphere demands a comparable adjustment of the formal laws regulating conduct to the character and circumstances of the individuals who make up a society. "Autonomy," or *Selbständigkeit*, is the name that Schiller gives to this balance between the universal or objective and the particular or subjective.[31]

Moving from morality to aesthetics and politics, Schiller's arguments also extend to the natural sciences. His first letter decries the narrow aridity of the sciences and the murderous consequences of their investigations: "truth is a paradox for the analytic thinker; analysis dissolves the very being of that which is analysed." The same is true for the analyst himself. Humanity pays a heavy price for the extreme specialization of the modern disciplines: "Once the increase of empirical knowledge, and more exact modes of thought, made sharper divisions between the sciences inevitable, and once the increasingly complex machinery of State necessitated a more rigorous separation of ranks and occupations, then the inner unity of human nature was severed too, and a disastrous conflict set its harmonious powers at variance."[32] Here we see science linked to the "machinery" of the State, the "objective form" of human interactions. Yet the solution is not an escape from all social bonds, nor a flight from science or technology into an idealized state of nature or a subjective state of reverie. Instead the aesthetic state will take science out of its austere and self-enclosed abstraction; science must humanize itself. Though Schiller chooses to retain the term "objectivity" to refer to pure formal knowledge, in fact the model of knowledge that he implies is one that places this pure formalism in suspension with the senses. Analytic a priori judgments and synthetic a priori judgments are placed in a continuum; the fusion of sense and category now takes place in the world of objects and communal activity.[33] As in his treatment of politics, Schiller continues to use "objective" in a sense like Kant's, as validation by an abstract, formal law. But the "free" practice of science, like that of politics, demands a mixture between the pure formalism of Kantian objectivity and the variable, sensuous "life" with which it engages. Once more this state of interdependence between two poles is called autonomy.

Scientific specialists must learn to share their knowledge; with diffusion comes transformation. "From within the Mysteries of Science, taste leads knowledge out into the broad daylight of Common Sense, and transforms a monopoly of the Schools into the common possession of Human Society as a whole"; by means of the aesthetic, freedom will enter the realm of science as surely as it did the realm of politics. The result will be a transformation of the lifeless machinery of the state, and of the mere instruments of art, into a free, cosmic republic: "At the touch of the wand [of taste], the fetters of serfdom fall away from the lifeless and the living alike. In the Aesthetic State, everything—even the tool which serves [*auch das dienende Werkzeug*]—is a free citizen, having equal rights with the noblest; and the mind, which would force the patient mass beneath the yoke of its purposes, must here first obtain its assent."[34]

The mind, whose province is truth, "object, pure and simple," can only rule if it has the assent of its subjects. All individuals over which the laws of form hold sway—the laws of the objective polity, the objective law of morality, or the objective categories of science—must be liberated, allowed to choose to obey the law, all the way down to the "lifeless" instruments which were previously only means to an end. The result is a kingdom in which the means (the instruments) are also ends; this kingdom includes all of nature, along with the material mediations—art, tools, language—that shape and articulate the relations among its members.

In Schiller's aesthetic state, knowledge is no longer the relation between a transcendental subject and sense data subsumed under universal categories; it is now a liberating relation of mutual respect among users, tools, and their objects. For the pedagogic or the political artist, "man is at once the material on which he works and the goal toward which he strives. In this case the end turns back upon itself and becomes identical with the medium; and it is only inasmuch as the whole serves the parts that the parts are in any way bound to submit to the whole." We move from a vertical relation to a horizontal one, from a hierarchical to an egalitarian model, from a linear, mechanical causality to a reflexive, organic one. The subjective individuality of all entities must be preserved within the beautiful appearance of the work of art, the logical interconnections established in the scientific "tableau," the objective apparatus of the state.[35]

While Schiller's main concern was the political consequences of the aesthetic state, his arguments, as we have seen, had direct implications for natural science. The *Aesthetic Education* presented the theory of how to get from an intellectual, immaterial, and individual view of objectivity and "autonomy" to one in which objectivity and autonomy are external, embodied, and collectively validated. The work of Alexander von Humboldt, culminating in his *Cosmos*, offered the practice.

A Cosmic Polity of Free Instruments

Cosmos was a true "Victorian sensation," one of the best-selling books of its time. As with François Arago's popular astronomy lectures at the Observatory of Paris, the renown of this work was to no small extent due to its pleasing style, vivid descriptions of natural phenomena, and liberal references to literature. Early on, for instance, Humboldt writes, quoting "the immortal poet" Schiller, that mankind, "amid ceaseless change [,] seeks the unchanging pole." But the "aesthetic" concerns that underwrote Humboldt's project were not just window dressing.[36] The task that Schiller announces of fusing particular, "ceaselessly changing" sensory givens with "the unchanging pole" of abstract principles is of a piece with Humboldt's overall project. Humboldt consciously worked in the "intermediate zone" balanced between extremes that Schiller opened up, with much of the same theoretical apparatus. While he set himself the task of actively submitting observations of nature "to the test of reason and intellect," he also paused to note how a "romantic landscape" can be "a source of enjoyment to man, by opening a wide field to the creative powers of his imagination." Just as Schiller found in the Juno Ludovisi a combination of melting and energizing beauty, so did Humboldt praise the "soothing yet strengthening influence" of observing nature. The guiding idea of *Cosmos*, to present the universe as a law-bound, unified whole while respecting the specificity and "freedom" of each individual part, was a scientific realization of Schiller's reconfiguration of Kantian autonomy.[37]

Cosmos made clear Humboldt's acceptance of post-Kantian epistemology: "Science is the labor of mind applied to nature, but the external world has no real existence for us beyond the image reflected within ourselves through the medium of the senses." The inclusion of the term "labor" to what might otherwise be construed as an expression of idealist faith goes beyond the Kantian assumption of a universality of perception and cognition; some kind of *activity* is demanded to shape what passes through the "medium" of the senses. On the next page, discussing the progress made in the theory of matter, Humboldt notes improvements in natural philosophy over earlier speculations and haphazard observations, "by the ingenious application of atomic suppositions, by the more general and intimate study of phenomena, and by the improved construction of new apparatus." Contemporary science is marked by its refinements in method and, crucially, in apparatus: the technical arrangements in the middle ground between mind and nature.[38]

Like other scientists of this period, Humboldt placed a heavy emphasis on the development of new instruments and observational apparatus to measure and to reduce differences in perception. His research depended on an embarrassment of devices for registering a huge range of phenomena: chronometers, telescopes, quadrants, sextants, repeating circles, dip needles, magnetic compasses, thermometers, hygrometers (built around a strand of human hair which grew longer or shorter depending on the moisture of the air), barometers, electrometers, and eudiometers (to measure the air's chemical composition).[39] These instruments were not understood as transparent means of registering nature "in itself." Like Schiller's "fine art," Humboldt's instruments were the concrete *media* occupying the milieu, the "halfway-place" between humans and the world: the concrete locus for the fusion of sense and intellect.

In other words, Humboldt's regime of instrumentation externalized, temporalized, and "communalized" Kant's categories. Jonathan Crary has argued that in this period perception was increasingly theorized as a phenomenon of particular bodies disciplined by external devices and practices. Humboldt's work with instruments shows that in the same moment that perception was somatized and particularized, thereby dismantling the Kantian notion that the categories of experience were universal properties of the transcendental ego, these categories were *made into universals in practice*, in the technical apparatus of the observatory sciences. The processes of instrumental calibration, standardization, and coordination externalized the process of the understanding described in the *Critique of Pure Reason*. By ensuring that instruments shared the same scale of measures for quantities of time, space, and degrees of magnitude, that they were subject to identical thresholds for determining the presence and composition of substances, and that they possessed the same sensitivity to specific causal relations, the work done in observatories to bring instruments into agreement literally built the concepts of the pure understanding—the basis for the communicability of knowledge—into the physical apparatus.[40]

The new instruments of nineteenth-century science were often seen to embody the qualities of the ideal human subject.[41] For Humboldt this symbolic identification was rooted in *interchangeability* and *intimacy*. In his extensive research on galvanism and animal magnetism, he constructed elaborate circuits for galvanic electricity in which different metals, chemical solutions, and frogs had equal status as instruments; in several experiments the main site of inscription and observation, another link in the chain, was his own body.[42] Furthermore, his correspondence repeatedly testified to his extraordinary care for his instruments: he typically identified them by the patronymic of their makers, went to great lengths to assure their well-being, and discussed his most cherished compasses, barometers, and sextants with the same enthusiastic affection as his dearest friends. Letters of introduction for human protégés followed equally solicitous letters on behalf of instruments. Gleefully describing a phase of the expedition in which all of his "instruments were in action," he followed an apparently rhetorical question, "But how can I tell you about that?," with thousands of words describing the behavior of each member of his brood. Observations, he said, must be made "with exactitude and 'con amore'" when tropical heat makes instruments burn one's hands. Instruments which made for good traveling companions—those which were small, light, and versatile—were favored, like the portable barometer which could be fitted onto the head of Humboldt's walking stick. The tools arrayed in the famous painting of Humboldt and Bonpland during their voyage—visually uniting the two researchers at the center of the painting—were totems and extensions of the researcher's self.[43] They not only extended his senses, heightening his perceptual faculties and submitting sensory phenomena to mathematical scaling: they were embodiments of his relations with others and his place in the natural and social world.[44]

This mode of sociability was central to Humboldt's conception of scientific knowledge. As in much late Enlightenment republican thought, including Schiller's, scientific "objectivity" was linked to ideals of moral and political freedom. But Schiller's assimilation of laws of nature to laws of freedom in the aesthetic state points us toward a distinct moral meaning in Humboldtian instrumentation. In their discussions of objectivity, Daston and Galison consistently stress its *negative* aspect, likening objectivity to the hollow remainder of wax impressed by the more robust and positively defined seal of subjectivity, and defining it as an "escape from perspective" and from intention: "Instead of freedom of will, machines offered freedom from will." Elsewhere Daston explains the moral economy of science in the mid-nineteenth century in terms recalling Kant's second critique: "The self-restraining and self-effacing counsels of mechanical and aperspectival objectivity reverberate with the stern voice of moral duty: the self-command required in both cases to suppress the merely personal is indeed the very essence of the moral." In these arguments freedom is associated with escape, suppression, and denial. Yet for Humboldt it was not simply the negative virtues of tirelessness, restraint from intervention, and lack of bias that commended machines as observers and scribes for natural observation. The freedom of Humboldtian instruments was a positive and active virtue.[45]

The well-tempered instrument, like a reliable but spontaneous human, oscillates within a specific range of values, passive in receiving its phenomena, active in transmitting them. The process of standardization and calibration builds an a priori principle into the instrument, a categorical imperative ruling over its "desire"—like Schiller's "objective man" who is "ennobled into participation with the law." Each instrument responds "freely" to its milieu and its particular circumstances; but like human laws and the regularities of human language and practice, the agreement produced between the field instrument and the master instrument, often located in the observatory, fixes the instrument's action within a defined range of values, providing the shared and stable background needed to make local difference communicable.[46] This is not, however, an automatic process or unilateral application of force. Just as freedom for Schiller can only emerge in reciprocal exchange with other beings, so does the objectivity of the Humboldtian tool demand cooperation with a highly skilled and patient human. The observer must gain the instrument's assent by entering into a dialogue, "playing" with it, becoming familiar with its limits and habits. Humboldt's letters and travel reports are filled with accounts of awkward moments at the beginning of his relationship with an instrument and his joy at learning to cooperate successfully with it. Making good measurements means knowing and adjusting to an instrument's particularities. For instance, he preferred a chronometer which lost time gradually at a regular rate to one which kept time perfectly yet was subject to unexpected stops. Precisely the same logic underwrote the device that François Arago introduced at the Paris Observatory to measure an observer's "personal equation," or regular lag time in registering a star moving across a fixed space. The law may be the same for all, yet as Schiller argued, the law is only a law of freedom if it adjusts to the living particularity of the individual, whether a person or a machine.[47]

These individuals form a cosmopolitan society. All instruments of a single type constitute a globally distributed "clan" of instruments registering the same phenomenon; this is one stratum of a larger society into which humans, machines, and certain privileged sites—notably observatories—

are woven. Humboldt played a preeminent role in the internationalization and institutionalization of science. He encouraged younger scientists, giving advice and training and supporting their candidacies; he facilitated contacts among scientists and secured government support. He organized the first meeting of the Society of Naturalists and Natural Philosophers in Berlin; his opening speech inspired Charles Babbage, in attendance, to form the similar British Association for the Advancement of Science.[48] Even more, his incessant correspondence helped him to form a network of natural scientists. The global network of investigators that Humboldt helped to coordinate directly recalled his early experiments in animal magnetism, in which animals, metals, batteries, and the experimenter formed an energetic circuit; even in the midst of inquiries about friends and their families, there is hardly a single letter in Humboldt's correspondence which does not mention an instrument or a meteorological observation.

The aim of Humboldt's physics of the earth was to map the patterns of global systems of natural forces, charting in its local detail and its particular interactions "a general equilibrium which reigns among disturbances and apparent turmoil," the result of an infinity of mechanical forces and chemical attractions balancing each other out. When he charted the values registered by the instruments distributed around the globe, global patterns became manifest. The clan of thermometers and their readers, for instance, made it possible to trace *isothermal lines*, regions of shared average temperature in bands across the earth. An early article that Humboldt wrote for the Society of Arcueil contains a tableau with the position in latitude and longitude, the average temperature over the year, and maximum and minimum temperatures taken from reports at forty-eight locations around the globe. The table is a who's who of observers, field stations, and observatories: Humboldt in Curana, Saussure in Geneva, Dalton in Kendall, Arago in Paris, Euler in Petersburg, Young in London, and Playfair in Edinburgh.[49] This study, in which half of the sites' averages were calculated on the basis of around eight thousand observations, prepared the way for the "magnetic crusade" which traced the distribution of magnetic intensity, inclination, and declination around the globe. As John Cawood has shown, this campaign—for which Arago's Observatory of Paris was briefly the clearinghouse—was a major step in the establishment of a global network of observational science, highlighting the need for communication and shared standards; the multiple readings taken in observatories in both metropolitan centers and colonial outposts played a key role in maintaining and implementing standards for time, space, and other measures.[50]

The range of local variations of phenomena could be depicted across the planet in the synoptic tableaux that Humboldt invented. These were maps of the distribution and relative change of numerous phenomena across a single limited region. In the painted tableau that accompanied his *Essai sur la géographie des plantes*, each instrument and the phenomenon that it registers belongs to one vertical column; the y-axis represents altitude (figure 41). For example, for each stratum the temperature, air pressure, amount of magnetic phenomena, light quality, blueness of the sky, moisture, and boiling temperature of water could thus be seen and compared at once. In a given tableau each instrument (like each citizen) performs its Kantian duty, but not in isolation, and not in pure, abstract relation to the law. When each does its duty, the whole system is described — a balancing act among opposed forces.

In this tableau Humboldt has rendered the image of a vast natural chorus, expressing itself freely through its liberated (and at the same time law-bound) instruments. Rotate it ninety degrees to the right, so that it stands on its side, with the columns reading across horizontally, and each column traces the part played by each phenomenon, isolated and recorded by its respective instrument; the tableau follows the structure of an orchestral score, with progression over time replaced by ascension in altitude. The observers, instruments, and phenomena united in Humboldt's tableau sing the same song, one variation of which may be expressed, again, in the "Ode to Joy" / "Ode to Freedom":

> All the world's creatures
> Draw joy from nature's breast;
> Both the good and the evil
> Follow her rose-strewn path.
>
> Be embraced, Millions! . . .
> Can you sense the Creator, world?
> Seek him above the starry canopy.

As in Kant, a view of the agreement of empirical laws within the realm of appearances suggests a supersensible basis for this harmony and splendor. The shimmering intermediate zone of appearances — with each phenomenon isolated, brought into a form that allows for comparison and joined with the others in a global system of energetic forces in constant interplay — gave testimony to a higher order of eternal things. The numerous connections that Humboldt wove between the starry canopy and the sub-

GÉOGRAPHIE DES PLANTES ÉQUINOXIALES.

Tableau physique des Andes et Pays voisins

Dressé d'après des Observations & des Mesures prises sur les lieux depuis le 10.e degré de latitude boréale jusqu'au 10.e de latitude australe en 1799, 1800, 1801, 1802 et 1803.

41. "Geography of Equinoxal Plants," plate accompanying Alexander von Humboldt and Aimé Bonpland, *Essai sur la géographie des plantes accompagné d'un tableau physique des régions équinoxiales* (Paris: Fr. Schoell, 1807). The image encapsulates the book's sections as well as the process of its construction: vertical columns correspond to chapter headings, which in turn correspond to specific observations and in most cases to the parts played by specific instruments. Vertical columns include: Height; Temperature Scale; Barometric Scale; Hygrometric Scale; Blueness of the Sky; Horizontal Refractions; Chemical Composition of the Atmosphere; Temperature of Boiling Water; Soil Culture. The instruments analyze and the image synthesizes, in intellectual and sensuous form, tropical milieus at all elevations.

lunary globe were indices of a transcendent dimension which he sought to bring as near as possible to presence. Each particular element and each sweeping line that joined it to wider phenomenal currents, each note, each melody, each movement pointed to the dynamic *whole* that it was part of and the sublime principle of order beyond it. A "symphony" is so named because it contains a range of sounds working together at once. Humboldt spent his life composing something altogether new, combining the surge and shades of feeling tapped by romanticism with the mechanical detail and social coordination of the new worldwide regime of the precision sciences. Humboldt was composing a cosmic *symphenomenony*—and helping to assemble the worldwide orchestra needed to perform it.[51]

The notion of a polity of scientific instruments and users inevitably raises the question of who makes the laws, who enforces them, and how. Such concerns recall the enduring political issue of the Enlightenment: how to determine and enforce the laws of a polity whose legislators are as exactly as numerous as its subjects. As other chapters in this book show, it would be difficult to identify one single mode of arbitrating over such questions in the physical sciences. Instead, diverse solutions emerged for establishing the "accuracy," the "reliability," the "objectivity," or even the "truth" of claims about physical phenomena: the construction of increasingly precise instruments, relying on trust and experience in makers and users; instruments and methods to account for and to correct individual error; Gauss's mathematical method of least squares to reduce the errors in a large number of observations; the creation and maintenance of standard values. From the late 1790s to the 1830s Humboldt was involved in attempts to perfect observation, measurement, and representation according to every one of these methods. While diverse, a common thread runs through these modes of verification. In each case objectivity was the product of communal activity, a process of exchange and coordination taking place within a single sphere—a common ground between mind and nature, shaped by the social instruments that articulated it.[52]

At all levels Humboldt's works played a key role in the shifting meaning of "objectivity": from an internal and rationalist model in Kant to an external, communal, and emergent model which relied increasingly on machines.[53] The development of scientific associations, journals, and vast collective research projects in the nineteenth century which Daston presents as evidence of an ascendant ideal of "aperspectival objectivity" might thus in Humboldt's case be thought of as an ideal of *multiperspectival* objectivity. While Humboldt's science relied thoroughly on mechanical devices, these were not seen as the negation of individual perspective and the embodiment of values of restraint and denial; they aspired to an ideal of communal, active, productive, and spontaneous mediation. Humboldt's instruments were free citizens in a cosmic polity: flexible individuals which nevertheless obeyed laws. As long as their "users"—whose actions they regulated in turn—understood and adjusted to their individual qualities and temperaments, they served as autonomous go-betweens, fulfilling their duty in the liberated universe of humans and nature depicted and, Humboldt hoped, realized in *Cosmos*.

Reframing the Modern World Picture

Humboldtian science had a generative effect on the laboratory science of the following century. Du Bois-Reymond wrote in the mid-1800s that "every industrious and ambitious man of science . . . is Humboldt's son"; and Helmholtz clearly modeled his persona as a humanist spokesman for German science on Humboldt.[54] Much suggests that after the middle of the century, however, a new image of science and of instrumentation prevailed. As noted by Daston and Galison, the image of scientific machines became increasingly inflexible, lifeless, and inhuman; Humboldt's regime was reconceptualized and tied to laboratories, with its instruments often described and developed in the direction of automatic action and self-inscription.[55] The delicacy with which such apparatus still had to be constructed and maintained was cloaked by a new rhetoric of objectivity, for which the daguerreotype has frequently been cited as the model.

Yet when it first appeared, photography, "essence and emblem of mechanical objectivity," was one more citizen in Humboldt's polity. Arago, Humboldt's close friend, introduced Daguerre's and Niepce's invention to the public in 1839 and secured a lifetime salary to its inventors. Throughout his physical and astronomical researches Arago had shown the same passion for instruments as Humboldt.[56] Obsessed with the effect of different interfering media on the observation of light, Arago did not believe that the daguerreotype's use of the "automatic" action of light implied any exceptional ontological or epistemological claim for its images. In his announcement of the discovery, he suggests a use for the technique that is completely indifferent to the content of the image. That the image develops at different rates depending on the season and weather suggests that "the meteorologist would have one more element to include in his *tableaux*, and to the former observations of the state of the thermometer, the barometer, the hygrometer and of the transparency of the air, he will have to add an element that the other instruments do not grasp."[57] Here Arago is not interested in the object depicted on the silver plate, but rather in what the process of its development tells us about invisible atmospheric phenomena. Along similar lines, he suggests that photography will allow for improvements in photometry, the comparison of intensities of light; other suggestions are firmly set among the field sciences of Humboldt, including a map of the moon and topographical applications. The daguerreotype was presented by its first public supporter as another member of the family of geophysical instruments—an interesting addition, no doubt, but for Humboldtian instruments and the experimental apparatus that they begat, the story traced here suggests that "mechanical objectivity" was not so much a set of practices as a way of *talking* about practice which developed after the revolutions of 1848, especially to an audience that had once been invited into the halls of science—by Humboldt and Arago—and, after the middle of the century, was to be kept out. After 1848 photography staged a coup d'état in the polity of instruments: only those methods that appeared untouched by human agency, like photography and the automatic, self-inscribing instruments of the graphic method, would be allowed into the court—or at least into the court's official representations. This moment of the ascendancy of "mechanical objectivity" parallels the failure of the German revolution of 1848 and the coup d'état of Napoleon III in 1851. The valence of the term "freedom" shifted as well: these instruments were now considered "free" of human idiosyncrasy, perspective, will, and imagination, no longer seen as participating in the oscillation between active and passive, whole and part, individual and general which marked other Humboldtian instruments. At the same time late romantic critics such as Charles Baudelaire virulently attacked the machine as the antithesis of the human.

In response to the demiurgic powers released by the new technologies of early-nineteenth-century science and industry, we are most familiar with those texts which express what has proved in many ways to be a horror of mechanization. Humboldt's *Cosmos*, on the contrary, is a decidedly romantic attempt to harmonize these new forces; unlike gothic and antirationalist strands of romanticism, his work incorporates the natural sciences and their machines into a vision of natural balance. The stable yet dynamic universe of *Cosmos* is a *world-picture technology* intended to bring awareness of the interconnectedness of the universe and to place human endeavors within a wider frame of meaning. In an age when the recognition of human responsibility for both the health and the conceptual order of nature became increasingly widespread, *Cosmos* can be read as an argument against the madness of seeing human freedom as liberation from all restraint. Instead it advances a view of autonomy that applies to all beings, the effects of whose freely chosen actions reverberate throughout the system.[58]

The observatory sciences seem paradoxical. They require a huge range of "eyes"—human and artificial—distributed across the globe, each working independently and immersed in its specific circumstances; at the same time the many voices and individuals brought into this circle of exchanges must be calibrated within known parameters, measured by shared standards, and held to universal principles. How do we create a national or global community ordered by laws that preserve and respect phenomenal variety, individual spontaneity, and freedom? While it took a variety of forms, this problem was faced in the same terms in the sciences, arts, and politics of the first half of the nineteenth century. Humboldt wrote to Arago in May 1848 after the uprising for a constitutional monarchy in Berlin: "My ardent hopes for democratic institutions, hopes which date back to 1789, have been fulfilled." By the next year, after the kaiser's repudiation of the constitution, Humboldt was much less sanguine: "I am reduced to the banal hope that the noble and ardent desire for free institutions is maintained by the people and that, though from time to time it may appear to sleep, it is as eternal as the electromagnetic storm which sparkles in the sun."[59] For Humboldt the work of freedom—and the discipline that it demands—combine hope, effort, and artifice, to give shape to natural potentials which lie sleeping. It was to awaken, guide, and frame this unity, in the natural world and among humans, that Humboldt served as a charismatic and gregarious instrument.

Notes

1. See Eagleton, *The Ideology of the Aesthetic*; Bürger, *Theory of the Avant-Garde*; Herbert Marcuse, *The Aesthetic Dimension: Toward a Critique of Marxist Aesthetics* (Boston: Beacon, 1978).

2. See Wilhelm von Humboldt, *On Language: The Diversity of Human Language-Structure and Its Influence on the Mental Development of Mankind* (Cambridge: Cambridge University Press, 1988); the above quote is from Martin Heidegger, "The Way to Language," *Basic Writings from "Being and Time" (1927) to "The Task of Thinking" (1964)* (New York: Harper and Row, 1977), 405.

3. My conception of Humboldtian Science owes much to the thesis and articles of

Michael Dettelbach, including his introduction in Humboldt, *Cosmos* (1997 edn); see also Cannon, *Science in Culture*; Botting, *Humboldt and the Cosmos*; and Bourguet, "La république des instruments."

4. See Bourguet, "La république des instruments."

5. Notably in Daston and Galison, "The Image of Objectivity." See also Lorraine J. Daston, "Objectivity and the Escape from Perspective," *Social Studies of Science* 22 (1992): 597–618; and Peter Galison, "Objectivity Is Romantic," ACLS Occasional Paper 47 (2000).

6. See Langdon Winner, *Autonomous Technology: Technics-Out-of-Control as a Theme in Political Thought* (Cambridge: MIT Press, 1977).

7. See Lenoir, "The Göttingen School and the Development of Transcendental *Naturphilosophie* in the Romantic Era"; Phillip F. Rehbock, *The Philosophical Naturalists: Themes in Early Nineteenth Century British Biology* (Madison: University of Wisconsin Press, 1983); Judith Schlanger, *Les Métaphores de l'organisme* (Paris: Vrin, 1971); and Richards, *The Romantic Conception of Life*.

8. See Frederick Beiser, *The Fate of Reason: German Philosophy from Kant to Fichte* (Cambridge: Harvard University Press, 1987).

9. Kant, *Grounding for the Metaphysics of Morals*, 53. Referring to Kant, Foucault speaks of man as the "empirico-transcendental doublet" in *The Order of Things: An Archaeology of the Human Sciences* (New York: Vintage, 1973), 318.

10. See Immanuel Kant, *Critique of Judgement*, trans. James Creed Meredith (Oxford: Oxford University Press, 1952), Introduction, sec. IX, 36; Gilles Deleuze, *Kant's Critical Philosophy: The Doctrine of the Faculties*, trans. Hugh Tomlinson and Barbara Habberjam (Minneapolis: University of Minnesota Press, 1984), 3–10.

11. On Kant on the categories see Cassirer, *Kant's Life and Thought*, 170.

12. "For we are constrained to think the pure will as something bound by law and hence 'objective,' but this objectivity belongs to a sphere totally distinct from that which is expressed in the spatiotemporal phenomenon. It is not a world of things we are assured of here, but one of free personalities; not a set of causally related objects, but a republic of self-sufficient subjects purposively united." Cassirer, *Kant's Life and Thought*, 247, 154.

13. See Kant, *Grounding for the Metaphysics of Morals*, 49. The notion of autonomy ("self-law") can be traced to Rousseau's general will: "freedom is obeying the law one gives oneself." It has roots in stoicism, in Saint Paul (who speaks of the gentiles as "a law unto themselves"), and in Luther's "Freedom of a Christian Man," in which an inner law of faith overrides laws of the state. See J. B. Schneewind, *The Invention of Autonomy: A History of Modern Moral Philosophy* (Cambridge: Cambridge University Press, 1998).

14. Kant, *Critique of Judgement*, secs. 64–65.

15. Ibid., secs. 65, 22.

16. On Kant's idea of reason see Paul Clavier, *Kant: les idées cosmologiques* (Paris: Presses Universitaires de France, 1997).

17. Kant, *Critique of Judgement*, secs. 82, 91: "the reconciliation of the two modes of picturing the possibility of nature [either mechanical or teleological] might

easily lie in the supersensible principle of nature, both external and internal. For the mode of representation based on final causes is only a subjective condition of the exercise of our reason in cases where it is not seeking to know the proper estimate to form of objects arranged merely as phenomena, but is bent rather on referring these phenomena, principles and all, to their supersensible substrate, for the purpose of recognizing the possibility of certain laws of their unity, which are incapable of being figured by the mind otherwise than by means of ends (of which reason also possesses examples of the supersensuous type)."

18. Ibid., secs. 82, 88.

19. Ibid., 95.

20. Ibid., 96. See also Kant's related arguments in "To Perpetual Peace: A Philosophical Sketch," *Perpetual Peace and Other Essays*, trans. Ted Humphrey (Indianapolis: Hackett, 1983), 106–43.

21. Kant, *Critique of Judgement*, 97; see Cassirer, *Kant's Life and Thought*, 333.

22. See Michel Foucault, "What is Enlightenment?," *The Foucault Reader*, ed. Paul Rabinow (New York: Pantheon, 1984), 32–50 (italics mine).

23. In the next decades the blurring of boundaries and the exchange of properties between humans and mechanical devices became an increasingly important intellectual and moral resource for making sense of a new technical and political order. See Schaffer, "Enlightened Automata"; and Adelheid Voskuhl, "Motions and Passions: Music-Playing Women Automata and the Culture of Affect in late Eighteenth-Century Germany." *Genesis Redux: Essays in the History and Philosophy of Artificial Life*, ed. Jessica Riskin (Chicago: University of Chicago Press, 2008). For one example in the biological sciences see Lenoir, "The Göttingen School and the Development of Transcendental *Naturphilosophie* in the Romantic Era," esp. 148 on Blumenbach's solution in biology to the opposition between mechanical and teleological explanations.

24. Perpetuated by such posthumous accounts as Thomas de Quincey, *The Last Days of Immanuel Kant and Other Writings* (Edinburgh: Adam and Charles Black, 1862).

25. Schiller, *On the Aesthetic Education of Man, in a Series of Letters*, 3.

26. Willoughby's and Wilkinson's translation of the letters systematically renders *Selbständigkeit* as "autonomy." Critical discussions of Schiller frequently assimilate the notion to Kant's term "autonomy," at the cost of some confusion. See Bürger, *Theory of the Avant Garde*, 41–46 on Kant and Schiller, 6–14 on Adorno, Lukács, and Marcuse. Bürger's argument rests upon tracing back to Schiller the theoretical source for the functional differentiation of art as a separate sphere of activity (the "autonomy" championed not only by "bourgeois ideology" but by critical theorists such as Adorno).

27. Schiller, *On the Aesthetic Education of Man, in a Series of Letters*, 34, 17.

28. Ibid., 55, 109, 189.

29. Ibid., 21, 213. On the influence of Kant's critic Reinhold on Schiller's—at times quite varied—formulation of autonomy see Sabine Roehr, "Freedom and Autonomy in Schiller," *Journal of the History of Ideas* 64 (2003): 119–34.

30. Schiller, *On the Aesthetic Education of Man, in a Series of Letters*, 215.

31. Ibid., 19, 21.

32. Ibid., 33.

33. We have here a version of the central paradox of Schiller's work. The aesthetic is presented as a means of leading us upward to the world of form, but at the same time it appears as a corrective to an excess of form; similarly, Schiller's view of truth as both "pure object" and part of the world of sense may well appear contradictory. Lovejoy attributes these difficulties to Schiller's attempt to combine into a single system "the two Gods of Plato—the immutable and self-contained Perfection and the Creative Urge which makes for the unlimited realization in time of all the possible": "Since they are essentially antithetic, in any actual juncture in experience one of them must in some degree be sacrificed to the other." Schiller's wish to have it both ways, for Lovejoy, results in incoherence, though Lovejoy's sense is that for Schiller and subsequent romantics, "plenitude has the last word": the principles of perfect form must constantly be creatively realized within the world of sense. Arthur O. Lovejoy, *The Great Chain of Being: A Study of the History of an Idea* (Cambridge: Harvard University Press, 1964), 299–303.

34. Schiller, *On the Aesthetic Education of Man, in a Series of Letters*, 217, 219.

35. Schiller, *On the Aesthetic Education of Man, in a Series of Letters*, 21. Upon mention of the "political artist" we must note the influence of Schiller on Goebbels's "Führerfestschrift" *Michael*, though as I hope to have made clear, Schiller's utopia had nothing to do with fascism. Under the Third Reich Schiller's *Don Carlos* was banned, as its performances became the occasion for protests against the Nazi regime. Cf. Eagleton, *The Ideology of the Aesthetic*; Michael Jones, "Schiller, Goebbels, and Paul de Man: The Dangers of Comparative Study," *Mosaic* 32–34 (1999): 53–72.

36. See Finkelstein, "'Conquerors of the Künlün'?"; and Dettelbach, Introduction to Humboldt, *Cosmos*, vol. 2, on Humboldt's aesthetics. Cf. also Secord, *Victorian Sensation*; Arago, *Astronomie populaire*.

37. Quotes taken from Humboldt, *Cosmos*, 1:36, 26, 25.

38. Humboldt, *Cosmos* 1:76, 77. On the visibility of scientific labor see Cawood, "François Arago, savant de l'industrie"; Blondel, "Electrical Instruments in Nineteenth-Century France"; and Frederic L. Holmes and Kathryn M. Olesko, "The Images of Precision: Helmholtz and the Graphical Method in Physiology," *The Values of Precision*, ed. Wise, 198–221.

39. Dettelbach, "Humboldtian Science."

40. See Kant, *Critique of Pure Reason*, ed. and trans. Paul Guyer and Allen W. Wood (Cambridge: Cambridge University Press, 1998), 153–92, 219–66.

41. See Daston and Galison, "The Image of Objectivity."

42. See Humboldt, *Expériences sur le galvanisme, et en général sur l'irritation des fibres musculaires et nerveuses* (Paris: Didot Jeune, 1799); Dettelbach, "The Face of Nature"; Schaffer, "Self Evidence"; and Richards, *The Romantic Conception of Life*.

43. The picture adorns the cover of Andrew Cunningham and Nicholas Jardine, *Romanticism and the Sciences* (Cambridge: Cambridge University Press, 1990).

44. Letters from Humboldt to Pictet, 22 June 1798 and 7 November 1798, letter to Forell, 24 June 1799; repr. in *Lettres Américaines d'Alexandre de Humboldt, 1798–1807*, ed. E.T. Hamy (Paris: E. Guilmoto, 1904), 5, 7, 23, 39.

45. Peter Galison, "Objectivity Is Romantic," ACLS Occasional Paper 47 (2000), 83; and Daston, "The Moral Economy of Science."

46. Relevant discussions of the role of instruments in social and epistemological co-ordination include Crary, *Techniques of the Observer*; Sibum, "Reworking the Mechanical Value of Heat"; Galison, *Einstein's Clocks and Poincaré's Maps*. See also the various contributions to *Culture technique* 7 (1992) and *Osiris* 9 (1993).

47. Dettelbach, "Romanticism and Administration," 133; and Schaffer, "Astronomers Mark Time."

48. Morell and Thackray, *Gentlemen of Science*, 509–17; Cannon, *Science in Culture*, 181–96.

49. Humboldt, "Des lignes isothermes et de la distribution de la chaleur sur le globe."

50. Cawood, "Terrestrial Magnetism and the Development of International Collaboration in the Early Nineteenth Century"; Mary Louise Pratt in *Imperial Eyes* places Humboldt's natural history in the context of exploration and imperialism, as does Michael Dettelbach in "Romanticism and Administration." Unquestionably the projects of surveying, mapping, and artistically evoking the Americas were enticements for colonial and imperial adventures, many of which had dire consequences for the inhabitants of these lands. It is possible, however, to read an image like the frontispiece of *Views of the Andes*, in which a mythological figure representing western classical culture raises a fallen Aztec god to his feet, not as paternalistic bad faith but as a depiction of the late Enlightenment ideal of freedom as *reciprocity and interdependence*. The success of Humboldt in instituting this ideal is another question.

51. See Schiller to Körner, 23 February, 1793: "I know of no better image for the ideal of a beautiful society than a well executed English dance. . . . A spectator located on the balcony observes an infinite variety of criss-crossing motions which keep decisively but arbitrarily changing directions without ever colliding with each other. Everything has been arranged in such a manner that each dancer has already vacated his position by the time the other arrives. Everything fits *so skillfully, yet so spontaneously*, that everyone seems to be following his own lead, without ever getting in anyone's way. Such a dance is the perfect symbol of one's own individually asserted freedom as well as of one's respect for the freedom of the other." Schiller, *On the Aesthetic Education of Man, in a Series of Letters*, 300 (italics mine).

52. Humboldt frequently plays a role in accounts of the institution of standards and new instrumental verification; in his correspondence he served as a human relay at the intersection of multiple networks of scientific practitioners. On methods discussed here see transitional chapters by Wise in *The Values of Precision*; Kathryn M. Olesko, "The Meaning of Precision," ibid., 103–24; Schaffer, "Astronomers Mark Time"; Joseph O'Connell, "Metrology: The Creation of Universality by the Circulation of Particulars," *Social Studies of Science* 23 (1993): 129–73; and Alder, *The Measure of All Things*.

53. Humboldt rarely foregrounds the language of "objectivity" versus "subjectivity"; when he does so it is not to disparage the latter in favor of the former. In *Cosmos* he writes that "the objective world, conceived and reflected in us by thought, is subjected to the eternal and necessary conditions of our intellectual being." Humboldt, *Cosmos*, 76. Humboldt, like Schiller (and Hegel), saw the goal as an eventual fusion of the two terms. In the same paragraph Humboldt writes, "Science only begins when the spirit takes possession of substance, when the attempt is made to subject the mass of experience to rational knowledge; science is spirit turned towards nature. The external world exists for us only when we take it into ourselves and it forms itself into a view of nature." The existence of the "external" or "objective" world requires that a human being "takes hold" of or "overpowers" (*bemächtigt*) external substance (*Stoffe*) and then "forms" it into a view of nature (*Naturaanschauung gestaltet*)—the same dynamic interaction between "Stoffe" and "Gestalt" that Schiller calls the aesthetic state. Humboldt's "view of objectivity" should thus be understood, perhaps paradoxically, as a mixture between "objectivity" and "subjectivity," a tension analogous to the paradox of *autonomy* in Schiller (see note 33, above), in which the goal is the "moral state," itself a mixture of "morality" (form) and its opposite, "sense" (substance). In both, crucially, the two terms are balanced by concrete mediators: art, instruments. (Thanks to Aaron Davis for philological assistance.)

54. On Humboldt as transitional figure to nineteenth-century laboratory science see Finkelstein, "'Conquerors of the Künlün'?"; see also cameo appearances in Cahan, *Herman von Helmholtz and the Foundations of Nineteenth-Century Science*. Lenoir has argued that we see Helmholtz's findings as accounts of complex experimental arrangements of diverse interacting apparatuses, the kind of amalgamated system that Humboldt arranged on a global scale.

55. Note also the radical shift in modes of sociability, labor organization, and the image of the science that accompanied Le Verrier's replacement of Arago at the Paris Observatory; on factory discipline in the observatories of the 1860s, especially Airy, see Schaffer, "Astronomers Mark Time"; and Aubin, "The Fading Star of the Paris Observatory in the Nineteenth Century."

56. For recent work on Arago as a popularizer and politician see Aubin, "The Fading Star of the Paris Observatory in the Nineteenth Century"; Levitt, "Biot's Paper and Arago's Plates." Arago's interest in the labor and skill of science was part of a general rise before 1848 in reflection on "work" as scientific concept and "labor" as political and moral concern; key references on this development include William H. Sewell, *Work and Revolution in France: The Language of Labor from the Old Regime to 1848* (Cambridge: Cambridge University Press, 1980); Jacques Rancière, *The Nights of Labor: The Workers' Dream in Nineteenth-Century France*, trans. John Drury (Philadelphia: Temple University Press, 1989); Anson Rabinbach, *The Human Motor: Energy, Fatigue, and the Origins of Modernity* (New York: Basic, 1990); Robert Michael Brain, "The Graphic Method"; François Vatin, *Le travail: économie et physique, 1780–1830* (Paris: Presses Universitaires de France, 1993); and Hannah Arendt, *The Human Condition* (Chicago: University of Chicago Press,

1958). "Essence and emblem" is from Daston and Galison, "The Image of Objectivity," 123.

57. On Arago see M. Susan Barger and William B. White, *The Daguerreotype: Nineteenth Century Technology and Modern Science* (Baltimore: Johns Hopkins University Press, 2000), 27; F. Arago, "Le Daguerréotype," *Œuvres complètes*, 7:455–63, is commented upon by Benjamin, "A Small History of Photography," *One-Way Street and Other Writings*, trans. Edmund Jephcott and Kingsley Shorter (London: Verso, 1979), 240–57.

58. On Humboldt and the birth of environmentalism, see Aaron Sachs, *The Humboldt Current: A European Explorer and his American Disciples* (Oxford: Oxford University Press, 2007). Humboldt's deployment of instruments as an autonomous externalization of the categories is part of a more general movement of post-Kantian philosophy. While retaining the idea that knowledge is a function of the subject as much as of things, many displaced the categories from their seat in the transcendental ego: Schiller's view that the formal universals of art, politics, and science must emerge within collective material practices was one response to this crux; Schopenhauer relocated the constitutive ego in the will and physical drives, making representations of the world a function of physiology; Wilhelm von Humboldt located the structures of thought in culturally and historically variable, external, and shared languages. For a comparison of the energetic conceptions underwriting the work of the Humboldts see Peter Hans Reill, "Science and the Construction of the Cultural Sciences in Late Enlightenment Germany: The Case of Wilhelm von Humboldt," *History and Theory* 33 (1994): 345–66; on the stature of the Humboldts later in the century and their influence on Boas's concept of "culture," see George W. Stocking Jr., *The Shaping of American Anthropology, 1883–1911: A Franz Boas Reader* (New York: Basic, 1974), and Matti Bunzl, "Franz Boas and the Humboldtian Tradition: From *Volksgeist* and *Nationalcharakter* to an Anthropological Concept of Culture," *Volksgeist as Method and Ethic: Essays on Boasian Ethnography and the German Anthropological Tradition*, ed. George W. Stocking Jr. (Madison: University of Wisconsin Press, 1996).

59. Botting, *Humboldt and the Cosmos*, 268, 273.

Chapter 10

"I thought this might be of interest . . .":

The Observatory as Public Enterprise

THERESA LEVITT

In 1862 Victor Hugo set down to record what he called one of his "deepest memories."[1] Twenty-eight years earlier, in 1834, he had paid a visit to the Paris Observatory. The director, François Arago, was in, and led him to one of the large telescopes with the instruction to look through it. As Hugo remembered it, they then had the following conversation:

"I see nothing," I said.
Arago replied: "you see the moon."
I insisted: "I see nothing."
Arago maintained: "keep looking."

Arago then explained to Hugo that he had just undertaken a voyage. Where before he had been, like all inhabitants of the earth, 90,000 leagues from the moon, he was now, because of the enlarging power of the telescope, only 225 leagues away. Hugo again claimed that he saw nothing, and Arago again instructed him to keep looking. Then, said Hugo, "I followed the example of Dante with respect to Virgil. I obeyed." Hugo marveled at the vision of the moon before him. Suddenly a streak of light appeared on the dark surface. The sun was rising on the moon. As Hugo looked, Arago listed the lunar features as the light revealed them: the volcano Messala, the Promontorium Somnii, Mount Proclus, Mount Céomèdes, Mount Petavius. "There is no more mysterious spectacle than the irruption of dawn in a universe covered in obscurity," Hugo later wrote. "One seems to be witnessing the payment of a debt of infinity."

What does this episode add to our understanding of the observatory? Was Hugo simply an extraneous guest passing through without altering the observatory's fundamental structure? Was his turn at the telescope an ir-relevant diversion to fit in between the duties of the real astronomers? Per-haps. But it may be more fruitful to follow the lead of Camille Flammarion, who had no hesitations about assigning to Hugo the rank of astronomer in his publication, *Clairs de lune: Victor Hugo, astronome* (1894). Hugo had pro-fessed a lifelong passion for the study of the heavens, following closely the debates over the canals on Mars and looking into constructing a telescope of his own on his island exile of Guernsey.

If Hugo was an astronomer, we should be better off thinking of the Paris Observatory not as consisting strictly of trained scientists but as a more porous entity encompassing a wide range of people not on the observa-tory's payroll. The building on the Avenue de l'Observatoire was not a self-contained and inviolable machine works of observation but more like the center node of a far-flung network of informants. More generally, as is argued in the Introduction to this book, the observatory sciences relied in an essential way on similar networks. The republic of letters had long provided a model of coordination, and it remained so in extensive corre-spondences between observatory directors. More and more, though, the egalitarian model was replaced by hierarchies for which military organiza-tions provided not only the model but a principal resource (see for example Simon Werrett's and Martina Schiavon's contributions to this book). In the transitional period that followed the French Revolution, wide voluntary participation to observatory networks was both encouraged and channeled in specific ways for humans (as well as nonhumans, John Tresch reminds us). As is explored further in other chapters (especially by Charlotte Bigg and Ole Molvig), popularization therefore lay at the center of the scientific networking activities of some observatory directors.

At the center of the network dealt with here was the figure standing beside Hugo: the Paris Observatory director Arago. Arago's role in Hugo's story was twofold. He was chiefly responsible for allowing Hugo, whom he knew personally and politically, to have access to the space. But he also had a part to play in the act of observing itself. When Hugo saw nothing, Arago urged him to keep looking. When Hugo saw only indistinct shapes, Arago guided his sight to recognize the moon. It is useful, I think, to take these two things together: on the one hand Arago sought to open the activity of observation as wide as possible and make it a truly public enterprise, but he also tried to maintain control over the process and in so doing help the public, more generally, learn to see.

This chapter treats Arago's double agenda in reverse order. I start with his efforts at public education. Through his free astronomy course at the observatory and his widely read popular essays, he tried to mold an astronomically literate French public, with the study of the heavens as a model for rational debate. I next turn to Arago's efforts to include the public within the practice of astronomy itself. I examine several of his instructions for observing, which outlined what sort of data would be useful to the observatory, and how to go about collecting them. I also examine the role of amateur observation in one specific instance, the total eclipse of 1842.

Educating the Public

One of the first things we learn about Combeferre, the most philosophical of Victor Hugo's young revolutionaries in *Les misérables*, is that he loved to attend public science lectures, where he "learned from Arago the polarization of light."[2] This fact headed up a list of cultural markers intended to paint the portrait of a student revolutionary in Paris of the 1830s. We learn also that he read passionately the works of Claude Henri de Saint-Simon and Charles Fourier, although he could never keep the two apart in his mind, and that he believed in all the nineteenth-century dreams of progress: the railroad, the suppression of pain in surgical operations, the fixation of the image in the camera obscura, the electric telegraph.

The fictional Combeferre was not the only one to learn from Arago's public lectures. From 1813 to 1848 Arago's free astronomy course at the observatory was a local institution. Hugo and Auguste Comte spoke of going. George Sand wrote to Louis Blanc, begging him to accompany her when her usual escort was indisposed.[3] "All of Paris runs to hear them," an English paper wrote in 1840. The courses stood as a public testament to Arago's "obstinate ardor in the cultivation of what the learned in x and y call the subaltern interests of the country and humanity."[4]

Arago gave his lectures once a week before a consistently large crowd in a specially built auditorium at the observatory. Although he began teaching the course in 1813, he always traced it back to its revolutionary origins. The course was born, in principle at least, on 7 messidor Year III, or as everyone besides Arago put it, 25 June 1795.[5] The Convention created the Bureau des Longitudes on that date, and included as one of its duties the teaching of a public astronomy course. The bureau only decided to begin honoring the statute on 11 November 1812.[6] It appointed Arago to give the lectures and allotted fifteen hundred francs a year for the cause.[7] He delivered the first lecture in February 1813.

In addition to the lectures Arago also affected his campaign of public science through the bureau's publication the *Annuaire*. The bureau originally intended this publication for the use of sailors. But when Arago was put in charge he also began including general-interest articles, and its readership expanded to include a wide swath of the literate public. "The *Annuaire* of the Bureau of Longitudes," a commentator wrote, "is read throughout Europe, and the articles of Arago on lightning, steam and the most delicate questions of astronomy have given it immense vogue."[8] Stendhal counted himself "an assiduous reader," and wrote to Arago asking about his work on the various effects of moonlight.[9]

One of the crucial questions for both the *Annuaire* and the observatory lectures was the appropriate level of mathematics. Arago was adamant throughout the thirty years that he taught the course: no prior mathematical knowledge whatsoever was required of his audience. This policy became something of a war cry. Laplace, Lagrange, and most of the other first-generation members of the Bureau des Longitudes had turned astronomy into a showcase for the power of advanced analytical mathematics. Many of them claimed that the subject could not be taught without this mathematical apparatus. Arago treated his course as a proving ground against the claim of the necessity of analysis. "It is thus a great, a solemn experiment that we are undertaking together," Arago informed his audience at the opening of the course in 1846: subjecting the principles of science to the critical reason common to mankind.

In the course Arago laid out a template for reasoned discourse. Just because he assumed no prior knowledge did not mean that he would let up on the rigor. The first several lectures were spent establishing the principles of geometry, optics, and mechanics that he would use later. Arago emphasized that these principles were not part of the mathematical "obscurity" intended to "drive away the ignorant." Rather, the theorems that he used were the "the geometry, optics, and mechanics of common sense." The intent was to establish a common ground for discourse that placed every participant at an equal level. Then one could proceed according to the rules of reason. Arago acknowledged that his pedagogical stance was at heart an issue of sociability. "Clarity," he quoted from Fontenelle at the opening of his course, "is the etiquette of those who speak in public." Arago's requirement that his course be comprehensible to any intelligent listener was a statement about how people should interact as a public.[10]

The "public opinion" that Arago sought to shape with his astronomy lessons was a category that had crystallized in the late eighteenth century and the early nineteenth.[11] The term "opinion" shed its connotation of uncertainty and came to stand for the critical judgments of a public capable of relying on its own reasoning power. It had above all a political resonance. Public opinion would serve as a check on state power. In 1813, when Arago began his lectures, the public participating in civic discourse offered its critique from a position largely outside of state power. With the expansion of constitutional representation in 1830, this began to change.

The July Monarchy had opened as a great experiment in incorporating public opinion into the framework of the state. Within two years, however, a palpable unease had spread. "The year 1832," Hugo wrote, "began with an air of ominousness and looming danger."[12] The papers began warning of a comet whose scheduled return brought it dangerously close to the earth. The comet in question had first been spotted in late 1826 by the astronomer Wilhelm von Biela. Soon after, the director of the Marseilles Observatory, Gambert, calculated its trajectory and determined that it had a period of six years and nine months.[13] Although its trajectory could be calculated exactly using celestial mechanics, its presence came wrapped in a discourse of portentous destruction. A first pass at the calculations revealed that the comet would pass directly through the ecliptic of the earth's orbit, and the alarm went out that it might run into the earth. Even if there were no collision, the papers mentioned the possibility that the earth could pass through the nebulous material making up the tail of the comet.

These dire forecasts closely followed another comet scare. In January 1831 Arago had reported at the Academy of Sciences that one of his provincial correspondents had spotted a new comet around the constellation of Ophincus.[14] Shortly after, the opposition paper *Le National* picked up the story and continued tracking the comet's progress through the sky.[15] A nervous concern over the unknown consequences of the earth's passage through the nebulous tail of the comet coincided with reports of a mysterious dry fog appearing throughout Europe, a series of disastrous crop failures, and the worst outbreak of cholera ever seen.

Over the next twenty months France seemed under the influence of some dark star, as nature itself seemed to conspire against the fledgling regime. Poor weather led to crop failures. Wheat prices soared. Grain riots were endemic. Protests broke out against the forestry administration. The stock market collapsed with a spectacular crash. Things only went downhill from there, sliding into what became known as the great fear of 1832.[16] Agricultural and economic crisis became mixed with meteorological aberrations. Earthquakes in Italy were paired with political revolution there. Mysterious

fires broke out in western France and were never fully explained.[17] Sometimes farmers blamed possessed young girls. The papers in 1831 covered the
story of "la fille Choleau," a young girl blamed for starting at least two fires.
She had admitted to setting the fires, claiming that she had been bewitched
and pushed by an evil spirit to do the deed. Yet at the trial she remembered
neither the fires nor the confessions. These "bizarreries" were all the more
likely, the council pointed out, as the girl was several months pregnant.[18]

By far the biggest natural disaster of 1832 was the outbreak of cholera.
The epidemic claimed 18,000 people (12,733 in April 1832 alone).[19] It almost
seemed to target the liberal opposition. Casimir Périer, the prime minister
and one of the original deputies to support the July Revolution, died in
May. General Lamarque, an opposition leader hailed as the regime's greatest protector of the people, soon followed. "For the past month," wrote the
papers on 30 April, "[cholera] alone has governed."[20]

General Lamarque's funeral, on 5 June 1832, became the occasion for
Paris to again erupt into violence. The next two days, immortalized in Hugo's Les misérables, had some of the worst barricade fighting that France had
ever seen. In 1830 the Chamber of Deputies had used a similar eruption of
violence to enact a change in regime. In the first moments of the action in
1832, however, the Chamber vacillated on whether to embrace or repudiate
the revolutionaries of the street.

The question was particularly pointed for Arago. He had gained a seat in
the Chamber after 1830 and sat on the extreme left as one of the most radical of the opposition deputies. These deputies of the opposition, roughly
thirty in number, met at the house of Lafitte on 6 June 1832 to discuss what
position they would take with respect to the insurgency. They elected three
of their members, Laffitte, Odilon Barrot, and Arago, to go speak to the
king about ways to end the violence.[21] The three spent an hour and a half
with the king, during which they performed the acrobatic task of representing the people while rejecting their actions.

The riots of 1832 were a disappointment for Arago. Shortly after, he
wrote to John Herschel, asking forgiveness for his delays in correspondence: "my damnable functions as a Deputy imposed quite onerous duties
upon me during the deplorable events for which Paris has been the theater!"[22] The events of 1832 pointed to perhaps the greatest danger facing a
constitutional government: the possibility that public opinion would not
be sane and reasonable. To address this problem Arago turned to what he
knew best: astronomy.

In 1832 Arago began a massive campaign to debunk pernicious forms of ignorant superstition.[23] His favorite topic was comets. He first tackled the subject with an article of several hundred pages that dominated the *Annuaire* of the Bureau des Longitudes of 1832. The daily papers, he claimed, were spreading fear with their announcements that the comet would imminently strike the earth and smash it to bits.[24] He felt compelled to counter these claims with "everything that science has uncovered" about the trajectory of the comet, "whose proximity, we are assured, will surely be so fatal to Earth and its inhabitants."[25]

It was true, Arago admitted, that according to the equations of celestial mechanics the comet would pass directly through the ecliptic of the earth's orbit. Yet a more thorough investigation, he added, revealed that there was no cause for fear. One could calculate the point at which the comet would cross the ecliptic, which turned out to be a distance equal to four and a third earth radii away from the orbit of the earth. This distance was certainly close enough to allow for the possibility that the earth would at least pass through the nebulous portion of the comet. But, Arago pointed out, the comet was scheduled to pass by this spot on 29 October, while the Earth would not get there until 30 November. Given that the earth traveled at a speed of 674,000 leagues a day, one could see that it would never be less than twenty million leagues from the comet.[26]

And thus the comet's nebulosity would not transmit a physical effect to the earth. Was there any other way the comet might influence worldly events? Popular prejudice said there was, and a host of incidents were cited when notable harvests followed comet sightings. Arago set out to show the error in this view. First, he asked what kind of influence a distant comet could have on the earth. It was possible that it exerted a sensible gravitational pull as it passed by. Such an effect would have very limited consequences on earth, though. It could perhaps alter the tides a bit, but it would hardly bring about the change in temperature needed to affect crops. One might then think of the various radiations given off by the comet, and the additional heat they might provide. But here Arago pointed to the negative results of his research on the calorific effects of moon rays, adding that the light from a comet was much weaker in intensity than that of the moon.

A few years later the return of Halley's comet once again focused public attention on the sky. Historians of astronomy have assigned to this comet a particular place in a story linking the rise of celestial mechanics to a decline in celestial portentousness. Halley's comet was after all the first to have its orbit worked out according to Newton's laws of mechanics and gravitation. Eighteenth-century geometers hailed its return in 1759, within days of its predicted arrival, as the most convincing proof available of the application of mechanical law to the heavens. This triumph did not necessarily dampen the public's enthusiasm for speculating about comets. Simon Schaffer has pointed to the many ways in which the cometographers' authority as prognosticators increased after 1759.[27] Comets frequently retained their reputations as harbingers of doom. In France, Lalande published *Réflexions sur les comètes qui peuvent approcher de la Terre* in 1773 amid great interest and terror. Cometary orbits suffered marked variations, making it impossible to rule out collisions. The memoir was not read in the public meeting of the Academy of Sciences on 21 April 1773, and this caused even greater fear, contributing to the *grande peur* that spread through France that year.

On Halley's return in 1835 Arago rushed to satisfy the public's curiosity and calm its fears in both the *Annuaire* and the observatory lectures.[28] He denied the widespread rumor that the comet had a noticeable effect on the year's harvest. Farmers had pointed out that the months of October and November were particularly mild in 1835, just as Haley's comet was visible in the sky. As the comet disappeared in December, a bitter cold set in. Some drew the implication that the comet provided additional heat. And yet, Arago warned, this account was not complete. The comet was still quite close to the earth in December, even if the public could not see it. It would have just passed by its closest point to the sun and had therefore, according to comet theory, replenished its heat for the long voyage out. This would mean that the comet had somehow heated the earth when it itself was cool, and cooled the earth when it was hot. Furthermore, Arago presented the results of his polariscopic measurements to show that the light of comets was entirely reflected and thus originated in the sun.[29] He would have full confidence in the power of these arguments, he claimed, if it were not for the sad fact that when it came to meteorology, far too many chose simply not to listen to arguments. "Nearly everyone is a perfect mimic of the famous Abbé Vertot, with their siege irrevocably done." Arago referred here to a well-known anecdote about the eighteenth-century historian Vertot declining to incorporate new evidence about the siege of Rhodes into his history of the Order of Malta because, as he put it, "My siege is done."[30] Arago's complaints about *what* people said about comets were thus tied to the *way* people talked about them. Namely, that those who attributed mysterious effects to them did not subject their claims to rational debate.

The particularly dramatic comet of 1843 excited even more speculation. This comet fell in the "unexpected" category, as no one had predicted its arrival. It was moreover a phenomenon of striking brilliance, unmistak-

42. An ad for a pamphlet entitled "The Comet: Past, Present, Future." On the left is a comet carrying water buckets entitled "The Comet of the Flood." On the right is a comet pulling wine casks entitled "The Comet of 1811," presumably recalling the year's exceptional harvest. *L'Illustration* 1, no. 7 (15 April 1843): 111.

able in the night sky and even visible in daylight. On top of everything, it coincided with a devastating flood in the center of France, and an even deadlier earthquake in the French Antilles. Tracts such as *The Comet: Past, Present, Future* appeared for sale, promising "revelations, opinions, predictions, whims" (figure 42).[31] This pamphlet showed a coarse-faced comet in the dress of a peasant girl; behind her streamed unruly *chevelure* (the word for hair and a comet's tail was the same), and the caption read: "The comet of the flood." The girl held two buckets of water, presumably to dump upon central France.

More expensive journals such as *L'Illustration* treated the comet very differently. They printed a star map of a small section of the sky that would allow its readers to track down the comet for themselves. *L'Illustration* chastised the population for interrogating its astronomers "with an eagerness that has not always been enlightened." But it nonetheless praised members of the public for elevating their minds by contemplating the great laws of nature (figure 43).[32] The journal then presented Arago's report of his observations, including an account of the readings given by the polarimeter.

Arago also devoted an issue of the *Annuaire* to the comet of 1843. "The common people [*le vulgaire*]," he pointed out, blamed it for everything from floods in the Midi to earthquakes in Guadeloupe. He assured his public that those responsible for "inventing or propagating" these complaints were "certainly strangers to the most elementary notions of science." They were not (and this was their fatal flaw) able to "produce an argument either good or bad" for their claims. Arago compared them to the Abyssinians, reported to tremble in terror at the great length of the comet's tail, or the Mexicans, who he claimed saw a comet as portending an imminent bonanza. He told

43. A star map allowing readers to locate the comet of 1843. Attached was a description of Arago's observations, including his polarimetric measurements concluding that the comet shone with its own light, and not that reflected by the sun. *L'Illustration* 1, no. 4 (25 March 1843): 64.

of his own research on the subject. A battery of meteorological data from around France gave no indication of anything unusual in the weather. He had also used the observatory's most sensitive thermometric instruments to test the light from various parts of the comet, and found that none of them produced a sensible calorific effect.[33]

Also in 1832 Arago took up in both the *Annuaire* and his observatory lectures another issue that he felt was overly influenced by peasant superstition: the physical effects of moon rays on terrestrial phenomena. Arago spent a considerable time in his lectures demystifying the effects of the moon. He directed his efforts at aligning the opinions of the public and the scientific élites while not flat-out rejecting the claims of the general populace. He began by setting up the division of opinion that lay at the heart of the problem: "Astronomers, physicists, and meteorologists seem generally convinced that the moon exercises no appreciable influence on our atmosphere; but it must be said that they are alone in this opinion. The immense majority of the public believes firmly in a powerful action of our satellite."[34] Arago proceeded to list several of these "powerful actions" that the public attributed to the moon, and examined their reasonableness. He found that several of them were not without merit.

Arago's principal claim was that the light of the moon was composed strictly of visible rays. He performed "the most delicate experiments" possible to determine that this light had neither calorific nor chemical properties.[35] To test for the presence of calorific rays, he designed a special apparatus that collected the moon light with mirrors, which was sent down a curved tube with two glass balls at the end. One of these balls was transparent, the other black. In between them was a colored liquid. If there were any heat associated with the light rays, Arago reasoned, the black ball would warm up faster than the transparent one, the air inside would expand, and the colored liquid would be displaced in the direction of the transparent ball. The device could detect a change in temperature as small as a thousandth of a degree, Arago reported, and he took the negative results of the experiment as conclusive proof that the moon had no calorific effect. Equally conclusive were the experiments denying the moon's chemical effect. Here Arago reported that silver hydrochloride, a substance that darkened immediately in sunlight, underwent no change from the light of the moon.

"However," Arago continued, "credulity has attributed moonlight with a great effect on the products of agriculture." The red moon in particular had a certain "wretched celebrity." Farmers credited the light of this moon with freezing their early buds and killing off the tender sprouts. Yet this effect, Arago claimed, could be explained without recourse to celestial interference. The red moon began in April and ended in May. During this season temperatures were often only 4 or 5 degrees or so above freezing. During the night crops could easily radiate off enough heat to drop their temperature 7 or 8 degrees, which would leave them well below freezing. The trick was that this radiation only occurred on clear nights, as the clouds would otherwise keep the radiated caloric heat from escaping into the atmosphere. The moon was of course only visible on these cloudless nights, and thus was blamed for the damage. The precautions that the farmers took against the moon rays, such as covering their crops with hay or other materials, worked well because they happened to prevent the plants from radiating their heat away.

Arago did not intend for his public audiences and readership to remain in "the passive role of contemplators." He felt rather that the proper form of education could allow them to be useful and active participants in the observatory's operations. The observatory was responsible, after all, for a complete survey of the heavens and earth, requiring observations from all points on the globe. Arago relied at least in part on the efforts of a well-instructed public to provide these observations (figure 44).

Arago wrote up instructions outlining both what sort of observations he was interested in and how these observations should be made. At the beginning of one of these sets of instructions, Arago invoked the eighteenth-century monument to accumulated knowledge, the *Encyclopédie*. He recounted that someone had once complained to d'Alembert about its size, only to receive the reply that the reader was lucky he did not have an encyclopedia of everything that was *not* known in front of him, as that would no doubt run into the hundreds of volumes. This response, Arago felt, was "more clever than exact." Although there were no doubt many things that people didn't even know they didn't know, there were also "many important questions, well defined, well characterized, that one can, with confidence, recommend to observers." This was the project he undertook in 1835, when he wrote up a set of instructions for the captain of the *Bonite* to make meteorological measurements as the ship completed its circumnavigation of the globe. "It thus appeared to me," he wrote, "that this genre of publication could become extremely useful, that a *foule* of people *instruites and désœuvrées* would receive from it an excitation that would transform them from the passive role of contemplators to the top ranks of militant science."[36] For this reason Arago published the instructions, giving detailed directions on how to record data in such a way that they could be usefully compiled by the observatory.

The instructions ran to several hundred pages, detailing such things as when and how to take readings with thermometers and barometers, and which features to look for and record with phenomena such as meteor showers and the aurora borealis. Arago repeated this procedure several times, first writing up instructions for the captains of ships sailing around the world, then publishing the instructions with the thought that others might be able to contribute their observations as well.

Arago received hundreds of letters at the observatory from around the globe. Typical was the one sent by Prosper Mérimée in 1840. He had been

44. An example of public participation in astronomy. The child begs, "Papa, let me look." The father, looking at a kite through his telescope, responds, "Quiet, I see the nucleus! We beat the Observatory!" *L'Illustration* 1, no. 5 (1 April 1843): 77.

traveling in Corsica and came across a meteorological phenomenon that he judged "could be of some interest" to Arago.[37] He arrived in Sollacaro on 2 September 1839, just in time to experience a violent storm of rain and hail that lasted two hours. After the storm had ended, around 8:00 in the evening, he went out to the terrace of his friend's house to smoke a cigarette. He thought at first that the terrace abutted a giant white wall. Shortly he realized that the terrace actually looked out over a sizable valley and that what he was witnessing was a wall of very thick fog. Mérimée then proceeded to perform a series of experiments, placing his friends and the lamp in various positions and noting what the shadows looked like on the wall of fog. He drew up a little diagram that he included in his letter to Arago, along with a detailed description of the design, oil, and wick of the lamp he used. Nothing is said about how Mérimée expected these observations to be of use to Arago. It was simply assumed that merely logging the occurrence of this unusual situation was an important step in the comprehensive cataloguing of the workings of the globe.

To see how these events played out in a particular instance, I shall focus

45. Path of the eclipse of 1842 in southeast France. Arago, *Œuvres*, 9:140.

on the eclipse of 1842. A total eclipse of the sun was scheduled to pass through southern France, cutting a swath across the Midi from the Pyrenees to the Alps (figure 45).

The event was a convenient one for Arago. One of the cities on the path of the eclipse was Perpignan, his hometown and the district he had represented in the Chamber of Deputies from 1830. And 1842, it turned out, was an election year, with Arago's seat up for grabs.[38] The observing expedition that he arranged for himself and his élèves astronomes coincided with campaign celebrations.

Arago left Paris on 22 June, accompanied by his assistants Eugène Bouvard, Ernest Laugier, and Victor Mauvais.[39] Arago, Laugier, and Mauvais each set up a telescope on the commander's terrace of the citadel at Perpignan. Bouvard went to Digne to observe the eclipse from another location (on his return, it was discovered that he had made an error in calculation, rendering his observations worthless).

If Arago and his students had Perpignan covered, there was still the question of the rest of the eclipse's path. For this Arago turned once again to the French public. He wrote up a set of instructions to those viewing the eclipse on their own, which he published in the widely read *Annuaire*, as well as in the *Comptes rendus* of the Academy of Sciences and the *Annales de chimie et de physique*. Arago began with a description of the phenomenon of the total eclipse, and a list of known eclipses throughout history.[40] There had been only one total eclipse seen in France in the eighteenth century, and the one in 1842 was the only one due to occur in the nineteenth century, so Arago looked toward the opportunity for data collection with some excitement.

The bulk of Arago's article was a section entitled "Réflexions et recommandations soumises aux observateurs."[41] He listed dozens of questions regarding eclipses that were still unresolved by astronomers and could be fruitfully investigated by anyone on the path of the eclipse. For example, he stated, it was of central importance to know whether the corona was centered on the moon or on the sun. Past observers disagreed on this issue, which was crucial for deciding whether a solar atmosphere existed. The question could be resolved, Arago claimed, by carefully measuring the size of the corona at both the beginning and the end of an eclipse. He specified the instruments appropriate for these measurements: reflection telescopes, *lunettes de Rochon* with birefrigent prisms, or refraction telescopes of moderate enlarging power whose lenses have the minutes marked with fine wires. Each of these methods would have its advantages, and all would complement one another.

Arago recommended many other observations that could be made without any particular equipment. "It goes without saying," he wrote, "that in each location one will seek to determine the number and size of the stars that become visible to the naked eye during the total darkness."[42] This observation would be a means for determining just how dark the darkness was. Another observation of interest was the slight coloration that the atmosphere and terrestrial objects took on. Here Arago gave a special warning to amateurs to guard against the effects of color contrast, and to be very careful about any light sources that they might have in the vicinity.

Arago collected the diverse sets of observations and published the conclusions he drew from them in his popular textbook on astronomy, *Astronomie populaire*.[43] Typical of the dozens of letters received at the observatory regarding the eclipse was that of M. Laurens, from Montpellier, who wrote to report on his observations of the behavior of livestock during the eclipse (the possible effect of the eclipse on animals had been mentioned in Arago's initial instructions).[44] Arago also reported the account of M. de Costa, who performed an experiment with his dog: he refrained from feeding the dog on the eve and morning of the eclipse. Shortly before it was to start, he threw the dog a piece of bread. The dog, he reported, jumped on it hungrily but dropped it as soon as the sun disappeared. He began eating again as soon as the light reappeared.[45]

Arago also used the measurements of the time of the eclipse to correct

Prédiction des principales circonstances de l'éclipse totale de Soleil visible dans le midi de la France, dans la matinée du 8 juillet 1842.

	Lever du Soleil.	Commencement de l'éclipse.	Commencement de l'éclipse totale.	Fin de l'éclipse totale.	Fin de l'éclipse.	Plus courte distance des centres.
	h. m.	h. m. s.	h. m. s.	h. m. s.	h. m. s.	
Perpignan..	4 31	4 53 21	5 46 14	5 48 28	6 45 47	1″.8
Montpellier.	4 28	4 57 53	5 51 20	5 53 12	6 51 6	21 .5
Marseille...	4 29	5 3 4	5 56 50	5 58 50	6 57 12	18 .5
Digne.....	4 26	5 7 12	6 1 8	6 3 28	7 2 3	1 .4

46. Table of predicted times for the eclipse of 1842. F. Arago, *Œuvres*, 9:141.

his own moon tables. He had printed, in the *Éphémérides*, a prediction of the times when the eclipse would achieve totality (figure 46).

Arago found that the measured times differed systematically from the predicted times by about thirty or forty seconds. The effect was consistent enough that he felt it warranted a revision in the measure he had been using of the length of the moon's radius. There was, he reported, only one instance in which the reported measurements did not match up with the others. This was the report from Narbonne, provided by a group of "zealous amateurs."[46] In proposing possible reasons for this discrepancy, Arago settled upon the explanation that the problem had been with the synchronizing of the chronometer. The only instruments mentioned in the report, he stated, were reflecting and refracting telescopes, and polariscopes. Although Arago did not question the ability of the zealous amateurs to use these instruments, he did point out that none of the devices could serve to determine the time, and for this reason he felt comfortable disregarding the anomalous readings.

After Napoleon III assumed power after his coup, he went about completely restructuring the Paris Observatory. Long known as a nest of Republicanism, it would be remade into a properly imperial institution. The first step was to get rid of Arago, a step obviated in 1854 when Arago died after an extended illness. Also implicated in the observatory's republican character was the group of *élèves astronomes* responsible for recording astronomical data. They regularly attended reform banquets and radical political meetings. One of them, obligingly enough for future historians in

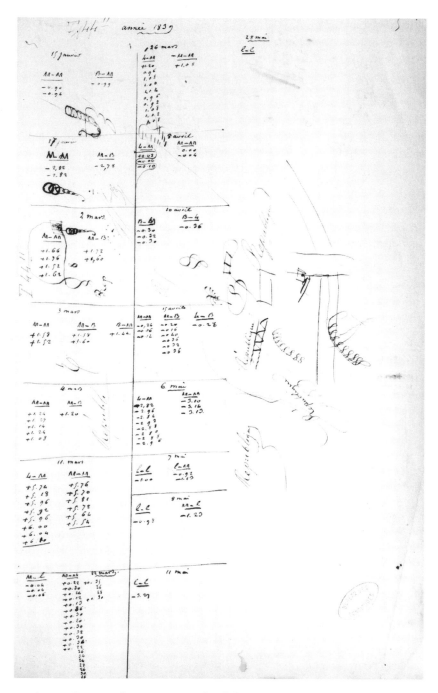

47. A page from an observatory notebook for 1839. By permission of the Bibliothèque de l'Observatoire de Paris. Note the doodles spelling "République" on the right-hand side.

search of smoking guns, seems to have passed the time when not looking through his telescope by executing doodles that betrayed his political allegiances (figure 47).

But is there a way to talk about the Republican observatory that does not simply make reference to the Republican politics of its employees? I think there is, but it involves looking beyond the observatory as a fixed spatial location with limited personnel. If I have chosen to concentrate on Arago's efforts to involve the public in the operations of the observatory, it is because I think that for him the public was the platform for both his astronomical and political visions. Arago saw his public instruction as a means of shaping the rational critical French public necessary for the proper functioning of a republic founded on public opinion. His willingness to allow this public into the activity of science, and his easy confidence in the reported observations of his distant correspondents, rested upon the possibility of universal communicability. Although both Auguste Comte and Adolphe Quetelet, two heralds of personal difference in observation, studied astronomy under Arago, Arago himself showed little interest in how one person might see differently from another. Transparency was the cornerstone for the observatory that Arago directed and the Universal Republic of 1848 that he helped to found. The rapid dismantling of both, in the middle of the nineteenth century, heralded the end of the dream of easy coordination.

Notes

1. Victor Hugo, *Promontorium somnii* (Paris: Les Belles Lettres, 1961).
2. Hugo, *Les misérables*, 664.
3. George Sand, *Correspondence*, ed. Georges Lubin (Paris: Garnier, 1964–95), 26:37. The letter, which has no heading, states: "Je voudrais votre bras si votre bras est libre, pour me conduire jeudi prochain au cours de Monsieur Arago." The attribution to Louis Blanc is by the editor, who also dates the letter to the end of 1844.
4. Both quotes above are from Loménie, *Sketches of Conspicuous Living Characters of France*, 281, 273. The preface notes that the sketches are translations of a series that appeared weekly in newspapers in Paris in 1840 and were signed "un homme de rien."
5. Arago, *Astronomie populaire*, 1:xi.
6. Procès-Verbaux of the Bureau des Longitudes, 11 November 1812, Archives et Bibliothèque de l'Observatoire de Paris, 87.
7. Archives Nationales, F^{17} 3704.
8. *Sketches of Conspicuous Living Characters of France*, 281.

9. Stendhal, *Correspondence*, ed. Henri Martineau (Paris: Le Divan, 1933–34), 3:200–201.

10. All quotes from this paragraph are from Arago, *Astronomie populaire*, 1:v, iii, x.

11. Jürgen Habermas, *The Structural Transformation of the Public Sphere*, trans. Thomas Burger (Cambridge: MIT Press, 1991), 89–102; the notion of an *opinion publique* has been particularly well studied in the pre-Revolutionary period. See for example *Habermas and the Public Sphere*, ed. Craig Calhoun (Cambridge: MIT Press, 1992); David A. Bell, *Lawyers and Citizens: The Making of a Political Elite in Old Regime France* (New York: Oxford University Press, 1994), Arlette Farge, *Dire et mal dire: l'opinion publique au XVIIIe siècle* (Paris: Le Seuil, 1992), Dena Goodman, *The Republic of Letters: A Cultural History of the French Enlightenment* (Ithaca: Cornell University Press, 1994).

12. Hugo, *Les misérables*, 859.

13. Arago named it after the man who calculated its return. Many people chose to call it Biela's comet, after the man who first spotted it. Arago, *Astronomie populaire*, 3:297.

14. "Académie des Sciences, Séance de 10 janvier," *Le National*, 12 January 1831, 1.

15. *Le National*, 14 January 1831, 3.

16. André Jardin and André-Jean Tudesq, *Restoration and Reaction, 1815–1848*, trans. Elborg Forster (Cambridge: Cambridge University Press, 1987), 106–7.

17. Paul Gonnet, "Esquisse de la crise économique en France de 1827 à 1832," *Revue d'histoire économique et sociale* 33 (1955): 249–92.

18. *Le National*, 11 January 1831. A quick look at how Arago handled another case of mysterious fires reveals his recurring strategy to use astronomy to dispel peasant superstition. In 1842 a justice of the peace in Montierender faced a similar problem when a number of grain stacks in his area burst into flames with no explanation. This outbreak coincided, he observed, with reports of fiery orbs falling from the sky. He concluded that these must be aerolithes or meteors. He wrote to Arago suggesting his hypothesis and providing several examples of its probable occurrence. Arago enthusiastically accepted this explanation. He read the judge's letter before the Academy of Sciences and urged all magistrates to consider this possibility before launching their proverbial or literal witch hunts. *L'Illustration* 1 (1843): 254.

19. M. Lucas-Dubreton, *La Grande Peur de 1832: le choléra et l'émeute* (Paris: N.R.F, 1932).

20. "Chronique de la Quinzaine, 30 avril 1832," *Revue des deux mondes* 6 (1832): 373–75, at 373.

21. Different "transcripts" exist of this interview. Arago provided one, which exists as a manuscript in his family papers. Arago, "Entretien avec le Roi, 6 juin 1832," Archives Nationales, Papiers de la famille Arago, MI 372. Étienne Cabet also provided a transcript that he based on the discussion of the three men after their interview. Étienne Cabet, *Faits préliminaires au procès devant la cour d'assises contre M. Cabet. 4ème partie: conférence du 6 juin 1832, entre S.M. Louis-Philippe et MM. Laffitte, Odilon-Barot, Arago* (Paris: Rouanet, 1833).

22. Arago to John Herschel (8 July 1832), Royal Society Archives, H.1.356.

23. His notebooks of that year contain reflections on the probability of a number of commonly believed superstitions, Archives et Bibliothèque de l'Observatoire de Paris, BL MSS z5(1).

24. Arago, "The Comet: Scientific Notices of Comets in General and in Particular the Comet of 1832," *New World* 65 (March 1843): 1.

25. Arago, "Les Comètes," *Astronomie populaire*, 3:292.

26. Ibid., 295.

27. Schaffer, "Authorized Prophets."

28. Arago gave reports on the comet's progress and properties at nearly every séance of the Academy of Sciences from August through December; *Comptes rendus des séances hebdomadaires de l'Académie des Sciences* 1 (1835): 40, 66, 87, 96, 129, 130, 235, 255, 256, 322. He also wrote an account in *L'Annuaire*, 1836, reproduced in Arago, *Œuvres*, 9:481.

29. Arago, *Leçons d'astronomie professées à l'Observatoire Royal, recueillies par un des ses élèves*, 4th edn (Paris: Chamerot, 1845), 352; *L'Annuaire*, 1836.

30. C. Volpilhac-Auger, *Histoire de l'ordre de Malte, ou histoire des chevaliers hospitaliers de St-Jean de Jérusalem, appelés depuis chevaliers de Rhodes* (Paris, 1726).

31. *L'Illustration* 1 (1843).

32. "La comète," *L'Illustration* 1 (1843): 64.

33. All quotes above are from Arago, *Leçons*, 335, 353–54; also in *L'Annuaire*, 1844, and Arago, *Astronomie populaire*.

34. Arago, "De l'influence de la lune sur les phénomènes terrestres," *Astronomie populaire*, 8:25.

35. All quotes in this paragraph and the next one are from Arago, *Leçons*, 253–54.

36. All quotes from Arago, *Œuvres*, 9:2–3.

37. Prosper Mérimée, *Correspondance générale*, ed. Maurice Parturier with Pierre Josserand and Jean Mallion (Paris: Le Divan, 1941–64), 2:309.

38. Daumas, *Arago*, 226.

39. Procès-verbaux of the Bureau des Longitudes, Archives et Bibliothèque de l'Observatoire de Paris, MS 1022, p. 182.

40. Arago, "Sur l'éclipse totale de Soleil du 8 juillet 1842," 843.

41. Ibid., 853–61.

42. Ibid., 857.

43. Arago, *Astronomie populaire* 3, 537–624.

44. Laurens to Arago, 8 July 1842, Archives et Bibliothèque de l'Observatoire de Paris, Bureau des Longitudes MSS, z12, Correspondance.

45. "7 juil. 1842," Archives et Bibliothèque de l'Observatoire de Paris, Bureau des Longitudes MSS, z12 (2), 52.

46. Arago, *Œuvres*, 9:157.

Chapter 11

Staging the Heavens: Astrophysics and Popular
Astronomy in the Late Nineteenth Century

CHARLOTTE BIGG

It now and then happens in the history of the human race upon this planet,
that one particular generation witnesses the most stupendous advancement
of knowledge, this advancement generally coming from what one might consider an exceeding small germ of thought. You will at once call to mind several such instances. You will recollect how once a Dutchman experimenting
with two spectacle-glasses produced the Telescope; and how the field of the
known and the knowable has been enlarged by the invention of that wonderful instrument. Again, you recollect how once Sir Isaac Newton was in a
garden and saw an apple fall, and how the germ of thought which was started
in his mind by that simple incident fructified into the theory of universal
gravitation. You will also acknowledge that each step of this kind has more
firmly knit the universe together, has welded it into a more and more perfect
whole, and has enhanced the marvellous beauty of its structure.

I think that future times will say that either this generation or perhaps the
next, is as favoured a one as that which saw the invention of the telescope or
the immortal discovery of Newton: for as by the invention of the telescope
the universe was almost indefinitely extended; as from Newton's discovery
we learned that like forces were acting in like manner everywhere; so in our
times does the wonderful instrument called the Spectroscope show us that
like matter is acting like matter everywhere; so that if matter and force be
not identical, then these two namely, matter and force, may be termed the
foundation stones of the universe in which we dwell.

—J. NORMAN LOCKYER (1874).[1]

Celestial Chemistry

On Tuesday, 16 December 1873, Joseph Norman Lockyer gave a public lecture on "Celestial Chemistry" at the Quebec Institute, London. Some of those present may have been attracted by the rising fame of the speaker, who had recently been made a fellow of the Royal Society and was the founder and editor of a fledgling but respected scientific journal, *Nature*. Others may have expected to hear of Lockyer's latest research in solar spectroscopy, since public lectures at the time were often used to announce new results. Some members of the audience were perhaps more interested in the pleasures of displaying and mingling which such an occasion provided, not unlike an evening at the theater. A number of listeners might have hoped to learn something of the new science of spectrum analysis which Lockyer, William Huggins, and Henry Enfield Roscoe had helped to import from Germany into Britain in the early 1860s, heralding a fashion for prisms and spectra which lasted well into the 1870s in polite and scientific circles alike.[2] If the speaker was right in suggesting that spectroscopes would contribute to a new understanding of the universe as momentous as that brought about by telescopes and Newton's theory of gravitation, then, some may have hoped, a corner of the veil of Nature might be raised on that evening.

The speaker obliged to the varied expectations of his audience. He balanced technical details with general metaphysical considerations; moved seamlessly from explanations of the basic principles of spectrum analysis to his own latest speculations on the dissociation of atoms in the sun, combining rhetorical *envolées* and direct appeals to his audience; and performed experiments and demonstrations.

The focus of this occasion was a new form of scientific investigation which Lockyer referred to indifferently as celestial chemistry, solar physics, and the new astronomy, and which would later be known as astrophysics. As Lockyer emphatically announced, spectroscopy had opened a new field of inquiry, the investigation of the physical and chemical properties of astronomical bodies. He was referring to Robert Bunsen's and Gustav Kirchhoff's elaboration of the principles of spectral analysis in 1859, conventionally acknowledged as the founding moment of astrophysics. Practiced in diverse ways and in different settings throughout Europe and the United States, this new science in the 1860s and 1870s was in the process of forging its own methods, institutions, and identity.

Early astrophysics occupied a transverse position between three types of spaces: the laboratory, the observatory, and the eclipse expedition.[3] To this trilogy should be added the lecture hall, together with such similarly public spaces as newspaper columns. These public spaces constituted essential sites for the establishment of the new astronomy, especially in Victorian Britain. They were neither used by early astrophysicists to simply disseminate a knowledge produced elsewhere, nor to attract support for the new science. Rather, they were integral to the practice and identity of astrophysics: this was where the new science was not only communicated but in part elaborated, articulated, and conceptualized. That astrophysics could take this form owed much to the existence of new urban, middle-class patrons, consumers and practitioners of science.

This chapter suggests that early astrophysics of the type practiced by Lockyer (one could also cite William Huggins, Jules Janssen, or Max Wolf) had much in common with popular astronomy of the late nineteenth century. The period from roughly 1860 to 1910 was one of profound transformation for the sciences of the observatory, one that also included the emergence of astrophysics and the establishment of a great number of popular observatories and Uranias. Popular astronomy can be conceived of as a matrix from which a number of specialized constituencies progressively emerged in the period: astrophysicists, science popularizers, and amateur astronomers.[4] As popular science became the domain of professional museologists and popularizers, and scientists themselves increasingly shut the public out of their laboratories and observatories, the fate of amateurs was sealed in the early twentieth century, as they were pushed to the margins of the scientific establishment.[5] In its most fertile decades, however, popular astronomy supplied an imaginative, playful, and participatory alternative to established science, with its own take on observatory techniques and narratives. Lockyer stands here for a wider class of popular scientists who appropriated observatory science for their own pursuits and ultimately contributed to change practices inside the observatory.

But let us first have a closer look at this new astronomy. For this we can turn to the opening words of Lockyer's lecture of December 1873, which point to some of the key themes in his conception of what he called celestial chemistry. Lockyer began with a quick survey of past achievements in astronomy that emphasized scientific progress. This was a common notion in the discourse of Victorian Britons, often expressed in conjunction with the new technologies that were transforming their daily lives. But progress also had religious connotations in a society that regarded self-improvement and the personal quest for knowledge as moral duties.

In linking scientific progress to the history of humanity, Lockyer was also drawing on the ideas of Charles Darwin, whose *Origin of Species* was published in the same year that Bunsen and Kirchhoff elaborated the principles of spectral analysis. Like other public exponents of science, including his friend T. H. Huxley, Lockyer appreciated Darwin's theory for its all-encompassing quality, and for proposing a general principle for understanding the organization of the world, a theology in which Nature had taken God's place. Subsequently Lockyer even developed a theory of "inorganic evolution" that generalized Darwin's scheme to the celestial ecosystem: "life in its various forms on this planet, now acknowledged to be the work of evolution, [is] an appendix, as it were, to the work of inorganic evolution."[6] Like early biologists, early astrophysicists explored the boundary between the inorganic and organic realms. They did not exclude the possibility that the sun itself might be part of the living, witness the controversy in which Lockyer, John Herschel, and others were involved in the 1860s, when they disputed whether the visible "willow-leaf" structures on the surface of the sun might be living organisms or patterns caused by physical phenomena.[7]

A related and major theme in Lockyer's lecture was the chemical uniformity of the universe as revealed by the spectroscope. Here Lockyer reasoned by analogy with notions recently introduced by physicists: the conservation of energy and the uniformity of a universe composed of matter in motion (Faraday's *Experimental Researches in Chemistry and Physics* also appeared in 1859). For Lockyer not only the laws of physics but also the laws of chemistry were uniform throughout the universe. He did not hesitate to extrapolate from his laboratory experiments about goings-on in the sun, just as he claimed that the sun was an ideal laboratory, giving him indications about the behavior of gases at temperatures and pressures impossible to achieve in the laboratory (figure 48).[8]

Borrowing from natural history and physics, Lockyer developed an astrophysics that sought to "knit the universe together," for instance by uncovering the connection between crop failure and the solar spot cycle and thus to prevent famine in India through the study of solar physics.[9] But like most early astrophysicists he adopted a qualitative, literary, and aesthetic approach rather than a quantitative, mathematical approach to phenomena. He mapped the surfaces of the moon, the sun, and Mars using telescopes, spectroscopes, and photographic cameras.

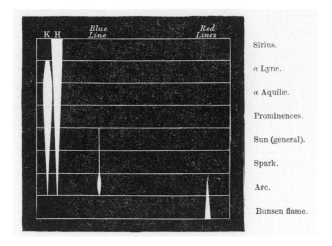

Sirius.

α Lyræ.

α Aquilæ.

Prominences.

Sun (general).

Spark.

Arc.

Bunsen flame.

48. The changes in the spectrum of calcium from the Bunsen flame to Sirius, exemplifying Lockyer's conviction that "to the spectroscope all of Nature is one." Lockyer believed in a continuous development from laboratory to stellar spectra by a gradual increase in temperature. J. Norman Lockyer, *The Chemistry of the Sun* (London: Macmillan, 1887), 250.

Popular Observatories

Lockyer's astrophysics was very different from the precise astronomy of the Astronomer Royal and head of Greenwich Observatory George Biddell Airy, or the concern for calculation expressed by the accountants who founded the Royal Astronomical Society in 1820[10]—though it must be noted that some of the early astrophysicists went down the road of institute building and precision measurement. Anders Ångström in Uppsala and Heinrich Kayser in Bonn, for instance, devoted their lives to the precise measurement and charting of the wavelengths of lines in terrestrial and astronomical spectra. In Britain this type of work was carried out mainly in established observatories, like Greenwich's program for the systematic measurement of stellar radial velocities.

Early astrophysics of the type practiced by Lockyer was closer, and in some cases grew out of the popular astronomy of the period. Many popular astronomers, such as Max Wilhelm Mayer, Max Wolf, Simon Archenhold, and Camille Flammarion, put forth their own visions of an astronomy very different from that practiced in existing observatories and shared the views articulated by Lockyer. For instance, they distanced themselves from the disciplines of precision, observation, and calculation characteristic of positional astronomy and quantitative astrophysics. They complained about the drudgery of positional astronomical work, of the unbearable hierarchies, of the astronomers' lack of imagination. The young Camille Flammarion, hired as a calculator in the Paris Observatory under LeVerrier, was bitterly disappointed about the astronomy he witnessed there and for the rest of his life criticized the authoritarian, mathematical approach to astronomy, preferring the new astrophysics and the speculations to which it could give rise:

> M. Leverrier . . . was essentially concerned about the publication of observations kept in old manuscripts and to complete the revision of Lalande's catalogue. . . . Neither of my five colleagues liked astronomy, neither was interested in celestial contemplation, neither asked himself what the other worlds were, neither travelled with their imagination in the infinite spaces of the sky. . . . Excellent office employees, careful calculators, they did not see beyond the columns of numbers. It was just like military service, the punctual execution of an administrative task. . . . Thus physical astronomy, living astronomy, which represented for me the admirable science of the heavens, the study of the conditions of life in the universe, was outside the programme of the Paris observatory! . . . For me, then as today, astronomy's mission was not to stop at the measurement of the positions of the stars, but to rise to the study of their nature.[11]

Instead the popular astronomers, much like the early astrophysicists, sought to convince their audiences by combining aesthetics and science, appealing to the senses and to reason in what they saw as the tradition of the great Humboldt.

These popular astronomers were central figures in the late-nineteenth-century movement that saw the creation of numerous popular observatories and Uranias throughout Europe. The Berlin Urania discussed by Ole Molvig in this book was perhaps the most famous and certainly the best endowed of these new institutions. Founded in 1888 by Wilhelm Förster (the director of the Berlin Observatory) and the popular astronomer Max Wilhelm Meyer, it had as its statutory aim "the dissemination of the enjoyment of the understanding of nature,"[12] and the very practical aim, from the point of view of Förster, of helping to channel and discipline public interest in astronomy (figures 49–50).

But most popular observatories were quite independent from official observatories: one can mention the Observatoire Populaire du Trocadéro

49–50. Disciplining public observation of the heavens in the nineteenth century. Popular reactions to the eclipse of 1820; peering at Venus through a telescope from the place du Châtelet, Paris, in 1889. Wilfred de Fonvielle, "Les astronomes de la place publique à Paris," *La Nature* 17, no. 1 (1889): 236–37.

51. Simon Archenhold's popular observatory in Berlin-Treptow, founded on the occasion of the *Gewerbeausstellung* of 1896. *Das Weltall* 1, no. 1 (1 October 1900), facing p. 2.

in Paris, founded in 1881 by Léon Jaubert; Camille Flammarion's observatory in Juvisy-sur-Orge outside Paris, founded in 1883; Louis Figuier's Théâtre Scientifique project; and the Archenhold observatory in Berlin, founded in 1896 (figure 51). In 1911 the existence of additional popular observatories was recorded in Paris (owned by the Société Astronomique de France), Brussels, St. Petersburg, and Zurich. One might even include the popular observatory and two planetariums of the Deutsches Museum, Munich (museum founded in 1904, observatory constructed in 1911, planetariums opened in 1925) as transitory configurations towards twentieth-century amateur observatories, planetariums, and space centers. Situated in the city center, open to the public, and available to society figures for parties and receptions as well as to visits from schoolchildren, workers' unions and teachers, popular observatories to a large extent took over the social and representational function of official observatories, which were concerned that admitting the public might disrupt their activities. The Universal Exhibition of 1900 provided an exemplary demonstration of the split between scientific and popular astronomy: while the public gathered at the Palais de l'Optique to witness the show put on by the popularizers, astronomers met separately at two conferences to discuss progress in measuring the earth's dimensions and unifying international standards of weights and measures.[13]

Popular observatories had much in common with the great exhibitions of the late nineteenth century. Interactive displays, an abundance of illustrations and projections, experiments and performances, were characteristic of popular observatories but also of exhibitions, and they were often set up by the same organizers, and viewed by the same publics. The astronomy section of the Universal Exhibition thus displayed "beautiful illustrations supplied by Flammarion, Antonialdi and others. . . . The relative sizes of the planets, remarkable comets and nebulae were skilfully and instructively exhibited. . . . projections of clouds, exposures of lightning flashes or experiments with Geissler tubes, spark inductors etc. were shown, always accompanied by a short lecture usually held by ladies. . . . The prehistory of the earth was shown in a series of dioramas of inferior standard. In an antechamber, next to the glass plates out of which the photographic objective was to be cut, several more astronomical photographs were exhibited, together with meteoritic stones and a famous golden statue representing an American singer. A few astronomical pictures were shown in a large projection room, in particular the magnificent exposures of the moon taken at the Paris observatory by Loewy and Puiseux."[14] The structure of popular observatories reflected their diverse origins and ambitions. They were hybrid spaces combining the museum, the observatory, the laboratory, and the theater. The Archenhold Observatory was described as comprising, in addition to a lecture room, "a large, valuable library, the [instrument] collections of the astronomical museum, a large refractor and the mechanical apparatus for moving it . . . , a valuable collection of meteoric stones, historical apparatus and unique, irreplaceable memorabilia from famous scholars, mostly presents."[15] Often, as with the Urania, the popular observatory's structure and organization mirrored those of the scientific institution, albeit in a playful, pedagogical, spectacular mode.

Astrophysics as Popular Astronomy

The early astrophysicists resembled the popular astronomers in shunning any kind of specialization and advocated an astronomy open to many other fields, including archaeology, meteorology, and Egyptology. Their forefather was William Herschel, the "natural historian of the heavens."[16] Like him they usually began as amateurs working in the back of their gardens or on terraces rather than as trainee calculators in an observatory; they learned

52. An eclipse of the sun by the earth, as seen from the moon, one of the many fantastic stage sets shown by popular astronomers in their public lectures destined to "exercise a stimulating influence on even the simplest mind, especially on the growing youth." This one was part of M. W. Meyer's show "From the Earth to the Moon" at the Urania in Berlin. Max Wilhelm Meyer, *Illustrirter Leitfaden der Astronomie, Physik und Mikroskopie in Form eines Führers durch die Urania zu Berlin* (Berlin: Paetel, 1892), 153, quotation at 152.

for themselves how to use a small telescope, spectroscope, and photographic camera. And like Herschel, they were less interested in the predictable, cyclical Laplacian time of positional astronomy, preferring the irreversible time of Darwin and Huxley, projecting their speculations into "inorganic evolutions" and "meteoritic hypotheses" to account for the life and death of stars (figure 52). Flammarion thus reported the words of an (ideal?) reader of his *Astronomie populaire*: "I want to know each star by its name, I want, when the sky shines with all its light, to be able to spell out these celestial hieroglyphs and to guess what mysteries operate up there; I want to know everything that is known about these distant lights analysed today using the chemistry of the heavens, to know their physical constitution, their worth, their power, and to appreciate the different richness of each constellation. . . . What splendours to behold, what riches to acquire! What pleasure to experience! What delicious hours to spend, telescope in hand, like true astronomers. . . . We want to distinguish ourselves from the common crowd; we do not waste our time reading novels; we are thirsty for science, we leave ignorance and its illusions to those who remain satisfied with it."[17]

While their peers became dependent for their living on institutions and addressed primarily their small community of specialists, they developed networks which were heterogeneous and generalist at once. Lockyer was rarely seen in universities or specialized societies, preferring the Royal Society and the British Association for the Advancement of Science as forums for presenting his work. He would give a lecture in a series which covered the breadth of contemporary knowledge. He exhibited his instruments and photographs at Royal Society *conversazioni* next to butterflies, mathematical curves, bacteria from the river Thames, sacred flutes, samples of steel, and the results of cathode ray experiments. The journal that he founded, *Nature*, was eminently generalist. It was aimed at the whole community of scientists as well as to an educated, urban public, which also made up the audiences of his lectures.

Lecture halls, scientific soirées, and popular journals were particularly important for this kind of popular astronomy and astrophysics in Britain, partly because neither succeeded in becoming institutionalized there to the same degree as on the continent. British astrophysicists were often led to invest in other venues for the practice and display of their work, as well as to earn their living. The lack of any great purpose-built astrophysical observatory in Britain, following the model of those built in France at Meudon and in Germany at Potsdam, not to mention the great American observatories, resulted in part from the resistance of positional astronomers: Airy opposed the creation of an astrophysical observatory that would not be under his direction. That British astrophysics failed to make a space for itself was also due to the British government's reluctance to subsidize science, a function considered outside the realm of its responsibilities. The ongoing implicit assumption in Britain that scientific investigation was an activity for men of independent means made it difficult for the rising generations of middle-class scientists such as Lockyer, who had to live from their work. Thus in the 1870s Lockyer could be simultaneously a fellow of the Royal Society and a third-class clerk of the War Office. While his peers abroad, most notably his friend Jules Janssen, often succeeded in becoming professional scientists, in Britain Lockyer and his science remained on the margins of the establishment.

While Lockyer's reputation enabled him to obtain one-off grants from various societies, he was not able to live from his research and had to rely in particular on lecture fees and journalism to supplement his civil servant's income and feed his large family. Lockyer was first introduced into most scientific circles as a journalist reporting on scientific events for the journal

53. Lockyer (with beard) arriving in Gibraltar during the eclipse expedition of 1905. Note what looks like a photographic camera held by Lockyer's interlocutor, probably his son W. J. S Lockyer. This image is a reproduction from a glass plate that could be inserted in a projector and shown to a large audience. Copyright the Science Museum, London.

the *Reporter*. Later on, his position as the editor of *Nature* enabled him to construct the networks essential to his continued presence in the scientific community, a network which he relied on to raise funding for eclipse expeditions, to borrow rooms for chemical experiments, or to obtain grants to purchase instruments. Deprived of a stable institutional setting, Lockyer was continuously on the move, packing and unpacking his instruments, photographs, and experiments. These always traveled with him, and were used and reused in diverse settings including borrowed laboratories, Indian mountains, impromptu lectures on steamships, soirees, and experiments reported in journals (figure 53). The early astrophysicists survived by continuously asserting themselves on a multitude of temporary stages. Navigating between different social and scientific milieus, they had a transient life and career that acquired their coherence through their personas and public expressions of their work.

Even the solar physics observatory founded by Lockyer in 1879 was a

54. Lockyer's Solar Physics Observatory remained a temporary structure in the South Kensington complex until it was removed to Cambridge in 1912. Lockyer's son, W. J. S. Lockyer, at South Kensington preparing to observe the Leonid meteor shower (1898–99). Copyright the Science Museum, London.

temporary structure made of wood and canvas. It subsisted on the edge of the new technical education complex in South Kensington until it was moved to Cambridge in the early twentieth century (figure 54). In the absence of permanent, secure institutional identities, Lockyer relied on activities such as lecturing, journalistic writing, and eclipse expeditions as important sources of income. It would be wrong, however, to ascribe solely to contingency this particular incarnation of astrophysics in the late nineteenth century. In many ways astrophysics in Britain and on the continent developed self-consciously in opposition to the astronomy carried out in established observatories.

Popular astronomers and astrophysicists can be seen as successors of the demonstrators of electrical experiments in the late eighteenth century such as Benjamin Martin, who toured the country putting on impressive shows that combined education, entertainment, and advertising for the apparatus they sold.[18] While popular astronomers and astrophysicists had only their stories to sell, they operated in a similarly commercial framework. In the context of the late-nineteenth-century expansion of the middle classes, the spread of literacy, and the related growth of the popular press, the astrophysicists participated in the commodification of writing by publishing serialized accounts alongside novels and other stories and reports. The activities of the astrophysicists were carried out in a public environment marked by fashion, that is, the regular appearance of products, always new but always similar. Thus over the last three decades of the nineteenth century Lockyer participated in eclipse expeditions approximately every other year. The expeditions always followed the same procedure: Lockyer drummed up support in public addresses and lobbied the relevant scientific bodies and government agencies, went off on his travels, sent telegraphs to *Nature* while away, returned triumphantly to tour the lecture halls, and wrote up his adventures and observations in a wide range of scientific and popular media before proceeding to launch his next project.

Techniques of Representing Nature and the Scientist

Eclipse expeditions are revealing of the importance of public display and representation for practitioners such as Lockyer. In 1871 the British Association for the Advancement of Science put Lockyer in charge of an expedition to India. There Lockyer and his colleagues busied themselves taking photographs. This was a central part of the work. Lockyer had refused the offer of the Royal Astronomical Society to support his expedition to make spectroscopic observations unless photography was included in his mission (figure 55).

The sun was of course photographed at different stages of the eclipse, directly or through a spectroscope. But the eclipse camp was also photographed, as were the travelers and their local helpers. Having successfully produced these images, Lockyer returned to London to tour the most prestigious and fashionable venues, where he would display himself and the photographs of his travels and observations, and report on his adventures. In the lecture hall Lockyer projected pictures of himself against different backgrounds, multiplying himself on the spot, asserting his presence and the importance of his work.

The activities of the astrophysicists were permeated by representation, but not the representation for a general public of what went on behind closed doors: Lockyer was forever performing before an audience, whether directly on stage or indirectly, as he photographed himself observing an

55. The spectroscope observatory at Baikal (India) in 1871. Lockyer is on the right, holding the telescope. This image shows the improvised (though well-equipped) nature of Lockyer's observatory during his eclipse expedition, but also his reliance on local helpers. It was printed several times, also as an engraving in his journal *Nature*. J. N. Lockyer, "The Approaching Total Eclipse of the Sun," reprinted by permission from Macmillan Publishers Ltd. *Nature* 56 (19 August 1897): 366, copyright 1897.

eclipse. Self-representation for the benefit of London's cultural and scientific crowds was a central component of his scientific work.[19] Lockyer further blurred the boundary between science and its popularization when he urged members of the public to experiment for themselves just as he had done when he bought his first telescope and set it up in his garden. Lockyer implied that he belonged to the public before him, to the wide population of those curious about nature.

The almost complete overlap of the production of representations *in* science and representations *of* science is particularly apparent when one examines the technologies that astrophysicists employed. The spectroscope, photographic camera, and photometer complemented the telescope as they each provided the technological basis of the new astronomy, its singularity, and what distinguished it from positional astronomy. The images produced by means of these new representational technologies were the hallmark of astrophysics as they were of popular astronomy. At Royal Society *Conversaziones*, the yearly evening events where scientists displayed objects representative of their research, Lockyer always brought either optical devices or photographs.

Lockyer first became known for his elaboration of a method of observing the spectrum of solar spots. Projecting the image of the sun on a screen, he punctured the screen at the position of a spot, and the light transmitted was then passed through a spectroscope. Lockyer was able to provide some evidence as to the nature of sun spots in the late 1860s. His setup employed elements which he later used frequently in his public lectures: screens, artificial light, lantern projectors (of which he was one of the early advocates), and spectroscopes (which were also advertised together in the back pages of popular scientific journals). In this way Lockyer was able to reproduce in the lecture hall, with suitable adaptations, the experiments and observations which he had devised in the field.

Upon his return from the eclipse expedition in India of 1871, Lockyer reported to the Royal Institution on his travels, describing the eclipse and his observations by means of a series of reconstructions. In particular he demonstrated the polarization of solar light in the following manner: "On the screen we now have a highly-magnified image of the hollow cone of iron which I am compelling to reflect the light from the lamp; and by inserting this Biquartz I throw various colours over different portions of that ring, which I beg you to consider for one moment as the solar corona, and the colours change as I rotate this prism. You will at once be able to explain the different actions of the Biquartz in this instance. The reflexion, and therefore the plane of polarisation, is no longer general, but varies from point to point of the reflecting surface. It is in fact radial, and hence the delicate arrangement of colour." The technologies of founding astrophysics had much in common with the technologies of representation in use in lecture halls in theaters: they were often produced by the same manufacturers. In particular Carl Zeiss, who was responsible for making photography available to a mass market, was also one of the most renowned makers of optical apparatus.

The history of projection techniques is closely linked to that of popular science: the first public projection on a screen of chemical and physical phenomena was alledgedly made by Abbé Moigno in 1862 using apparatus made by the optical instrument maker François Soleil with the explicit purpose of speaking to the senses.[20] Moigno wrote that "when instruction is addressed indistinctly to all classes of society, it will only have a chance of penetrating the masses if it seduces by the power of exposition and illustration. The word is always cold, and the greatest number does not understand the aridities of scientific language; but when the eyes are shown greatly enlarged and vividly lit objects, the intelligence grasp them instantly with minimal effort. The most ordinary minds accept abstract demonstrations more easily when speech is complemented by images, the sensible representations of spoken language."

Scientists, instrument makers, and popular scientists experimented for the rest of the century with different light sources and optical arrangements for projection as the technique became widespread in educational institutions and lecture theaters. Drawings, objects, and photographs were projected, stereoscopical and color projections were tried out.[21] Popular astronomers possessed collections of thousands of glass slides to enliven their lectures and shows. Among the different theatrical techniques, projection technologies were crucial for popular science because they enabled the presentation of objects to large audiences. The solution for displaying an experiment or astronomical observation to the greatest number and in the most vivid manner was projection: the large telescopes at the Trocadéro Observatory and at the Universal Exhibition of 1900 were fitted with a so-called celestial projector (*projecteur céleste*), which projected the image at the focus of the telescope onto a large screen for all to see. Observers equipped with additional microscopes or telescopes would be able to examine a sun with a diameter of 30, 50, 100, 150, or 300 meters (figure 56).[22]

The history of projection techniques testifies to the constant exchanges between the observatory, the lecture hall, and the instrument maker's workshop. It also shows the multitude of ways in which observatory optical techniques were appropriated in the social and cultural context of the late nineteenth century.

The composition and motivations of the audience at Lockyer's lecture of December 1873—the location of the lecture, the words of the speaker, indeed the very fact that this lecture took place—all spoke of the particular culture of science in Victorian Britain. The lecture was a part of the broader economy of public science, alongside the annual meetings of the British Association for the Advancement of Science, the Royal Society's *conversaziones*, and the expeditions by pioneering scientists, tourists, and mountaineers whose accounts continuously poured into the metropolis

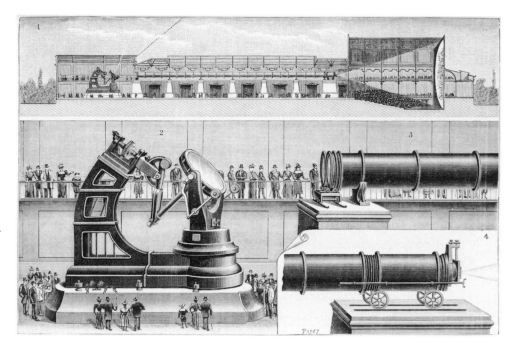

56. The great telescope shown at the Universal Exhibition of 1900 in Paris was fitted with a device for projecting views to a large public (1 and 4). L. Barré, "La grande lunette de 1900," *La Nature* 27 (1899): 169.

together with a stream of exotic objects, images, and stories which fed the growing popular press and helped sustain and display the unity and manageability of the immense and heterogeneous Empire.[23] The content and values that this astrophysics embodied, the specific technologies that it involved as well as the fragmented social and institutional situation of its practitioners, created a strong affinity with public spaces and public representations which became integral to its identity.

In its technologies and its objects of predilection, Lockyer's astrophysics indeed had much in common with the popular astronomy that developed simultaneously from mid-century onward, sharing with it not only a precarious institutional environment but also a preference for optical technologies (photography, spectroscopy), for representation rather than calculation, for speculation rather than standardization, for planetary rather than stellar astronomy. The images produced by the new representational technologies were the hallmark of astrophysics, and of popular astronomy. Like early astrophysicists, popular astronomers were interested in showing the unity of nature and of the universe, and frequently appealed to Arago and Humboldt as their models in their pursuit of science and its popularization. Astrophysics and popular astronomy had a similar outlook on science, which the astrophysicists saw as a democratic pursuit open to all, as a curiosity about the heavens and the earth, as a source of wonder and admiration. The ethos of early astrophysics was congenial to the modes of expressions and public, generalist spaces typical of popular astronomy.

This integrated astrophysics, practiced and performed before a broad and varied audience, provides a striking counterpoint to the narratives of institute building, specialization, professionalization, and precision measurement which characterize the establishment of some of the other physical sciences in this period. Lockyer's personal situation and that of early astrophysics testify to a different kind of establishment, one which confused the distinction between science and its popularization and between amateurs and professionals. If the late-nineteenth-century transformation of science into a profession was in part the expression of the wish of the expanding middle class to invest it, some early astrophysicists showed that this was not the only strategy. It was also possible to exploit the expansion of the middle classes as consumers of science.

Notes

1. J. Norman Lockyer, "Celestial Chemistry," *Nature* 9 (26 March 1874): 411.
2. Jim Bennett, "The Spectroscope's First Decade," *Bulletin of the Scientific Instrument Society* 4 (1984): 3–6.
3. See Aubin, "Orchestrating Observatory, Laboratory, and Field."
4. Methodological reflections on popular science include R. Cooter and S. Pumfrey, "Separate Spheres and Public Places: Reflections on the History of Science Popularization and Science in Popular Culture," *History of Science* 32 (1994): 237–67; Terry Shinn and Richard Whitley, *Expository Science: Forms and Functions of Popularisation*, Sociology of the Sciences, no. 9 (Dordrecht: Kluwer, 1989); Bernadette Bensaude-Vincent, *L'opinion publique et la science: à chacun son ignorance* (Paris: Sanofi-Synthélabo, 2000).
5. On the astrophysicist's persona see Aubin and Bigg, "Neither Genius nor Context Incarnate."
6. Lockyer, *Inorganic Evolution as Studied by Spectrum Analysis*, 168.
7. See Meadows, *Science and Controversy*, 44–47.
8. See Schaffer, "Where Experiments End."
9. On the realization of this program at the Kodaikanal Observatory in India see Simon Schaffer, "Laboratories Ashore," paper given at the conference La Science Fin de Siècle (1870–1910), Paris, 8–9 December 2000.
10. See Ashworth, "The Calculating Eye."

11. Flammarion, *Mémoires biographiques et philosophiques d'un astronome*, 144, 154–56.

12. Max Wilhelm Meyer, *Illustrierter Leitfaden der Astronomie, Physik und Mikroskopie in Form eines Führers durch die Urania zu Berlin* (Berlin: Hermann Paetel, 1892), 1. On the Urania see Ole Molvig's chapter in this book.

13. Wilhelm Förster, "Die Wissenschaft vom Weltall in den Pariser Herbsttagen von 1900," *Das Weltall* 1 (1900): 45–48 emphasizes extreme precision in measurement and international cooperation.

14. Simon Archenhold, "Die Astronomie auf der Pariser Weltausstellung 1900," *Das Weltall* 1, no. 16 (1901): 134–35.

15. "Aufruf zu Beiträgen für den Neubau der Treptow-Sternwarte," *Himmel und Erde* 5 (1893): 398.

16. Schaffer, "Herschel in Bedlam."

17. Camille Flammarion, *Les étoiles et les curiosités du ciel* (Paris: C. Marpon et E. Flammarion, 1882), 2–3.

18. Schaffer, "The Consuming Flame."

19. On eclipse expeditions, representation, and the role of the amateur see Pang, *Empire and the Sun*.

20. François (Abbé) Moigno, "Enseignement de tous, l'art et la pratique des projections," *Cosmos–Les mondes*, ser.3, no. 2 (6 May 1882): 2.

21. Gaston Tissandier, "L'art des projections, son histoire," *La Nature* 18, no. 2 (1890): 373–74; Flammarion recounts his experiments with oxyhydric light in Camille Flammarion, *Mémoires d'un astronome* (Paris: Ernest Flammarion, 1911), 345; Albert Londe, "Les projections et l'enseignement," *La Nature* 18, no. 2 (1890): 195–98; "Les projections stéréoscopiques," *La Nature* 18, no. 2 (1890): 218; L. Vidal, "Projections polychromes," *La Nature* 20, no. 1 (1892): 339–43.

22. "L'observatoire du Trocadéro à Paris," *La Nature* 9, no. 1 (1881): 13–14; L. Barré, "La grande lunette de 1900," *La Nature* 27 (1899): 169.

23. On the popular appeal of spectra at conversaziones see Susan Gamble, "An Appealing Case of Spectra: Photographs on Display at the Royal Society, London, 1891," "Spectroscope Histories," ed. Bigg and Staubermann, 17:635–51; see also Sam Alberti, "Conversaziones and the Experience of Science in Victorian England."

Chapter 12

The Berlin Urania, Humboldtian
Cosmology, and the Public

OLE MOLVIG

> "In the beginning it was formless and empty." That was the time when the
> atoms of the World-aether, which had torn themselves from the awakening
> body of the Sun, settled themselves along the Earth's orbit, in order to cre-
> ate there an independent, free body [*Weltkörper*]. The dense, whirling mists,
> which ran their course through the silence of dark space, were a myriad of
> atoms seeking one another out to form a more beautiful union. This process
> has now brought about such wonderful organization. These atoms now ac-
> tively work and create and flower and feel in millions of well developed cre-
> ations. The mist surged through the emptiness as vapor and steam. It surged!
> That means: it did not stagnate, did not stand still. Every atom was a heav-
> enly body and drew its neighbor to itself by its own strength. The Chaotic
> mass clenched itself into a sphere: A planet, the Earth, was born.
> —WILHELM MEYER (1889).[1]

These words, first read aloud late in the summer of 1889 in a booming,
theatrically trained voice, echoed throughout the main hall of the Berlin
Urania, the capital's magnificent, newly constructed institution for public
science. However, unlike at most scientific lectures, the speaker was nowhere to be seen. His disembodied voice was projected from behind a stage
filled by a cloud of steaming gas and undulating light, a stage where the very
actions he described were in process, for all of Berlin to see. One observer
of the drama described the remainder of the scene this way:

> The opening scene, denominated "Chaos," shows the stage filled with
> whirling and hissing clouds of steam, suffused with weird light that
> slowly changes color as the awful elemental battle accompanying the
> deposition of the first ocean upon the new formed and still heated crust

of the earth proceeds. Finally, the raging clouds are chased away, the commotion ceases, and the face of the earth gradually emerges to sight, covered by the sea.

By slow gradations the scene changes. The creative throes have been transferred to the interior of the planet, and the effect of the strain upon its crust from within, as the pent-up fires struggle to break forth, begins to be manifested. A huge black ridge of rock appears in the gloom, thrust up from the sea bottom, and representing the first land of the new planet. Then an angry red overspreads the sky; fierce and broken storm clouds stream across the scene; the threatening hue of the heavens deepens; blinding flashes of lightening illuminate for a moment the rising land, which has swelled up into a mountain; heavy, rolling thunder is heard, and presently there is a deafening crash, the summit of the mountain is rent open, and volcanic fires pour forth. From the ocean, thus assailed by floods of molten lava, clouds of steam again arise, and enveloped in tumbling vapors, the scene closes.[2]

This performance, entitled "Die Geschichte der Urwelt" (The history of the primeval world), continued for eleven more scenes, each dramatically representing a stage in the earth's development, from its chaotic origins through geological turmoil to a series of biological epochs (of which the Jurassic was reported as the most spectacular; figure 57). Humankind would arrive in the penultimate scene on the banks of Lake Geneva, where among the earliest known signs of organized settlement were found. The performance concluded slightly over an hour after it began, on the shores of a Greek Mediterranean dotted by cities and villas, with the narrator reminding his audience that the harmonious tone of Greek culture first found its pitch in the "music of the spheres."[3]

The show, written and performed by the Urania's director, Max Wilhelm Meyer, was a smash success, performed over two hundred times over the institute's first two years. Another dramatic performance, "Von der Erde bis zum Monde" (From the earth to the moon), reenacted how a highly anticipated total eclipse would have appeared had not Berlin's famously gray skies intervened.[4] This presentation was performed 164 times during the same two years, with both productions garnering popular and critical acclaim at home and abroad. After four years in Berlin the two productions, at the suggestion of Thomas Edison and at the expense of Andrew Carnegie, made a trip to America, performing at length in Carnegie Hall, as well as Philadelphia and Boston.[5]

57. A landscape from the Carboniferous age, shown during Meyer's show "History of the Primeval World." Max Wilhelm Meyer, *Illustrirter Leitfaden der Astronomie, Physik und Mikroskopie in Form eines Führers durch die Urania zu Berlin* (Berlin: Paetel, 1892), 155.

While the *"wissenschaftliches Theater"* was certainly the most dramatic and popular innovation of the Berlin Urania (figure 58), the institute offered far more comprehensive facilities to introduce the interested public to the recent developments of science and technology. Founded in 1888 as a public astronomical observatory, the Urania quickly expanded into a commercial institution for popular science, covering physics, meteorology, geology, biology, medicine, travel, and technology in addition to its humble astronomical origins. Buoyed by Berlin's success, cities across Europe, from London to Moscow, experimented with establishing their own "Uranias" before the outbreak of the First World War.[6]

This chapter traces the founding of the Berlin Urania along two interrelated fronts. The first concerns the imagery with which the Urania presented itself and science to the public. The second examines how, within the local context of the Berlin Observatory and debates over the Physikalisch-technische Reichsanstalt (PTR), the Urania come to represent this particular vision of the sciences.

58. The Urania's theater hall, where lectures were held daily on all areas of science, livened by the use of projections, experiments, etc. Max Wilhelm Meyer, *Illustrirter Leitfaden der Astronomie, Physik und Mikroskopie in Form eines Führers durch die Urania zu Berlin* (Berlin: Paetel, 1892), 148.

Creating The Berlin Urania

In 1888 the Prussian state, at the suggestion of the minister of culture Gustav von Gossler, donated a substantial property within the Landes-Ausstellungspark in Berlin to the recently inaugurated Gesellschaft Urania. The Gesellschaft, headed by the astronomer and director of the Berlin Observatory Wilhelm Foerster, was a commercial association formed on publicly traded stock. On 3 March 1888 the association held its first shareholders' meeting, having raised over 205,000 Marks.[7] Among its biggest investors were an impressive collection of Berlin's scientific, cultural and financial élite, including Werner von Siemens, Hermann von Helmholtz, the *Optiker* Carl Bamberg, the publisher Hermann Paetel, the industrialist

Louis Schwarzkopff, and the banker Adolf Salomonsohn, all of whom would serve on its executive committee. Shortly thereafter ground was broken on the donated parcel, located in eastern Moabit just off Invalidenstrasse, and by 1 July 1889 the public observatory was open (figure 59).[8]

The astronomer Wilhelm Foerster, the Urania's founder, was no stranger to the politics of science. Director of the Berlin Observatory since 1865, Foerster also served on the philosophical faculty of the Berlin University, the advisory board of the Royal Geodetic Institute (Königliches Geodätisches Institut), and the planning committee for the Physikalisch-technische Reichsanstalt before he developed his interest in the public observatory. In 1891 he was asked to serve as rector of the Berlin University, and in the same year was appointed chairman of the International Committee on Weights and Measures, a position that he retained until his death in 1921.[9]

As director of the Berlin Observatory Foerster was constantly reminded of the public's astronomical interest by the constant flow of curious visitors, who hoped to catch a glimpse of the moon through Fraunhofer's fa-

59. The façade of the Urania in the Ausstellungspark in Berlin, bearing close resemblance to the great astronomical observatories of the time. P. Schwann, "Zum 25-jährigen Bestehen der Gesellschaft Urania in Berlin," *Himmel und Erde 25* (1913): 341.

mous refractor. Foerster, long devoted to Alexander von Humboldt's suggestion of creating a public observatory in Berlin, was equally plagued by the idea's impracticalities. A proper observatory, functional only at night, would be further restricted to the civilized hours of the evening. Given the further complication of Berlin's persistent foul weather, Foerster and the cultural minister agreed that a public observatory would not be available with enough frequency to justify its cost.[10]

However, in 1886 Foerster met the astronomer and popular writer Max Wilhelm Meyer, who had recently relocated from Vienna after the failure of his first "dramatic lecture," *Bilder aus der Sternwelt*. Employed by the *Berliner Tagesblatt* to write five *Feuilletons* monthly, Meyer continued to seek support for his astronomical *Schauspiel*.[11] Upon meeting Foerster, Meyer pitched his scientific theater, requesting funds from either the Berlin Observatory or the Prussian State to rent for one night a week a small theater off Friedrichstrasse. It apparently did not take long for the two astronomers to realize the compatibility of their two projects: Meyer's theater, a production simulating the heavens, would not depend on observational conditions, and Foerster's public observatory could provide the theater with permanent housing. By the time the Gesellschaft was founded a little over a year later, Foerster's simple telescope had grown considerably. In addition to the performance hall and observational dome, it included a museum, a library, a journal, and several hands-on experiment halls (figure 60).[12]

Speaking just after the Gesellschaft's first meeting, Foerster outlined its mission: "The object of the society is to promote knowledge. In order to inculcate knowledge, it is necessary to educate man to use his mental powers. Therefore institutions for the diffusion of knowledge can only be successful when they try to teach one how to use one's mental powers. The society has limited its work principally to astronomical, geographical and physical phenomena, and for reaching its object has established a great popular observatory, which will be a model for all similar institutions, and publishes a journal, *Himmel und Erde* (Berlin, H. Paetel), which is beautifully printed and illustrated, and gives, in a popular form, reports on astronomical and geographical phenomena and questions."[13] For both Foerster and Meyer the Urania was a project of *Volksbildung*.[14] Foerster, as well as many of those sitting with him on the executive committee, predictably lamented the difficulty of properly enculturating the common Berliner. Members of the Gesellschaft acknowledged the prohibitive price of entrance to the Royal Theater and the Opera, and were disturbed by the cultural degeneration on offer in Berlin's crass but affordable popular theaters. It was this audience

60. The cover of the first issue of the Urania's periodical *Himmel und Erde*, edited by Max Wilhelm Meyer. *Himmel und Erde* 1 (1888), front cover.

that Foerster, by setting the price of admission between 50 Pfennig and 3 Marks, sought to inspire through the drama of the Urania's new theater. Well aware of a range of interests and abilities among his clientele, Foerster imagined the Urania as a funnel, with the theater attracting the masses, and the scientific curiosity thus instilled driving them into the institution's pinnacle, the great observational dome. "With a variety of offerings can the different levels of motivations and expectations of the population be met. The 'Scientific Theater' should attract his curiosity and rouse in him the pleasure of gaining knowledge of nature, which should encourage him in the museum to experiment on his own. Doing so will propel him into the observatory in search of more accurate observations, where instruction in the observatory will instill in him a deep admiration of the universe [*tiefe Bewunderung des Weltalls*]."[15] Foerster's rhetorical notion of funneling the audience from the most accessible level of presentation (*Angebotsebene*) to the most difficult was in fact built in stone. As shown by an early plan, traffic in the Urania was routed first through the theater, then into the experiment halls, filled with a selection of precision optical and electrical instruments, and finally up into the telescope's eight-meter dome.

Before entering the dome proper, in Wilhelm Meyer's graphic description, "we first arrive in a darkened area, where we are asked to take a place before those wonders of the sky, which, *in natura* generally only visible at night, are shown as true to nature as possible through impressive reproductions with the assistance of powerful lighting-works. Shining star clusters flicker and glimmer like a diamond; nebulae, whose diffused, mysterious forms at immeasurable distances, describe the world-creating strength of the universe; double stars . . . , in short, all that is beautiful and worthy of notice that the wide heavens offer, presents itself here for convenient inspection at any time of day."[16]

During the day these exhibits were accompanied by short lectures of explanation, along with demonstrations of the scientific instruments relevant to each object's study. At night, however, a visitor entered "the domed structure of the great telescope. Here our gaze rests on the powerful instrument, which can be guided with ease merely with a gentle touch. Likewise, with the mere press of an electrical button, the entire domed roof can be made to turn." Throughout the entire dome "the tik-tak of an accompanying pendulum clock" could be heard over a microphone, as the minutes of each observation were registered by a chronograph.[17] The telescope's twelve-inch refracting lens, ground by the craftsman Carl Bamberg in nearby Friedenau, was at the time of installation the largest in Berlin and

61. The Observation of the solar eclipse of 17 April 1912 attracted a great crowd to the Urania's telescope. P. Schwann, "Zum 25-jährigen Bestehen der Gesellschaft Urania in Berlin," *Himmel und Erde* 25 (1913): 363.

second only to one other in Germany. In addition to the main cupola the Urania was equipped with two smaller domes on either side of the building, one installed with a small refractor designed for lunar observation, the other with a passage instrument intended exclusively for training observers.[18] In short, the Berlin Urania provided its visitors with a range of attractions, from examples of scientific showmanship to an unusual glimpse into the workings of a professionally equipped observatory (figure 61).

Foerster's and Meyer's vision, as it turned out, proved remarkably successful. In the Urania's opening year 313 performances attracted more than 95,000 visitors, and in excess of 100,000 Marks was raised in ticket revenue. Furthermore, the Gesellschaft authorized the release of additional public stock, bringing its total offering to 500,000 Marks.[19] By 1896, with the number of annual visitors reaching 200,000, the popularity of the "scientific theater" had outgrown the four-hundred-seat venue used by the observatory. Construction soon began on Taubenstrasse, just off Friedrichstrasse in the heart of Mitte, on an impressive new building that would house the theater alone (figure 62).[20]

The elaborate façade of the new theater merits particular attention,

62. The façade of the Urania in the Taubenstrasse, with the busts of Copernicus, Siemens, Humboldt, Helmholtz, and Kepler. P. Schwann, "Zum 25-jährigen
Bestehen der Gesellschaft Urania in Berlin," *Himmel und Erde* 25 (1913): 339.

especially the five busts overlooking the street from above the second-floor windows. The five figures, from left to right, were Copernicus, Siemens, Humboldt, Helmholtz, and Kepler. It is to the three middle figures, the technologist Werner von Siemens, the naturalist Alexander von Humboldt, and the physicist Hermann von Helmholtz, that we now turn, examining both their specific role in the Urania's founding and their relationship to the vision of science that it promoted.

Built in their Image: Werner Siemens,
Alexander von Humboldt, Hermann von Helmholtz

We have seen that the Urania used novel stage techniques to demonstrate to its visitors the extraordinary wonders of nature and to inculcate in them an edifying desire for self-improvement through continuing education. The initiative of Wilhelm Foerster and the theatrical flair of Wilhelm Meyer convinced the state, their financial supporters, and their visitors that an institution for public science was socially desirable, fiscally viable, and popularly engaging. While Foerster's and Meyer's actions were obviously critical to the story, and their socially progressive intentions genuine, another, parallel history of the Urania can be told, one in which the role of *das Publikum* is shifted slightly: from satisfying the need of the masses for self-betterment to using the masses (and their thirst for entertainment) as a means of solving a variety of disparate problems besetting members of Berlin's scientific and financial elite.

Foerster claimed later in life that Alexander von Humboldt had been the real "Father of the Urania" (*Vater der Urania*).[21] Foerster was referring to Humboldt's critical role in integrating the natural sciences into university education, to his successful plea to Friedrich Wilhelm III for construction of the new Berlin Observatory, and most significantly to his status as the icon for popular science. Humboldt's immensely popular work *Cosmos: A Sketch of the Physical Description of the Universe* left a lasting impression on science writing until the First World War. In it Humboldt described his "wild [*tollen*] idea, that the entire material world, all that we now know of the phenomena of the universe and terrestrial life, from the nebulae to the geography of moss on granite rock,"[22] was of a single piece and should be understood as a complete work. While it was surely in this vein that Foerster declared Humboldt the Urania's true father, the statement is, as we shall see, true in a more practical sense as well.

At Humboldt's request Wilhelm III commissioned the new Berlin Observatory. It was completed in 1835 by the architect Friedrich Schinkel, with the provision that twice monthly the observatory's primary telescope was to be made available to the public. Foerster, who directed the observatory from 1865 to 1903, was acutely aware of this responsibility. While Foerster was certainly dedicated to his role as a science educator, it was in his capacity as director of the observatory that he began a campaign to dedicate solely for public use a separate telescope, designed primarily for lunar observation, to free his observatory's equipment and assistants for strictly scientific research.

It was fortunate for Foerster and Wilhelm Meyer that their meeting occurred when it did, in 1887, for the timing helped each to advance his own goal: for Foerster, an independent, public telescope; for Meyer, a place to house the science theater. But these two men's intentions are insufficient to explain fully why the Urania took the form and scale that it did. To answer this question we must turn to the two men flanking Humboldt on the institute's façade, Werner von Siemens and Hermann von Helmholtz, as well as Berlin's other large-scale scientific construction project, the hotly debated PTR, the so-called Institute for an Empire.[23]

Werner von Siemens, the industrial magnate and accomplished scientist who built his fortune first in telegraphy and later in electric power, was the driving force behind the establishment of the PTR and its eventual success as the world's premier institution for precision measurement and technical standards. Siemens, who eventually donated or sold nearly the entirety of the land required for the PTR's campus, had perhaps more to gain than any other German from the proposed state-sponsored institution responsible for creating, maintaining, and testing national, standardized electrical and scientific units as well as the calibrating and assessing precision measurements and instrumentation. But as the historian David Cahan has demonstrated, Siemens desired a very specific *kind* of institute, one designed not for maximum industrial or technical payoff but explicitly to advance his vision of pure science.[24]

In the form envisioned by Siemens as early as 1884 (and in which it was eventually constructed), the institute would be composed of a scientific section and a technical section, both to be under the direct control of the institute's president, whom Siemens had selected long before the institute received governmental approval: Germany's best-known scientist and Siemens's close friend, Helmholtz. Central to Siemens's vision for the PTR, a vision shared by a tight-knit circle of academic supporters including Helmholtz, du Bois-Reymond, and Foerster, was the scientific section. It was to be dedicated solely to the pursuit of pure physical research, a haven for "great scientists who are forced to waste [time] in instruction and administration as well as for talented younger scientists . . . where, undisturbed by other duties, they can serve science alone."[25] This was appropriate in an institution designed to ensure for the Reich technical and industrial superiority over its neighbors, because, in Siemens's opinion, precision technology was only improved through advances in "deeper scientific foundations."

Besides providing for precision technology, it was seen as even more necessary that Prussia have, through the establishment of a physical laboratory, at least the nucleus of an institute for the "entire field of exact natural [scientific] research. . . . [before] the higher development of precision technology begins to fail."[26] However, to the surprise of Siemens and his supporters, both this vision of the institute and this understanding of pure science in relation to practical science encountered heavy opposition from a broad spectrum of opponents.

Foerster, Siemens's key ally on the PTR's planning commission, was asked to present the committee's report to the Reichstag's budget commission in the autumn of 1886. "In all my life I had never encountered a thornier obligation. All parties, with exception of the Social-Democrats, expressed their energetic dislike of the proposal and would hardly listen to what I had to say. . . . The keywords? 'State management of trade' or 'interference of the free market' by the liberal side and 'excessive pampering of industry' on the conservative side. And among other things, the so-called civil parties [bürgerlichen Parteien] sharply criticized the proposed official residence of the institute's president, a position for which Helmholtz was being considered."[27] In addition to such diverse opposition questioning the institute's political utility, a consortium of engineers, instrument makers, and industrialists organized a campaign against the proposed structure of the Reichsanstalt itself. The "technologists," led by the instrument maker Rudolf Fuess and the optical instrument maker Carl Bamberg, decried the emphasis on pure science at the expense of practical results. From their point of view, an institute structured as Siemens suggested would merely duplicate the activities of university science: they recommended instead that the scientific section be canceled altogether and that the technical section be attached to the Technical University in Berlin-Charlottenburg as a mechanics institute for the sole purpose of industrial testing.[28]

While Siemens, Foerster, and Helmholtz were taken aback by the initial success of the technologists' campaign, opposition was not entirely unexpected. Siemens had long ago enlisted his friends in an intensive lobbying effort. In 1886 he appointed Foerster the acting commissioner of the technical section; Foerster, at the suggestion of Siemens and the crown prince Friedrich Wilhelm (who would ennoble Siemens in 1888), published a pamphlet emphasizing his section's need for the existence of a strong, independent scientific section.[29] The tactic was successful: by the middle of 1887 the Reichstag officially allocated funds for the institute, including 480,000 Marks for the construction of the scientific section.[30] Despite the apparent victory, Siemens and Helmholtz thought it temporary. The Reich and its citizens would need to be further convinced of the practical utility of the section if they were to commit themselves to staffing and maintaining what was to become Germany's most costly research institution, and thus guarantee its success. As early as 1884 Siemens was aware that many did not share his vision of the sciences. In a letter to the mayor of Cologne he wrote: "I have strived to set up an imperial institute dedicated exclusively to natural scientific work, not to teaching. I shall make considerable personal sacrifice for it. *Bismarck, however, currently still holds science for a type of sport without practical meaning; and this view is still rather widespread.* . . . I intend to introduce an effective journalistic propaganda [campaign] in order to spread knowledge of the great social meaning of natural scientific research."[31] Foerster, Siemens, Helmholtz, and Du Bois-Reymond continued to lobby not just for the Reichsanstalt but for a specific form of the Reichsanstalt, one predicated on a particular understanding of the natural sciences, and their relationship to technology, precision mechanics, and society.

It is therefore notable that Wilhelm Foerster's Urania should have begun to gain momentum at precisely the point when the political machinations concerning the formation of Siemens's institute for the advancement of precision mechanics were in highest gear. Foerster, nearly as committed as Siemens to the success and value of the scientific section, served on the PTR's planning committee, executive board, and committee on weights and measures, wrote essays, and testified before the Reichstag in support of the scientific section. Likewise, by giving over 100,000 Marks,[32] Siemens was the single largest donor to the Urania, helping to equip its observatory with precision instrumentation;[33] four years after his death in 1892, his name was carved in stone over the new theater's entrance.

Early in his promotion of the Urania, after the formation of the Gesellschaft in 1887 but before Siemens's donation in 1888 and the construction of the Urania, Foerster had stressed exclusively that "the object of the society is to promote knowledge, . . . [and] the society has limited its work principally to astronomical, geographical and physical phenomena."[34] However, describing his initial involvement with the Urania years later, Foerster wrote: "Through my experiences on behalf of the Berlin Observatory grew a lively desire to create, for the astronomical interests of the larger public, a more complete and suitable mechanism for astronomical and scientific instruction than the imperfect performance of astronomical episodes the observatory was capable of. I then had the thought to connect this with the already mentioned Protectorate for the advancement of precision technologies [the PTR], so as to advance the interests of the broad social classes and the development of scientific research, namely through the presentation of astronomical, physical and chemical precision and experimental techniques not just through demonstration, but also through their handling and testing."[35] By 1889, after construction had begun on the observatory itself, Foerster's struggles over and support for the PTR were built into the organizational structure of the Urania. In its final form the public observatory was divided into five departments: Astronomy, Physics, Biology (Microscopy), Precision Mechanics, and the Scientific Theater. In the space devoted to the fourth *Abteilung*, one could experiment with a variety of precision optical equipment, test new electrical communication devices, or examine the internal workings of the phonogram. If one were so fortunate, the majority of displays were even available for purchase, from the very firms debating Foerster and Siemens, on both sides, over the mission of the PTR.[36] Even the type of lens for the central telescope, a twelve-inch Gaussian refractor, was attributable to Foerster's participation in the PTR. It was ground by Carl Bamberg, one of the scientific section's staunchest critics.[37] Throughout the 1890s the Gesellschaft erected a number of "Urania Towers," including one on Unter den Linden in front of the Culture Ministry. The towers would provide the time, meteorological information, and soon traffic reports in addition to advertisements promoting forthcoming events at the Urania.

In short, the construction of the Berlin Urania gave Foerster and his supporters the opportunity to present before the public their specific version of *die gesammten Wissenschaften*. The observatory's dome was large enough to house not just Humboldt's cosmology, from "the nebulae to moss on granite rock," but their own expanded version, from "the nebulae to the micrometer."

We have seen that there are two narratives for the history of the Berlin Urania. The first focuses on the *explicit* imagery employed by the Urania for its public image—the wonders of nature, the majesty of the universe, the complexity of life on earth—and the way it promoted a Humboldtian cosmology, emphasizing the harmony of nature and the unity of the sciences. The second narrative focuses on the specific, local circumstances out of which the Urania was constructed and pays more attention to the Urania's *implicit* imagery. By including electrical appliances and precision instruments under the same roof that housed Humboldt's *Cosmos*, Foerster was able, by association, to promote both the "deeper scientific foundations" of precision technologies and the practical payoff of pure science.

In conclusion, I would like to suggest a few ways in which these two narratives—the public presentation of science and the private motivations underlying it—played out during the remainder of the nineteenth century. Before the establishment of the Urania the enterprise of popularizing and disseminating science was a diverse one. It could include sporadic presentations and book-length treatises by professional scientists (such as those by Foerster at the Berlin Observatory and Humboldt's *Cosmos*), journal articles by professionals and interested amateurs, newspaper columns, and participation in the countless scientific associations that dotted the German social landscape.[38]

It was within this framework that Foerster, as late as 1886, imagined his popular observatory, "limited principally to astronomical, geographical, and physical phenomena."[39] What was remarkable about the Urania, as its ambitions increased and its founders encountered new challenges in the PTR, was its attempt to consolidate nearly all fields and methods of science into one enterprise. In addition to its novel theater, the Urania published its performances as short books, circulated a journal containing articles as diverse as its exhibitions, and presented occasional specialist lectures of unmatched caliber.[40] Although the full significance of the Berlin Urania for the structure of German scientific popularization is unknown, it appears that many large-scale popular science efforts followed its lead. As the historian Andreas Daum has demonstrated, by 1900 scientific periodical publications were consolidating swiftly.[41] In 1904 the largest of these consolidators, the Gesellschaft der Naturfreunde in Stuttgart, reintroduced its popular nineteenth-century magazine *Kosmos*. The publication, originally intended to promote the study of nature through evolutionary theory, was retooled to include well-illustrated articles across the spectrum of scientific and technical interests. As Daum notes: "scientific journalism over the preceding four decades [1860–1900] had been pluralizing and specializing, [but the modern] magazine once again bound the complete spectrum of themes and popular-scientific interests together."[42]

But perhaps the most significant outcome of the Berlin Urania's construction was somewhat paradoxical. The observatory, originally intended to bridge the divide between research science and its popularization, would in effect push the two further apart. As the term "Urania" became synonymous with "scientific popularization,"[43] Berlin's scientific institutions were relieved of their obligations to the public. The majority of university popular science lectures were held at the Urania; the PTR was released from a proposal requiring modest public outreach; and the Berlin Observatory, no longer responsible for public access to astronomical equipment, was moved, in a bid by Foerster, outside the city limits, to a remote hilltop outside Babelsberg. Thus as the sciences continued to specialize and their popularization to generalize, the distance between the practice and presentation of the sciences, like that between the Urania and the Berlin Observatory, would only increase.

Notes

1. Meyer, *Die Geschichte der Urwelt*, 7. The first quotation is a paraphrase of Genesis 1:1. In Luther's Bible (1535) it reads: "Im anfang schuff Gott himel und erden. Und die erde war wuest und leer, und es war finster auff der tieffe, und der Geist Gottes schwebet auff dem wasser. Und Gott sprach. Es werde liecht. Und es ward liecht." The author of this passage purposefully transfers the chaos and void from the "earth" to that which came before it.
2. Garrett Serviss, "From Chaos to Man," *Scientific American*, 25 June 1892, 405.
3. Meyer, *Die Geschichte der Urwelt*, 36.
4. M. Wilhelm Meyer, *Von der Erde bis zum Monde: ein astronomischer Gedankenausflug* (Berlin: Gronau, 1889). Reactions to this performance were similarly evocative: "The curtain rises on a scene near Berlin on the morning of the last great eclipse of the sun, 19 August 1887. Morning twilight comes on. The world begins to stir in anticipation of the usual sunrise, when lo! in the place of the usual sun, up comes a blood red sickle, which soon disappears, and weird lights appear around a black disk. Nature is shrouded in a veil worse than pitch darkness. Animals feel the terror which men uninstructed to look for such a phenomenon used to feel." Rupus Richardson, *Independent*, 1889, Nachlaß Förster 64, BBAW.
5. Upon returning to Berlin Meyer created a new theatrical production based on the experiences of this tour. See Meyer, *Eine Amerikafahrt 1492 und 1892* (Berlin, 1893). For information on attendance in Berlin see Meyer, *Die Urania zu Berlin*.
6. Ebel and Lührs, "Urania," esp. 36.
7. See Schwahn, "Zum 25-jährigen Bestehen der Gesellschaft Urania in Berlin."

8. Ibid., 360.

9. Foerster's own life is best recorded in his memoir *Lebenserinnerungen und Lebens-hoffnungen*, written in his seventy-eighth year. After the founding of the Urania, Foerster became publicly active, serving on the following commissions until his death: Kaiserliche Normaleichungskommission, Kuratorium der physikalisch-technischen Reichsanstalt, Permanente Kommission der internationalen Erd-messung, Executivkomitee des internationalen ethischen Bundes, Vereinigung der Freunde der Astronomie und kosmischen Physik, Deutsche Vereinigung für Chronometrie, Gesellschaft Urania, Gesellschaft Schillertheater, and Abwehr-verein gegen den Antisemitismus.

10. Harro Hess, *Aus der Geschichte der Berliner Gesellschaft Urania, 1888–1927* (Berlin: Archenhold-Sternwarte, 1979).

11. Mathias Iven, *Von den Sternen auf die Erde* (Potsdam: Urania, 1988).

12. Ebel and Lührs, "Urania," 32.

13. Foerster as quoted in Anon., "A Popular Observatory," 109.

14. Daum, *Wissenschaftspopularisierung im 19. Jahrhundert*, esp. 168–84.

15. Wilhlem Foerster, "Vorschläge betreffend die Begründung einer öffentlichen teleskopischen, spektroscopischen und mikroscopischen Schaustätte zugleich zur Vorführung optischer und elektrischer Experimente sowie zu mannigfachen naturwissenschäftlichen Erläuterungen durch Wort und Bild, endlich als Aus-stellungsort für einschlägige Instrumente und Apparate dienend" (Berlin, 1887), as quoted in Ebel und Lührs, "Urania," 21.

16. Meyer, "Was soll die 'Urania' dem Publikum bieten, und was darf sie von der Betheiligung desselben erwarten?" (Berlin, 1887), exerpted in Ebel and Lührs, "Urania," 22.

17. Ibid.

18. Schwahn, "Zum 25-jährigen Bestehen der Gesellschaft Urania in Berlin," 360.

19. *Bericht der Gesellschaft Urania* (1889). In addition see "Science on the Stage," *New York Times*, 28 December 1890. Of those 95,000, 11,000 were schoolchildren, 6,000 members of workingmen's societies, and 8,000 members of other scientific societies, all of whom received reduced admission.

20. The observatory continued to serve in its capacity as a public observatory until 1913, when the Prussian government reclaimed the property to begin construc-tion on the Lehrter Bahnhof.

21. Ebel und Lührs, "Urania," 17.

22. Alexander von Humboldt, *Kosmos: Entwurf einer physischen Weltbeschreibung* (Stuttgart: Cotta, 1845–62), 179.

23. Easily the most comprehensive study remains Cahan, *An Institute for an Empire*.

24. Ibid., esp. 29–53.

25. Werner Siemens, "Conferenz zur Berathung von Vorschlägen auf Förderung der wissenschaftlichen Mechanik im Sitzungszimmer des Kultusministieriums," 2nd session, as quoted in Cahan, *An Institute for an Empire*, 30.

26. "Denkschrift betreffend die Begründung eines Instituts . . . vom 16. Juni 1883," 2–3, as quoted in Cahan, *An Institute for an Empire*, 31.

27. Foerster, *Lebenserinnerungen und Lebenshoffnungen*, 192. The planned construction of the presidential residence and Helmholtz's own demands for the extraordinarily high salary of 24,000 Marks would continued to plague and delay the institute's approval for years. See in particular chapter 3 of Cahan, *An Institute for an Empire*, 59-121.

28. Cahan, *An Institute for an Empire*, 30-31.

29. See Foerster, *Lebenserinnerungen und Lebenshoffnungen*, 192-96, and Cahan, *An Institute for an Empire*, 48-49.

30. Cahan, *An Institute for an Empire*, 51.

31. Werner Siemens to Mayor Rosenthal, 17 September 1884, as quoted in Cahan, *An Institute for an Empire*, 45; emphasis added.

32. Foerster, *Lebenserinnerungen und Lebenshoffnungen*, 200.

33. The firm of Siemens & Halske would also provide the electrical equipment needed for the building and demonstrations. Meyer, *Die Urania zu Berlin*, 15.

34. Anon., "A Popular Observatory," 109.

35. Foerster, *Lebenserinnerungen und Lebenshoffnungen*, 199.

36. Meyer, *Die Urania zu Berlin*, 20.

37. It remains unclear whether the lens was donated at cost to the Urania or the Bamberg firm was compensated for its work, and if so at what cost.

38. A few of these associations in Berlin alone included Gesellschaft für Mikroscopie, Entomologischer Verein, and Vereinigung von Freunden der Astronomie und kosmischen Physik. For a complete list of German *Naturvereine* see Daum, *Wissenschaftspopularisierung im 19. Jahrhundert*, 91-95.

39. Foerster in Anon., "A Popular Observatory," 109.

40. The speakers included Thomas Edison, Hermann Helmholtz, Emil Du Bois-Reymond, Walter Nernst, Fridtjof Nansen, and Michael Faraday. After 1900 they included Henri Poincaré, Heinrich Hertz, Conrad Röntgen, Max Planck, Karl Schwarzschild, and Albert Einstein.

41. Daum, *Wissenschaftspopularisierung im 19. Jahrhundert*, 370-76.

42. Ibid., 372.

43. Ebel and Lührs, "Urania," 15.

BIBLIOGRAPHY

Aasen, Clarence. *Architecture of Siam: A Cultural History Interpretation*. Kuala Lumpur: Oxford University Press, 1998.

Abetti, Giorgio. *Padre Angelo Secchi, il pioniere dell'astrofisica*. Milan: Agnelli, 1928.

Adas, Michael. *Machines as the Measure of Men: Science, Technology, and Ideologies of Western Dominance*. Ithaca: Cornell University Press, 1989.

Airy, George Biddell. "Schreiben an den Herausgeber." *Astronomische Nachrichten* 26 (1848): 353–60.

Akrell, Carl. "Tal hållet den 12 Nov. 1841 . . . vid nedläggandet af Styresmans-Embetet i Kongl. Krigs-Vetenskaps-Akademien." *Krigsvetenskapsakademiens Handlingar*, 1841, 220–34.

———. *Minnen från Carl XIV's, Oscar I's och Carl XV's dagar*. Stockholm, 1884.

Alabaster, Henry. *The Wheel of the Law: Buddhism Illustrated from Siamese Sources by the Modern Buddhist, a Life of Buddha and an Account of the Phrabat*. London: Trübner, 1871.

Alberti, Samuel. "Placing Nature: Natural History Collections and Their Owners in Nineteenth-Century Provincial England." *British Journal for the History of Science* 35 (2002): 291–311.

———. "Conversaziones and the Experience of Science in Victorian England." *Journal of Victorian Culture* 8 (2003): 208–30.

Alder, Ken. *The Measure of All Things: The Seven-Year Odyssey and Hidden Error That Transformed the World*. New York: Free Press, 2002.

Anderson, Katharine. *Predicting the Weather: Victorians and the Science of Meteorology*. Chicago: University of Chicago Press, 2005.

Andrewes, William J. H., ed. *The Quest for Longitude*. Cambridge: Harvard University Press, 1996.

Anon. Review of *Astronomical Observations Made at the Royal Observatory Greenwich in the Year 1847*, ed. George Biddell Airy, and six other books. *Living Age* 25 (1850): 433–57.

———. *Histoire de l'École Navale et des institutions qui l'ont précédée, par un ancien officier*. Paris: Quantin, 1888.

————. "A Popular Observatory." *Science* 13 (1889): 109–10.

Arago, François. "Sur l'éclipse totale de Soleil du 8 juillet 1842; sur les phénomènes qui devront plus particulièrement fixer l'attention des astronomes; sur les questions de physique céleste dont la solution semble devoir être liée aux observations qui pourront être faites pendant les éclipses totales du soleil." *Comptes rendus des séances hebdomadaires de l'Académie des Sciences* 14 (1842): 843–61.

————. *Leçons d'astronomie professées à l'Observatoire Royal, recueillies par un des ses élèves*, 4th edn. Paris: Chamerot, 1845.

————. *Œuvres Complètes*, ed. Jean-Augustin Barral. Paris: Gide, 1854–62.

————. *Astronomie populaire*. Paris: Baudry, 1855–59.

Arendt, Hannah. *The Human Condition*. Chicago: University of Chicago Press, 1958.

Arosenius, J. F. N. "Årsberättelse för åren 1846–1848 af föredraganden för topografien." *Krigsvetenskapsakademiens Handlingar*, 1849, 107–30.

Ashworth, William J. "The Calculating Eye: Baily, Herschel, Babbage and the Business of Astronomy." *British Journal for the History of Science* 27 (1994): 409–41.

————. "'Labour Harder Than Thrashing': John Flamsteed, Property and Intellectual Labour in Nineteenth-Century England." *Flamsteed's Stars: New Perspectives on the Life and Work of the First Astronomer Royal, 1646–1719*, ed. F. Willmoth, 199–216. Rochester, N.Y.: Boydell, 1997.

————. "John Herschel, George Airy and the Roaming Eye of the State." *History of Science* 36 (1998): 151–78.

Aubin, David. "La métamorphose des éclipses de soleil." *La Recherche* 321 (1999): 78–83.

————. "Orchestrating Observatory, Laboratory, and Field: Jules Janssen, the Spectroscope, and Travel." *Nuncius* 17 (2002): 143–62.

————. "The Fading Star of the Paris Observatory in the Nineteenth Century: Astronomers' Urban Culture of Circulation and Observation." *Osiris* 18 (2003): 79–100.

————. "Un passage de Vénus en politique." *La Recherche*, hors série, 15 (2004): 85–89.

————, ed. "L'événement astronomique du siècle? Une histoire sociale des passages de Vénus, 1874–1882." *Cahiers François Viète*, 2007.

Aubin, David, and Charlotte Bigg. "Neither Genius nor Context Incarnate: Lockyer, Janssen and the Astrophysical Self." *The Poetics of Biography in Science, Technology and Medicine*, ed. T. Söderqvist, 51–70. Adelshot, Hampshire: Ashgate, 2007.

Axelson, G. J. von. "Årsberättelse för åren 1857, 58 och 59 af föredragande för topografien." *Krigsvetenskapsakademiens Handlingar*, 1860, 81–128.

Baeyer, J. J. *Über die Grösse und Figur der Erde: eine Denkschrift zur Begründung einer mittel-europäischen Gradmessung*. Berlin: Reimer, 1861.

Bartky, Ian R. *Selling the True Time: Nineteenth-Century Timekeeping in America.* Stanford: Stanford University Press, 2000.

Batten, Alan H. *Resolute and Undertaking Characters: The Lives of Wilhelm and Otto Struve.* Dordrecht: D. Reidel, 1988.

Beauvoir, Ludovic de. *Java, Siam, Canton: voyage autour du monde.* Paris: Plon, 1874.

Beltrano, Maria Carmen, ed. *Presenze scientifiche illustri al Collegio Romano.* Rome: UCEA, 2001.

Benjamin, Walter. "The Work of Art in the Age of Mechanical Reproduction." *Illuminations,* ed. Hannah Arendt, 217–51. New York: Schocken, 1968.

Bennett, James A. *The Celebrated Phaenomena of Colours.* Cambridge: Whipple Museum, 1984.

———. *The Divided Circle: A History of Instruments for Astronomy, Navigation and Surveying.* Oxford: Phaidon, 1987.

Bergman, George. "Christian Carl Ludwig Rümker: Australia's First Government Astronomer." *Royal Australian Historical Society Journal* 46 (1960): 247–89.

Berthaut, Henri Marie Auguste. *La carte de la France, 1750–1898: étude historique.* Paris: Service Géographique, 1898–99.

Bhumichitr, Teerawat. *Phra Chomklao, roi du Siam: étude de l'émergence de l'anglophilie et de la francophobie au Siam du XIXe siècle.* Bern: P. Lang, 1993.

Bialas, Volker. *Erdgestalt, Kosmologie und Weltanschauung: die Geschichte der Geodäsie als Teil der Kulturgeschichte der Menschheit.* Stuttgart: Wittwer, 1982.

Bigg, Charlotte. "Photography and the Labour History of Astronomy: The 'Carte du Ciel.'" *The Role of Visual Representations in Astronomy: History and Research Practice,* ed. Klaus Hentschel and Axel D. Wittmann, 90–106. Thun: Harri Deutsch, 2000.

———. "Behind the Lines: Spectroscopic Enterprises in Early Twentieth-Century Europe." Ph.D. diss., Cambridge University, 2002.

———. "Spectroscopic Metrologies."

Bigg, Charlotte, and Klaus Staubermann, eds. "Spectroscopic Histories." *Nuncius* 17 (2002): 583–690, 18 (2003): 732–852.

Bigourdan, Guillaume. *Histoire de l'astronomie d'observation et des observatoires en France.* Paris: Gauthier-Villars, 1930.

Biot, Jean-Baptiste. "Compte rendu sur le traité de navigation de Jean-Baptiste du Bourguet." *Annales maritimes et coloniales* 2 (1817): 216–25.

Blondel, Christine. "Electrical Instruments in Nineteenth-Century France, between Makers and Users." *History and Technology* 13 (1997): 157–82.

———. "Les physiciens français et l'électricité industrielle à la fin du XIXe siècle." *Physis: rivista internazionale di storia della scienza* 35 (1998): 245–71.

Boistel, Guy. "L'astronomie nautique au XVIIIe siècle en France: tables de la Lune et longitudes en mer." Doctoral thesis, Université de Nantes, 2001.

———. "Les longitudes en mer au XVIIIe siècle sous le regard critique du père Pezenas." *Le calcul des longitudes: un enjeu pour les mathématiques, l'astronomie, la mesure du temps et la navigation*, ed. Vincent Jullien, 101–21. Rennes: Presses Universitaires de Rennes, 2002.

———. "Inventaire chronologique des œuvres imprimées et manuscrites du père Esprit Pezenas (1692–1776), jésuite, astronome et hydrographe marseillais." *Revue d'histoire des sciences* 56, no. 1 (2003): 221–45.

———. "De quelle précision a-t-on réellement besoin en mer? Quelques aspects de la diffusion des méthodes de détermination astronomique et chronométrique des longitudes en mer en France, de Lacaille à Mouchez (1750–1880)." *Histoire et mesure* 21 (2006): 121–56.

———, ed. *Observatoire et patrimoine astronomique français*. Cahiers d'histoire et de philosophie des sciences, 54. Lyon: SFHST-ENS, 2005.

Borda, Jean-Charles de, and Pierre Lévêque. "Rapport sur le mémoire et la carte trigonométrique présentés par le citoyen Maingon, lieutenant de Vaisseau." *Procès-verbaux de l'Académie des Sciences de l'Institut de France* 1 (1798): 465–73.

Borheck, Georg Heinrich. *Grundsätze über die Anlage neuer Sternwarten mit Beziehung auf die Sternwarte der Universität Göttingen*, ed. Klaus Beuermann. Göttingen: Universitätsverlag, 2005.

Botting, Douglas. *Humboldt and the Cosmos*. New York: Harper and Row, 1973.

Bourguet, Marie-Noëlle. "La république des instruments: voyage, mesure et science de la nature chez Alexandre de Humboldt." *Marianne–Germania: deutsch-französischer Kulturtransfer in europäischen Kontext*, ed. M. C. Hoock-Demarle, É. François, and M. Werner, 405–36. Leipzig: Leipziger Universitätsverlag, 1998.

Bourguet, Marie-Noëlle, Bernard Lepetit, Marcel Nordmann, et al., eds. *L'invention scientifique de la Méditerranée: Egypte, Morée, Algérie*. Paris: École des Hautes Études en Sciences Sociales, 1998.

Bourguet, Marie-Noëlle, Christian Licoppe, and H. Otto Sibum, eds. *Instruments, Travel and Science*. London: Routledge, 2002.

Brain, Robert Michael. "The Graphic Method: Inscription, Visualization, and Measurement in Nineteenth-Century Science and Culture." Ph.D. diss., UCLA, 1996.

Bratt, Einar. *En krönika om kartor över Sverige*. Stockholm: Generalstabens Litografiska Anstalt, 1958.

Brenni, Paolo. "Il meteorografo di padre Angelo Secchi." *Nuncius* 8 (1993): 197–247.

———. "19th-Century French Scientific Instrument Makers." *Bulletin of the Scientific Instrument Society* [series of 13 articles, 1993–96; see esp. 38 (1993): 11–3, 40 (1994): 3–6, 49 (1996): 3–8, 50 (1996): 19–24, and 51 (1996): 7–16].

Bret, Patrice. "Le dépôt général de la guerre et la formation scientifique des ingénieurs-géographes militaires en France (1789–1830)." *Annals of Science* 41 (1991): 113–57.

Brisbane, Thomas. *Reminiscences of General Sir Thomas Makdougall Brisbane of Brisbane and Makerstoun*. Edinburgh: Constable, 1860.

Brück, Mary T. *Agnes Mary Clerke and the Rise of Astrophysics*. Cambridge: Cambridge University Press, 2002.

Brush, Stephen G. "Looking Up: The Rise of Astronomy in America." *American Studies* 20 (1979): 41–67.

Bürger, Peter. *Theory of the Avant Garde*, trans. Michael Shaw. Minneapolis: University of Minnesota Press, 1983.

Cacciatore, Giuseppe, ed. *Rapporti sulle osservazioni dell'ecclisse totale di sole del dicembre 1870 eseguite in Sicilia dalla commissione italiana, pubblicati a spese del Regio Governo*. Palermo: Stab. Tip. Lao, 1872.

Cahan, David. *An Institute for an Empire: The Physikalisch-Technische Reichsanstalt, 1871–1918*. Cambridge: Cambridge University Press, 1989.

———, ed. *Herman von Helmholtz and the Foundations of Nineteenth-Century Science*. Berkeley: University of California Press, 1993.

Calhoun, Craig, ed. *Habermas and the Public Sphere*. Cambridge: MIT Press, 1992.

Canales, Jimena. "Exit the Frog: Physiology and Experimental Psychology in Nineteenth-Century Astronomy." *British Journal for the History of Science* 34 (2001): 173–97.

———. "Photogenic Venus: The 'Cinematographic Turn' and Its Alternatives in Nineteenth-Century France." *Isis* 93 (2002): 585–613.

Cannon, Susan Faye. *Science in Culture: The Early Victorian Period*. New York: Dawson and Science History Publications, 1976.

Cantor, Geoffrey. *Michael Faraday, Sandemanian and Scientist: A Study of Science and Religion in the Nineteenth Century*. London: Macmillan, 1991.

Cassirer, Ernst. *Kant's Life and Thought*, trans. James Haden, introd. by Stephan Körner. New Haven: Yale University Press, 1981.

Cawood, John. "Terrestrial Magnetism and the Development of International Collaboration in the Early Nineteenth Century." *Annals of Science* 34 (1977): 551–88.

———. "The Magnetic Crusade: Science and Politics in Early Victorian Science." *Isis* 70 (1979): 492–518.

———. "François Arago, savant de l'industrie." *La Recherche* 172 (1980): 1464–72.

Certeau, Michel de. *The Practice of Everyday Life*, trans. Steven Rendall. Berkeley: University of California Press, 1984.

Chabirand. "Étude sur l'éclipse totale de soleil du 18 août 1868, suivie de l'exposé d'une théorie nouvelle sur la constitution physique du soleil." *Revue maritime et coloniale* 25 (1869): 378–401.

Chapin, Seymour L. "P. J. C. Janssen and the Advent of the Spectroscope into Astronomical Prominence." *Griffith Observer* 48 (July 1984): 2–15.

Chapman, Allan. "Private Research and Public Duty: George Biddell Airy and the Search for Neptune." *Journal for the History of Astronomy* 19 (1988): 121–39.

———. "The Astronomical Revolution." *Möbius and His Band: Mathematics and*

Astronomy in Nineteenth-Century Germany, ed. John Fauvel, Raymond Flood, and Robin Wilson, 32–77. Oxford: Oxford University Press, 1993.

———. *Dividing the Circle: The Development of Critical Angular Measurement in Astronomy, 1500–1850*, 2nd edn. Chichester: John Wiley and Sons, 1995.

———. *The Victorian Amateur Astronomer: Independent Astronomical Research in Britain, 1820–1920*. Chichester: Praxis, 1999.

Chinnici, Ileana. "Nineteenth-Century Spectroscopic Instruments in Italian Astronomical Observatories." *Nuncius* 15 (2000): 671–80.

———, ed. *La carte du ciel: correspondance inédite conservée dans les archives de l'Observatoire de Paris*. Paris: Observatoire de Paris, 1999.

Clerke, Agnes M. *A Popular History of Astronomy during the Nineteenth Century*. London: Adam and Charles Black, 1887.

———. *A Popular History of Astronomy during the Nineteenth Century*, 4th edn. London: Adam and Charles Black, 1908.

Cook, Nerida. "A Tale of Two City Pillars: Mongkut and Thai Astrology on the Eve of Modernization." *Patterns and Illusions: Thai History and Thought*, ed. Gehan Wijeyewardene and E. C. Chapman, 276–309. Canberra: Richard Davis Fund / Australian National University, 1993.

Crary, Jonathan. *Techniques of the Observer: On Vision and Modernity in the Nineteenth Century*. Cambridge: MIT Press, 1990.

Crowe, Michael J. *The Extraterrestrial Life Debate, 1750–1900: The Idea of a Plurality of Worlds from Kant to Lowell*. Cambridge: Cambridge University Press, 1986.

Curtiss, John Shelton. *The Russian Army under Nicholas I, 1825–1855*. Durham: Duke University Press, 1965.

Darrigol, Olivier. *Electrodynamics from Ampère to Einstein*. Oxford: Oxford University Press, 2000.

Daston, Lorraine J. "The Physicalist Tradition in Early Nineteenth-Century French Geometry." *Studies in History and Philosophy of Science* 17 (1986): 269–95.

———. "Objectivity and the Escape from Perspective." *Social Studies of Science* 22 (1992): 597–618.

———. "The Moral Economy of Science." *Osiris* 10 (1995): 3–24.

Daston, Lorraine J., and Peter Galison. "The Image of Objectivity." *Representations* 40 (1992): 81–128.

Daum, Andreas. *Wissenschaftspopularisierung im 19. Jahrhundert: bürglische Kultur, naturwissenschaftliche Bildung und die deutsche Öffentlichkeit, 1848–1914*. Munich: R. Oldenburg, 2002.

Daumas, Maurice. *Arago: la jeunesse de la science*, 2nd edn. Paris: Belin, 1987.

Débarbat, Suzanne, Solange Grillot, and Jacques Lévy. *L'Observatoire de Paris: son histoire, 1667–1963*. Paris: Observatoire de Paris, 1984.

Dening, Greg. *The Death of William Gooch: A History's Anthropology*. Honolulu: University of Hawai'i Press, 1995.

————. *Performances*. Melbourne: Melbourne University Press, 1996.

Dettelbach, Michael. "Romanticism and Administration: Mining, Galvanism and Oversight in Alexander von Humboldt's Global Physics." Ph.D. diss., Cambridge University, 1992.

————. "Global Physics and Aesthetic Empire: Humboldt's Physical Portrait of the Tropics." *Visions of Empire: Voyages, Botany, and Representations of Nature*, ed. David Phillip Miller and Peter Hanns Reill, 258–92. Cambridge: Cambridge University Press, 1996.

————. "Humboldtian Science." *Cultures of Natural History*, ed. Nicholas Jardine and James A. Secord, 287–304. Cambridge: Cambridge University Press, 1996.

————. "The Face of Nature: Precise Measurement, Mapping, and Sensibility in the Work of Alexander von Humboldt." *Studies in History and Philosophy of Biological and Biomedical Sciences* 30C (1999): 473–504.

DeVorkin, David H. *Henry Norris Russell, Dean of American Astronomers*. Princeton: Princeton University Press, 2000.

————, ed. *The American Astronomical Society's First Century*. Washington: American Astronomical Society, 1999.

Dewhirst, David. "Meridian Astronomy in the Private and University Observatories of the United Kingdom." *Vistas in Astronomy* 28 (1985): 147–58.

Dick, Steven J. "Pulkovo Observatory and the National Observatory Movement: An Historical Overview." *Inertial Coordinate Systems on the Sky: Proceedings of the IAU Symposium 141, Leningrad, October 1989*, ed. J. H. Lieske and V. K. Abalakin, 19–39. Dordrecht: Kluwer Academic, 1990.

————. *Sky and Ocean Joined: The U.S. Naval Observatory, 1830–2000*. Cambridge: Cambridge University Press, 2003.

Dick, Wolfgang R., and Jürgen Hamel, eds. *Astronomie von Olbers bis Schwarzschild: nationale Entwicklungen und internationale Beziehungen im 19. Jahrhundert*. Acta Historia Astronomiae, 14. Thun: Deutsch, 2002.

Dickens, Charles. *Hard Times: For These Times*. London: Penguin, 1995 [1854].

Dörries, Matthias. "Balances, Spectroscopes, and the Reflexive Nature of Experiment." *Studies in the History and Philosophy of Science* 25 (1994): 1–36.

Drayton, Richard. *Nature's Government*. New Haven: Yale University Press, 2000.

Dreyer, J. L. E. "Observatory." *Encyclopaedia Britannica*, 11th edn, 19 (1910–11): 953–61.

Dvoichenko-Markov, Eufrosina. "The Pulkovo Observatory and Some American Astronomers of the 19th Century." *Isis* 43 (1952): 243–46.

Eagleton, Terry. *The Ideology of the Aesthetic*. Oxford: Basil Blackwell, 1990.

Ebel, Gerhard, and Otto Lührs. "Urania: eine Idee, eine Bewegung, eine Institution wird 100 Jahre alt!" *100 Jahre Urania Berlin, Festschrift: Wissenschaft heute für Morgen*, 15–74. Berlin: Urania, 1988.

Evans, Chris J., and Deborah Jean Warner. "Precision Engineering and Experimental Physics: William A. Rogers, the First Academic Mechanician in the U.S." *The Michelson Era in American Science: 1870–1930*, ed. Stanley Goldberg and Roger H. Stuewer, 2–12. New York: American Institute of Physics, 1988.

Faye, Hervé. "Sur l'état actuel de la géodésie et sur les travaux à entreprendre par le Bureau des Longitudes, de concert avec le Dépôt de la Guerre, pour compléter la partie astronomique du réseau français." *Connaissance des temps pour 1864* (Paris, 1863), 1–20.

Feltus, George Haws, ed. *Abstract of the Journal of Rev. Dan Beach Bradley, M.D., Medical Missionary in Siam, 1835–1873.* Cleveland: Multigraph Department of Pilgrim Church, 1936.

Feudge, Fannie Roper. "An Oriental Monarchy." *Galaxy* 18 (1874): 795–804.

Finkelstein, Gabriel. "'Conquerors of the Künlün'? The Schlaginweit Mission to High Asia, 1854–57." *History of Science* 38 (2000): 179–218.

Flammarion, Camille. *Pluralité des mondes habités.* Paris: Flammarion, 1862.

———. *Astronomie populaire: description générale du ciel.* Paris: Marpon et Flammarion, 1880.

———. *Mémoires biographiques et philosophiques d'un astronome.* Paris: Ernest Flammarion, 1911.

Fleck, Ludwik. "Schauen, sehen, wissen." *Ludwik Fleck: Erfahrung und Tatsache: gesammelte Aufsätze,* ed. Lothar Schäfer and Thomas Schnelle, 147–74. Frankfurt am Main: Suhrkamp, 1983.

Fleming, James Rodger. *Meteorology in America, 1800–1870.* Baltimore: Johns Hopkins University Press, 1990.

Foerster, Wilhelm. "Abendbesuch auf einer Sternwarte." *Sammlung von Vorträgen und Abhandlungen,* vol. 2. Berlin: Reimer, 1887.

———. *Lebenserinnerungen und Lebenshoffnungen (1832 bis 1910).* Berlin: Reimer, 1911.

Forbes, Eric G. *The Birth of Scientific Navigation: The Solving in the 18th Century of the Problem of Finding Longitude at Sea.* Maritime Monographs and Reports, 10. London: National Maritime Museum, 1974.

Forgan, Sophie. "The Architecture of Display: Museums, Universities and Objects in Nineteenth-Century Britain." *History of Science* 32 (1994): 139–62.

Foucault, Michel. *The Order of Things: An Archaeology of the Human Sciences.* New York: Vintage, 1973.

———. "What Is Enlightenment?" *The Foucault Reader,* ed. Paul Rabinow, 32–50. New York: Pantheon, 1984.

———. *Discipline and Punish: The Birth of the Prison,* trans. Alan Sheridan. London: Penguin, 1991.

Fox, Robert. "The Savant Confronts His Peers: Scientific Societies in France, 1815–1914." *The Organization of Science and Technology in France, 1808–1914,* ed. Robert Fox and George Weisz, 241–82. Paris: Maison des Sciences de l'Homme, 1980.

Frängsmyr, Tore, John L. Heilbron, and Robin E. Rider, eds. *The Quantifying Spirit in the 18th Century*. Berkeley: University of California Press, 1990.

Freiesleben, Hans Christian. *Max Wolf: der Bahnbrecher der Himmelsbotographie, 1863–1932*. Stuttgart: Wissenschaftliche Verlagsgesellschaft, 1962.

Frercks, Jan. "Creativity and Technology in Experimentation: Fizeau's Terrestrial Determination of the Speed of Light," *Centaurus* 42 (2000): 249–87.

Galison, Peter. *Einstein's Clocks and Poincaré's Maps: Empires of Time*. New York: W. W. Norton, 2003.

Galison, Peter, and David J. Stump, eds. *The Disunity of Science: Boundaries, Contexts, and Power*. Stanford: Stanford University Press, 1996.

Gingerich, Owen, ed. *Astrophysics and Twentieth-Century Astronomy to 1950*. Cambridge: Cambridge University Press, 1984–.

Gispert, Hélène. "The Effect of War on France's International Role in Mathematics, 1870–1914." *Mathematics Unbound: The Evolution of an International Mathematical Research Community, 1800–1945*, ed. Karen Hunger Parshall and Adrian Rice, 105–21. Providence: American Mathematical Society, 2002.

Godlewska, A., and M. Smith, eds. *Geography and Empire: Critical Studies on the History of Geographical Thought*. Oxford: Basil Blackwell, 1993.

Goldberg, Stanley, and Roger H. Stuewer, eds. *The Michelson Era in American Science, 1870–1930*. New York: American Institute of Physics, 1988.

Gooday, Graeme. "Precision Measurement and the Genesis of Teaching Laboratories in Victorian Britain." *British Journal for the History of Science* 23 (1990): 25–51.

Gould, Benjamin A. "The Observatory at Pulkowa." *North American Review* 144 (July 1849): 143–62.

Hacking, Ian. "Biopower and the Avalanche of Printed Numbers." *Humanities in Society* 5 (1982): 279–95.

———. *The Taming of Chance*. Cambridge: Cambridge University Press, 1990.

Hahn, Roger. "Les observatoires en France au XVIIIe siècle." *La curiosité scientifique au XVIIIe siècle: cabinets et observatoires*, 653–59. Paris: Hermann, 1986.

———. *Pierre Simon Laplace, 1749–1827: A Determined Scientist*. Cambridge: Harvard University Press, 2004.

Hannaway, Owen. "Laboratory Design and the Aim of Science: Andreas Libavius versus Tycho Brahe." *Isis* 77 (1986): 585–610.

Haynes, Raymond, ed. *Explorers of the Southern Sky: A History of Australian Astronomy*. Cambridge: Cambridge University Press, 1996.

Haywood, Richard M. *The Beginnings of Railway Development in Russia in the Reign of Nicholas I, 1835–42*. Durham: Duke University Press, 1969.

Hazelius, J. A. "Berättelse om Fältmätnings-corpsen och dess arbeten, från corpsens ställande under gemensamt befäl med Kongl. Fortifikation 1811, intill 1832 års slut." *Krigsvetenskapsakademiens Handlingar*, 1832, 248–306.

Headrick, Daniel. *The Tools of Empire: Technology and European Imperialism in the Nineteenth Century*. New York: Oxford University Press, 1981.

————. *The Tentacles of Progress: Technology Transfer in the Age of Imperialism, 1850–1940.* Oxford: Oxford University Press, 1988.

Hearnshaw, J. B. *The Analysis of Starlight: One Hundred and Fifty Years of Astronomical Spectroscopy.* Cambridge: Cambridge University Press, 1986.

Hentschel, Klaus. *Mapping the Spectrum: Techniques of Visual Representation in Research and Teaching.* Oxford: Oxford University Press, 2002.

Hentschel, Klaus, and Axel D. Wittmann, eds. *The Role of Visual Representations in Astronomy: History and Research Practice.* Acta Historica Astronomiae, 9. Thun: Harri Deutsch, 2000.

Herrmann, Dieter B. "An Exponential Law for the Establishment of Observatories in the Nineteenth Century." *Journal for the History of Astronomy* 4 (1973): 57–58.

————. *The History of Astronomy from Herschel to Hertzsprung,* rev. and trans. Kevin Krisciunas. Cambridge: Cambridge University Press, 1984.

Herschel, John. *Essays from the Edinburgh and London Quarterly Reviews with Addresses and Other Pieces.* London: Longman, Brown, Green, Longmans, and Roberts, 1857.

Hess, Haro. *Aus der Geschichte der Berliner Urania (1887–1927).* Berlin: Archenhold Sternwarte, 1979.

Hetherington, Norris. "Amateur versus Professional: The British Astronomical Association and the Controversy over Canals on Mars." *British Astronomical Association Journal* 86 (1976): 303–8.

Höjer, Henrik. *Svenska siffror: nationell integration och identifikation genom statistik, 1800–1870.* Hedemora: Gidlunds, 2001.

Hoskin, Michael. "Astronomers at War: South versus Sheepshanks." *Journal for the History of Astronomy* 20 (1980): 175–212.

————, ed. *The General History of Astronomy.* Cambridge: Cambridge University Press, 1984.

Howse, Derek. *Greenwich Time and the Discovery of the Longitude.* Oxford: Oxford University Press, 1980.

————. "The Greenwich List of Observatories: A World List of Astronomical Observatories, Instruments and Clocks." *Journal for the History of Astronomy* 17 (1986): 1–89; and "Amendment List No. 1," *Journal for the History of Astronomy* 25 (1994): 205–18.

Hufbauer, Karl. *Exploring the Sun: Solar Physics since Galileo.* Baltimore: Johns Hopkins University Press, 1991.

Hugo, Victor. *Les misérables.* Paris: Gallimard, 1995.

Humboldt, Alexander von. "Des lignes isothermes et de la distribution de la chaleur sur le globe." *Mémoires de physique et de chimie de la Société d'Arcueil.* Paris, 1817. Repr. New York: Johnson Reprint, 1967.

————. *Lettres américaines d'Alexandre de Humboldt, 1798–1807,* ed. E. T. Hamy. Paris: E. Guilmoto, 1904.

————. *Cosmos: A Sketch of the Physical Description of the Universe,* ed. Michael

Dettelbach and E. C. Otté. Baltimore: Johns Hopkins University Press, 1997.

Hutchins, Roger. "John Philips, 'Geologist-Astronomer,' and the Origins of the Oxford University Observatory, 1853-1875." *History of Universities* 13 (1991): 193-249.

Jackson, Myles W. *Spectrum of Belief: Joseph von Fraunhofer and the Craft of Precision Optics*. Cambridge: MIT Press, 2000.

Jullien, Vincent, ed. *Le calcul des longitudes: un enjeu pour les mathématiques, l'astronomie, la mesure du temps et la navigation*. Rennes: Presses Universitaires de Rennes, 2002.

Jungnickel, Christa, and Russell McCormmarch. *Intellectual Mastery of Nature: Theoretical Physics from Ohm to Einstein*. Chicago: University of Chicago Press, 1986.

Kant, Immanuel. *Critique of Judgement*, trans. James Creed Meredith. Oxford: Oxford University Press, 1952.

———. *Critique of Pure Reason*, trans. Paul Guyer and Allen W. Wood. Cambridge: Cambridge University Press, 1988.

———. *Grounding for the Metaphysics of Morals, with On a Supposed Right to Lie because of Philanthropic Concerns*, 3rd edn, trans. James Ellington. Indianapolis: Hackett, 1993.

Kern, Stephen. *The Culture of Time and Space, 1880-1918*. Cambridge: Harvard University Press, 1983.

Kevles, Daniel J. *The Physicists: The History of a Scientific Community in Modern America*. Cambridge: Harvard University Press, 1995.

King, Henry C. *The History of the Telescope*. Mincola, N.Y.: Dover, 2003 [1955].

Konvitz, Josef W. *Cartography in France, 1660-1848: Science, Engineering, and Statecraft*. Chicago: Chicago University Press, 1987.

Krisciunas, Kevin. *Astronomical Centers of the World*. Cambridge: Cambridge University Press, 1988.

Krüger, Lorenz, Lorraine J. Daston, and Michael Heidelberger, eds. *The Probabilistic Revolution*. Cambridge: MIT Press, 1987.

Kuklick, Henrika, and Robert E. Kohler, eds. "Science in the Field." *Osiris* 11 (1996) [special issue].

Lacombe, Henri, and Pierre Costabel, eds. *La figure de la terre du XVIIIᵉ siècle à l'ère spatiale*. Paris: Gauthier-Villars, 1988.

Lalande, Jérôme. *Bibliographie astronomique, avec l'histoire de l'astronomie depuis 1781 jusqu'à 1802*. Paris: Imprimerie de la République, 1803.

Lamy, Jérôme. *L'observatoire de Toulouse aux XVIIIᵉ et XIXᵉ siècles: archéologie d'un espace savant*. Rennes: Presses Universitaires de Rennes, 2007.

———. *La carte du ciel: histoire et actualité d'un projet scientifique international*. Paris, EDP Sciences, 2008.

Lankford, John, with Ricky L. Slavings. *American Astronomy: Community, Careers, and Power, 1859-1940*. Chicago: University of Chicago Press, 1997.

Lankford, John, and Ricky L. Slavings. "The Industrialization of American Astronomy, 1880–1940." *Physics Today* 49, no. 1 (1996): 34–40.

Latour, Bruno. *Science in Action: How to Follow Scientists and Engineers through Society*. Cambridge: Harvard University Press, 1987.

Launay, Françoise. *Un globe-trotter de la physique céleste: l'astronome Jules Jannssen*. Paris: Vuibert / Observatoire de Paris, 2008.

Laurie, P. S. "The Board of Visitors of the Royal Observatory." *Quarterly Journal of the Royal Astronomical Society* 7 (1966): 169–85, 8 (1967): 334–53.

Lenoir, Timothy. "The Göttingen School and the Development of Transcendental *Naturphilosophie* in the Romantic Era." *Studies in History of Biology* 5 (1981): 111–205.

Leonowens, Anna. *The English Governess at the Siamese Court, Being Recollections of Six Years in the Royal Palace at Bangkok*. Boston: Fields, Osgood, 1870.

———. *The Romance of the Siamese Harem Life*. London: Trübner, 1873.

Lesté-Lasserre, Nicolas. "Le journal d'observations astronomiques au XVIIIᵉ siècle: entre autocritique raisonnée et affirmation d'un art," mémoire de Diplôme d'Études Approfondies. École des Hautes Études en Sciences Sociales, 2004.

Levitt, Theresa. "Editing Out Caloric: Fresnel, Arago and the Meaning of Light." *British Journal for the History of Science* 33 (2000): 49–65.

———. "The Shadow of Enlightenment: Practices of Representation in France, 1789–1848." Ph.D. diss., Harvard University, 2002.

———. "Biot's Paper and Arago's Plates: Photographic Practice and the Transparency of Representation." *Isis* 94 (2003): 456–76.

Livingston, Dorothy Michelson. *The Master of Light: A Biography of Albert A. Michelson*. New York: Charles Scribner's Sons, 1973.

Locher, Fabien. "Le nombre et le temps: la météorologie en France (1830–1880)." Doctoral thesis, École des Hautes Études en Sciences Sociales, 2004.

———. "The Observatory, the Land-Based Ship and the Crusades: Earth Sciences in European Context, 1830–50." *British Journal for the History of Science* 40 (2007): 491–504.

———. *Le savant et la tempête: étudier l'atmosphère et prévoir le temps au XIXᵉ siècle*. Rennes: Presses Universitaires de Rennes, 2008.

Lockyer, Joseph Norman. *Contributions to Solar Physics*. London: Macmillan, 1874.

———. *Inorganic Evolution as Studied by Spectrum Analysis*. London: Macmillan, 1900.

Lockyer, T. Mary, and Winifred L. Lockyer. *Life and Work of Sir Norman Lockyer*. London: Macmillan, 1928.

Loménie, Louis de. *Sketches of Conspicuous Living Characters of France*, trans. Robert M. Walsh. Philadelphia: Lea and Blanchard, 1841.

Loomis, Elias. "Astronomical Observatories in the United States." *Harper's New Monthly Magazine* 13 (1856): 25–52.

Mädler, Johann Heinrich. *Populäre Astronomie*. Berlin: Heymann, 1841.

Maffeo, Sabino, S. J. *The Vatican Observatory in the Service of Nine Popes*. Vatican City: Vatican Observatory Publications, 2001.

Mahmoud-Bey. *Rapport à son altesse Mahammed Saïd, vice-roi d'Égypte sur l'éclipse totale de soleil observée à Dongalah (Nubie) le 18 juillet 1860*. Paris: Mallet-Bachelier, 1861.

Maison, Laetitia. "La fondation et les premiers travaux de l'observatoire astronomique de Bordeaux (1871–1906): histoire d'une réorientation scientifique." Doctoral thesis, Université de Bordeaux-I, 2004.

Marcuse, Herbert. *The Aesthetic Dimension: Toward a Critique of Marxist Aesthetics*. Boston: Beacon, 1978.

Marguet, Frédéric. *Histoire générale de la navigation du XVᵉ au XXᵉ siècle*. Paris: Société d'Éditions Géographiques, Maritimes et Coloniales, 1931.

Martina, Giacomo. *Pio IX*. Rome: Studium, 1973-91.

Mascart, Eleuthère. "Sur la réfraction des gaz." *Comptes rendus des séances hebdomadaires de l'Académie des Sciences* 78 (1874): 617-21.

Maunder, E. Walter. *The Royal Observatory, Greenwich: A Glance at Its History and Work*. London: Religious Tract Society, 1900.

Mauss, Marcel. "Les techniques et la technologie." Mauss, *Œuvres*, ed. Victor Karady, 3:250-6. Paris: Éditions de Minuit, 1969 [1941].

Mazzotti, Massimo. "For Science and for the Pope-King: Writing the History of the Exact Sciences in Nineteenth-Century Rome." *British Journal for the History of Science* 33 (2000): 257-82.

McCrea, W. H. *The Royal Greenwich Observatory*. London: Her Majesty's Stationery Office, 1975.

McGucken, William. *Nineteenth-Century Spectroscopy: Development of the Understanding of Spectra, 1802-1897*. Baltimore: Johns Hopkins University Press, 1969.

McNeill, William H. *The Pursuit of Power: Technology, Armed Force, and Society since A.D. 1000*. Oxford: Blackwell, 1983.

Meadows, A. J. *Early Solar Physics*. Oxford: Pergamon, 1970.

———. *Science and Controversy: A Biography of Sir Norman Lockyer*. Cambridge: MIT Press, 1972.

———. *Greenwich Observatory: Three Volumes by Different Authors Telling the Story of Britain's oldest Scientific Institution, the Royal Observatory at Greenwich and Herstmonceux, 1675-1975*, vol. 2, *Recent History (1836-1975)*. London: Taylor and Francis, 1975.

Mérimée, Prosper. *Correspondance générale*, ed. Maurice Parturier with Pierre Josserand and Jean Mallion. Paris: Le Divan, Privat, 1941-64.

Meyer, Max Wilhelm. *Die Geschichte der Urwelt: eine Wanderung durch vorsintflutliche Landscjhaften: Vorgetragen im wissenschaftlichen Theater der Urania zu Berlin*, 3rd. edn. Berlin: Wilhelm Gronau, 1890.

———. *Die Urania zu Berlin: Bericht des Direktors*. Berlin: Paetel, 1891.

Michelson, Albert A. "Experimental Determination of the Velocity of Light." *Proceedings of the American Association for the Advancement of Science* 27 (1878): 71–77.

———. "Experimental Determination of the Velocity of Light made at the U.S. Naval Academy, Annapolis." *U.S. Nautical Almanac Office Astronomical Papers* 1, no. 3 (1882): 109–45.

———. "A Plea for Light Waves." *Proceedings of the American Association for the Advancement of Science* 37 (1889): 67–78.

Miller, David Philip. "The Revival of the Physical Sciences in Britain, 1815–1840." *Osiris* 2 (1986): 107–34.

Mitchell, S. A. *Eclipses of the Sun*. New York: Columbia University Press, 1923.

Moffat, Abbot Low. *Mongkut, the King of Siam*. Ithaca: Cornell University Press, 1961.

Morley, Edward Williams. "Biographical Memoir of William Augustus Rogers, 1832–1898." *Biographical Memoirs of the National Academy of Sciences* 4 (1902): 187–99.

Morrell, Jack, and Arnold Thackray. *Gentlemen of Science: Early Years of the British Association for the Advancement of Science*. Oxford: Oxford University Press, 1981.

Morton-Gledhill, Rowan I. "The Architecture of Astronomy in the British Isles: A General Study." *Vistas in Astronomy* 32 (1989): 235–83.

Mouchez, Ernest. "Observation du passage de Vénus effectuée à l'île Saint-Paul: phénomènes optiques observés aux environs des contacts." *Comptes rendus des séances hebdomadaires de l'Académie des Sciences* 80 (1875): 611–18.

———. "Création d'un observatoire astronomique d'étude, dans le parc de Montsouris, par le Bureau des Longitudes." *Annuaire pour l'an 1876 publié par le Bureau des Longitudes* (Paris, 1876), 449–56.

———. "Observations relatives à l'ouvrage présenté à l'Académie par M. Yvon Villarceau, sous le titre de *Nouvelle Navigation*." *Comptes rendus des séances hebdomadaires de l'Académie des Sciences* 84 (1877): 1207–11.

Moyer, Albert E. *A Scientist's Voice in American Culture: Simon Newcomb and the Rhetoric of Scientific Method*. Berkeley: University of California Press, 1992.

Müller, P. *Sternwarten in Bildern: Architektur und Geschichte der Sternwarten von den Anfangen bis ca. 1950*. Berlin: Springer, 1992.

Norberg, Arthur L. "Simon Newcomb's Early Astronomical Career." *Isis* 69 (1978): 209–25.

Novokshanova (Sokolovskaia), Zinaida Kuz'minichna. *Vasilii Iakovlevich Struve*. Moscow, 1964.

O'Connell, Joseph. "Metrology: The Creation of Universality by the Circulation of Particulars." *Social Studies of Science* 23 (1993): 129–73.

Olesko, Kathryn M. "Michelson and the Reform of Physics Instruction at the Naval Academy in the 1870s." *The Michelson Era in American Science, 1870–*

1930, ed. Stanley Goldberg and Roger H. Stuewer, 111–32. New York: American Institute of Physics, 1988.

Ophir, Adi, and Steven Shapin. "The Place of Knowledge: A Methodological Survey." *Science in Context* 4 (1991): 3–21.

Ord, Harry St. George. "An Account of the Visit to the Late King of Siam at Hua Wan in August 1868." *Bangkok Calendar for the Year of Our Lord 1870, Corresponding to the Siamese Civil Era 1231–2, and Nearly to the Chinese Cycle Era 4507, Being the 7th Year of the 76th Chinese Cycle of 60*, comp. D. B. B[radley], 118–31. Bangkok: American Missionary Association, 1870.

Palladino, Paolo, and Michael Worboys. "Science and Imperialism." *Isis* 84 (1993): 91–102.

Pang, Alex Soojung-Kim. *Empire and the Sun: Victorian Solar Eclipse Expeditions*. Stanford: Stanford University Press, 2002.

Pansini, Valeria. "L'œil du topographe et la science de la guerre: travail scientifique et perception militaire (1760–1820)." Doctoral diss., École des Hautes Études en Sciences Sociales, 2002.

Pápay, Gyula. "Studien zur Herausbildung der Wissenschaftsdisziplin Kartographie." Ph.D. diss., Willhelm-Pieck-Universität, Rostock, 1988.

Perkins, Maureen. *Visions of the Future: Almanacs, Time, and Cultural Change, 1775–1870*. Oxford: Clarendon, 1996.

Perrier, François. "Nouvelle méridienne de France." *Mémorial du dépôt de la guerre*, 1885, vol. 1, no. 1.

Petit, Annie. "L'esprit de la science anglaise et les français au XIXᵉ siècle." *British Journal for the History of Science* 17 (1984): 273–83.

Peytier, Jean-Pierrc. "Nouvelle description géométrique de la France, 3ᵉ partie." *Mémorial du dépôt de la guerre* (Paris, 1853), 9.

Pizzamiglio, Pierluigi, and Giorgio Tabarroni. *Pio IX e le macchine ottiche*. Bologna: Clueb, 1981.

Poincaré, Henri. *The Value of Science*, trans. George Bruce Halsted. New York: Dover, 1958.

Porter, Theodore M. *The Rise of Statistical Thinking, 1820–1900*. Princeton: Princeton University Press, 1986.

———. *Trust in Numbers: The Pursuit of Objectivity in Science and Public Life*. Princeton: Princeton University Press, 1995.

Portolano, Marlana. "John Quincy Adams's Rhetorical Crusade for Astronomy." *Isis* 91 (2000): 480–503.

Prakash, Gyan, ed. *After Colonialism: Imperial Histories and Postcolonial Displacements*. Princeton: Princeton University Press, 1995.

Pratt, Mary Louise. *Imperial Eyes: Travel Writings and Transculturation*. London: Routledge, 1992.

Pyenson, Lewis. *Cultural Imperialism and Exact Science: German Expansion Overseas, 1900–1930*. New York: Peter Lang, 1983.

————. *Civilizing Mission: Exact Sciences and French Overseas Expansion, 1830–1940.* Baltimore: Johns Hopkins University Press, 1993.

————. "On the Military and Exact Sciences in France." *National Military Establishments and the Advancement of Science and Technology: Studies in Twentieth-Century History*, ed. Paul Forman and Jose M. Sánchez-Ron, 135–52. Boston: Kluwer Academic, 1996.

Rabinbach, Anson. *The Human Motor: Energy, Fatigue, and the Origins of Modernity.* New York: Basic, 1990.

Raj, Kapil. *Relocating Modern Science: Circulation and the Construction of Knowledge in South Asia and Europe, 1650–1900.* Houndmills, Basingstoke, Hampshire: Palgrave Macmillan, 2007.

Recht, Roland. *La lettre de Humboldt: du jardin paysager au daguerréotype.* Paris: C. Bourgois, 1989.

Redondi, Pietro. "Cultura e scienza dall'illuminismo al positivismo." *Storia d'Italia. Annali*, ed. Gianni Micheli, 3 (1980): 797–811. Turin: Einaudi.

Rehbock, Phillip F. *The Philosophical Naturalists: Themes in Early Nineteenth-Century British Biology.* Madison: University of Wisconsin Press, 1983.

Reingold, Nathan, ed. *Science in Nineteenth-Century America: A Documentary History.* New York: Hill and Wang, 1964.

Renan, Ernest. *L'avenir de la science*, 2nd edn. Paris: Calman-Lévy, 1910.

Riasanovsky, Nicholas V. *Nicholas I and Official Nationality in Russia, 1825–1855.* Berkeley: University of California Press, 1967.

Richards, Robert. *The Romantic Conception of Life: Science and Philosophy in the Age of Goethe.* Chicago: University of Chicago Press, 2002.

Richardson, William. *A Catalogue of 7385 Stars Chiefly in the Southern Hemisphere.* London: Clowes, 1835.

Rock, Allan J. *Nationalizing Science: Adolphe Wurtz and the Battle for French Chemistry.* Cambridge: MIT Press, 2001.

Rondeau Jozeau, Marie Françoise. "Géodésie au XIXème siècle: de l'hégémonie française à l'hégémonie allemande: regards belges: compensation et méthode des moindres carrés." Doctoral thesis, Université Paris VII–René Diderot, 1997.

Rowland, Henry Augustus. "A Plea for Pure Science." *Proceedings of the American Association for the Advancement of Science* 32 (1883): 105–26.

Rümker, Christian Carl Ludwig. *On the Most Effectual Means of Encouraging Scientific Undertakings.* Hamburg: Nestler, 1832.

Russell, H. C. "Astronomical and Meteorological Workers in New South Wales, 1778–1860." *Australasian Association for the Advancement of Science, Reports* 1 (1888): 45–94.

Said, Edward W. *Culture and Imperialism.* London: Vintage, 1993.

Sauzereau, Olivier. 2000. *Nantes au temps de ses observatoires.* Nantes: Coiffard, 2000.

Schaffer, Simon. "Herschel in Bedlam: Natural History and Stellar Astronomy." *British Journal for the History of Science* 13 (1980): 211–39.

———. "Authorized Prophets: Comets and Astronomers after 1759." *Studies in Eighteenth-Century Culture* 17 (1987): 45–74.

———. "Astronomers Mark Time: Discipline and the Personal Equation." *Science in Context* 2 (1988): 115–46.

———. "Self Evidence." *Critical Inquiry* 18 (1992): 327–62.

———. "The Consuming Flame: Electric Showmen and Tory Mystics in the World of Goods." *Consumption and the World of Goods*, ed. John Brewer and Roy S. Porter, 489–526. London: Routledge, 1993.

———. "Where Experiments End: Tabletop Trials in Victorian Astronomy." *Scientific Practice: Theories and Stories of Doing Physics*, ed. Jed Z. Buchwald, 257–99. Chicago: University of Chicago Press, 1995.

———. "On Astronomical Drawing." *Picturing Science, Producing Art*, ed. Caroline A. Jones and Peter Galison, 441–74. New York: Routledge, 1998.

———. "Enlightened Automata." *The Sciences in Enlightened Europe*, ed. William Clark, Jan Golinski, and Simon Schaffer, 126–65. Chicago: Chicago University Press, 1999.

———. "Metrology, Metrication and Victorian Values." *Victorian Science in Context*, ed. Bernard Ligthman, 438–74. Chicago: University of Chicago Press, 1999.

Schiavon, Martina. "François Perrier et la Nouvelle Méridienne de France (1870–1895): un savant officier rénovateur de la géodésie militaire française." *Revue scientifique et technique de la défense* 55 (2002): 65–78.

———. "Itinéraires de la précision: géodésiens, artilleurs, savants et fabricants d'instruments en France, 1870–1930." Doctoral diss., École des Hautes Études en Sciences Sociales, 2002.

———. "Des savants-officiers entre science, armée, état et industrie de précision: les géodésiens du Service Géographique de l'Armée, 1887–1920." *Le sabre et l'éprouvette: l'invention d'une science de guerre, 1914–1939*, ed. David Aubin and Patrice Bret, 60–73. Paris: Agnès Viénot, 2003 [no. 6 in series "14–18 aujourd'hui"].

———. "Astronomie de terrain, entre monde académique et armée." *Cahiers François Viète*, 2007.

Schiller, Friedrich. *On the Aesthetic Education of Man, in a Series of Letters*, ed. and trans. Elizabeth M. Wilkinson and L. A. Willoughby. Oxford: Clarendon, 1986.

Schivelbush, Wolfgang. *The Railway Journey: The Industrialization of Time and Space in the Nineteenth Century*, trans. Anselm Hollo. Berkeley: University of California Press, 1981.

———. *Disenchanted Night: The Industrialization of Light in the Nineteenth Century*. Berkeley: University of California Press, 1988.

Schmidgen, Henning. "Time and Noise: The Stable Surroundings of Reaction

Experiments, 1860–1890." *Studies in History and Philosophy of Biological and Bio-medical Sciences* 34 (2003): 237–75.

Schumacher, Hans Christian."Nachrichten über die Instrumente der kaiserlichen Hauptsternwarte Pulkowa." *Astronomische Nachrichten* 18, no. 411 (1840): 33–44.

Schwann, P. "Zum 25-jährigen Bestehen der Gesellschaft Urania in Berlin." *Himmel und Erde* 25 (1913): 337–69.

Secchi, Angelo. "Sugli spettri prismatici della luce de' corpi celesti." *Memorie dell'Osservatorio del Collegio Romano* 2 (1860–63): 121–28.

——. 1867–69. "Sugli spettri prismatici delle stelle fisse." *Memorie della Società Italiana delle Scienze* 1 (1867): 71–78, 2 (1869): 73–133.

——. *Descrizione del meteorografo dell'osservatorio del Collegio Romano*. Rome: Tip. delle Belle Arti, 1870.

——. *L'unità delle forze fisiche*. Milan: Treves, 1874 [1864].

Secchi, Angelo, and Giovanni Virginio Schiaparelli. *Corrispondenza (1861–1878)*, ed. L. Buffoni, A. Manara, and P. Tucci. Milan: Artes, 1991.

Secord, James A. *Victorian Sensation: The Extraordinary Publication, Reception, and Secret Authorship of "Vestiges of the Natural History of Creation."* Chicago: University of Chicago Press, 2000.

Service, John. *Thir notandums, Being the Literary Recreations of Laird Canticarl of Mongrynen (of Kittle Memory), to Which Is Appended a Biographical Sketch of James Dunlop, Esq.* Edinburgh: Pentland, 1890.

Sewell, William. *Work and Revolution in France: The Language of Labor from the Old Regime to 1848*. Cambridge: Cambridge University Press, 1980.

Shackleford, Jole. "Tycho Brahe, Laboratory Design, and the Aim of Science: Reading Plans in Context." *Isis* 84 (1993): 211–30.

Shapin, Steven. "The House of Experiment in Seventeenth-Century England." *Isis* 79 (1988): 373–404.

——. "'The Mind Is Its Own Place': Science and Solitude in Seventeenth-Century England." *Science in Context* 4 (1991): 191–218.

Shinn, Terry, and Bernward Joerges, eds. *Instrumentation between Science, State, and Industry*. Dordrecht: Kluwer Academic, 2001.

Sibum, H. Otto. "Reworking the Mechanical Value of Heat: Instruments of Precision and Gestures of Accuracy in Early Victorian England." *Studies in the History and Philosophy of Science* 26 (1995): 73–106.

——. "Experimentalists in the Republic of Letters." *Science in Context* 16 (2003): 89–120.

——. "Narrating by Numbers: Keeping an Account of Early Nineteenth-Century Laboratory Experiences." *Reworking the Bench: Research Notebooks in the History of Science*, ed. Frederic Lawrence Holmes, Jürgen Renn, and Hans-Jörg Rheinberger, 141–58. Dordrecht: Kluwer Academic, 2003.

Siestrunck, René. "La carte militaire." *Cartes et figures de la terre*, ed. Jean-Loup Rivière, 363–74. Paris: Centre Georges Pompidou, 1980.

Smith, Malcolm. *A Physician at the Court of Siam*. London: Country Life Limited, 1947.

Smith, Merritt Roe, ed. *Military Enterprise and Technological Change: Perspectives on the American Experience*. Cambridge: MIT Press, 1985.

Smith, Robert W. "A National Observatory Transformed: Greenwich in the 19th Century." *Journal for the History of Astronomy* 45 (1991): 5–20.

Smyth, Charles Piazzi. *Three Cities in Russia*. London: Rovell Reeve, 1862.

Sörbom, Per. *Läsning för folket: studier i tidig svensk folkbildningshistoria*. Stockholm: Norstedts, 1972.

Spens, Carl Gustaf. "Jemnförelse emellan de metoder at förfärdiga kartor, som vid svenska och utländska topografiska corpsernas arbeten blfvit begagnade." *Krigsvetenskapsakademiens Handlingar*, 1837, 269–324.

Stål, C. *Lärobok i topografien för linie-officerare*. Stockholm, 1835.

Staley, Richard. "Michelson's Interferometer: Experiment or Instrument?" *Instrument–Experiment: historische Studien*, ed. Christoph Meinel, 192–200. Berlin: Verlag für Geschichte der Naturwissenschaften und der Technik, 2000.

———. "The Interferometer and the Spectroscope: Michelson's Standards and the Spectroscopic Community." *Nuncius* 18 (2003): 779–801.

Stanley, Matthew. "'An Expedition to Heal the Wounds of War': The 1919 Eclipse and Eddington as Quaker Adventurer." *Isis* 94 (2003): 56–89.

Stéphan, Édouard. "Rapport sur l'observation de l'éclipse de soleil du 18 août 1868." *Archives des missions scientifiques et littéraires*, 2nd ser., 5 (1868): 534–75.

———. "Voyage de la commission française envoyée par M. le ministre de l'instruction publique sur la côte orientale de la presqu'île de Malacca, pour y observer l'éclipse totale de soleil du 18 août 1868." *Annales scientifiques de l'École Normale Supérieure* 7 (1870): 99–162.

Struve, Friedrich Georg Wilhelm. *Expédition chronométrique exécutée par ordre de Sa Majesté l'Empereur Nicolas Ier, entre Altona et Greenwich pour la détermination de la longitude géographique relative de l'Observatoire Central de Russie*. St. Petersburg: Académie Impériale des Sciences, 1844.

———. *Description de l'observatoire astronomique central de Poulkova*. St. Petersburg: Académie Impériale des Sciences, 1845.

———. *Arc du méridien de 25°20′ entre le Danube et la Mer Glaciale, mesuré, depuis 1816 jusqu'en 1855, sous la direction C. de Tenner, Chr. Hansteen, N. H. Selander et F. G. W. Struve*. St. Petersburg: Académie Impériale des Sciences, 1860.

———. *Wilhelm Struve: zur Erinnerung an den Vater den Geschwistern dargebracht*. Karlsruhe: Braun, 1895.

Susalla, Peter. "The Old School in a Progressive Science: George Cary Comstock and the Middle Ground between 'Old' and 'New' Astronomy, 1879–1922." M.A. thesis, University of Wisconsin, 2006.

Svanberg, Jöns. *Exposition des opérations faites en Laponie*. Stockholm: J. P. Lindh, 1805.

Sweetnam, George K. *The Command of Light: Rowland's School of Physics and the Spectrum*. Philadelphia: American Philosophical Society, 2000.

Tacchini, Pietro. "Sulla nuova società degli spettroscopisti italiani." *Memorie della società degli spettroscopisti italiani* 1 (1872): 3–6.

Thiphakorawong, Chaophraya. *The Dynastic Chronicles: Bangkok Era: The Fourth Reign*, B.E. 2394–2411 (A.D. 1851–1868), trans. Chadin Flood. Tokyo: Centre for East Asian Cultural Studies, 1966.

Thongchai Winichakul. *Siam Mapped: A History of the Geo-Body of a Nation*. Honolulu: University of Hawaii Press, 1994.

———. "The Quest for '*Siwilai*': A Geographical Discourse of Civilized Thinking in the Late Nineteenth Century and Early Twentieth Century Siam." *Journal of Asian Studies* 59 (2000): 528–49.

Tibell, Gustaf Wilhelm. "Om militäriska undervisnings-verk i Sverige." *Krigsmannasällskapets handlingar*, 1797, 72–104.

———. "Om Kongl. Fältmätnings-corpsens inrättning och arbeten." *Krigsvetenskapsakademiens Handlingar*, 1826, 59–97.

———, ed. *Samling af Kongl. Maj[estä]ts nådiga bref och instruktioner rörande Kongl. Fältmätnings-corpsens organisation och tjenstgöring i fred och fält*. Stockholm, 1810.

Tingsten, Lars. *Gustaf Wilhelm af Tibell: huvuddragen av hans liv, hans verksamhet såsom generaladjutant för armén och hans avskedand, tillika en studie av Sveriges krigföring år 1808*. Stockholm: Norstedts, 1924.

Tobin, William. *The Life and Science of Léon Foucault: The Man Who Proved the Earth Rotates*. Cambridge: Cambridge University Press, 2003.

Turner, Gerard L'E. *Nineteenth-Century Scientific Instruments*. Berkeley: University of California Press, 1983.

Udías, Augustin. *Searching the Heavens and the Earth: The History of Jesuit Observatories*. Astrophysics and Space Science Library, 286. Dordrecht: Kluwer Academic, 2003.

Vagts, Alfred. *A History of Militarism, Civilian and Military*, 2nd edn. New York: Free Press, 1967 [1959].

Van Helden, Albert, and Thomas L. Hankins, eds. "Instruments." *Osiris* 9 (1994): 1–242 [special issue].

Vorontsov-Vel'iaminov, Boris Aleksandrovich. *Ocherki istorii astronomii v Rossii*. Moscow: Gos. izd-vo tekhniko-teoret. lit-ry, 1956.

Warner, Brian, and Nancy Warner. *MacLear and Herschel: Letters and Diaries at the Cape of Good Hope, 1834–1838*. Cape Town: Balkema, 1984.

Wasana Obayawath. "L'emprise thaïe sur le Cambodge et l'établissement du protectorat français." *Études indochinoises: frontières et contacts dans la péninsule indochinoise (XVIᵉ–XIXᵉ s.)*. Études et documents 13:48–73. Aix en Provence: Institut d'Histoire des Pays d'Outre-Mer, 1981.

Werrett, Simon. "An Odd Sort of Exhibition: The St. Petersburg Academy of Sciences in Enlightened Russia." Ph.D. diss., Cambridge University, 2000.

Widmalm, Sven. *Mellan kartan och verkligheten: geodesi och kartläggning, 1695–1860.* Uppsala: Inst. för idé- och lärdomshistoria, 1990.

————. "A Commerce of Letters: Astronomical Communication in the Eighteenth Century." *Science Studies* 2 (1992): 43–58.

Williams, Mari E. "Astronomical Observatories as Practical Space: The Case of Pulkowa." *The Development of the Laboratory: Essays on the Place of Experiment in Industrial Civilization,* ed. Frank A. J. L. James, 118–36. Basingstoke, Hampshire: Macmillan, 1989.

Wilson, Constance M. "State and Society in the Reign of Mongkut, 1851–1868: Thailand on the Eve of Modernization," Ph.D. diss., Cornell University, 1970.

Wise, M. Norton, ed. *The Values of Precision.* Princeton: Princeton University Press, 1995.

ABOUT THE CONTRIBUTORS

David Aubin is a professor at the Université Pierre et Marie Curie–Paris 6 and a member of the Institut de Mathématiques de Jussieu.

Charlotte Bigg is a research scientist at the Centre National de la Recherche Scientifique/Centre Alexandre Koyré, Paris.

Guy Boistel is a physics teacher and a member of the Centre François Viète for the history of science and technology at the University of Nantes.

Massimo Mazzotti teaches the history and sociology of science at the University of California, Berkeley.

Theresa Levitt is an assistant professor in the Department of History at the University of Mississippi.

Ole Molvig is an assistant professor in the Department of History at Vanderbilt University.

Simon Schaffer is a professor of the history of science at the University of Cambridge.

Martina Schiavon is a postdoctoral fellow at the Archives Henri Poincaré, University of Nancy 2, France.

H. Otto Sibum is the Hans Rausing Professor of History of Science and director of the Office for History of Science at Uppsala University, Sweden.

Richard Staley is an associate professor of the history of science at the University of Wisconsin, Madison.

John Tresch is an assistant professor of the history and sociology of science at the University of Pennsylvania.

Simon Werrett is an assistant professor in the Department of History at the University of Washington.

Sven Widmalm is a professor at the Department of Technology and Social Change, Linköping University, Sweden.

INDEX

Nicholas I, Tsar, 33–52

Night, 1, 4, 10, 22, 26, 49, 87, 90, 120, 160, 213–14, 293, 295, 330–32

Nikolaev Observatory, 38–42, 51

North Pole, 150, 201

Numbers, 7, 9–16, 24–26, 50, 63–66, 93–94, 125–29, 133–35, 160, 175, 183, 204–5, 213, 226, 230, 255–56, 269, 300, 308, 310. *See also* Mathematics

Objectivity, 66, 76, 205, 257–69, 271; aperspectival, 256, 270, 275; mechanical, 6, 14, 256, 270, 276–77

Obscurity, 87–88, 285, 288, 299, 325, 332

Observation, 11, 73, 86–90, 100–101, 108, 120–22, 129–40, 163–68, 180, 202–4, 210, 213, 215–17, 229–30, 234–35, 240–46, 256, 259, 268–76, 285–87, 293–302; directions for making, 49, 106–7, 296; with naked eye, 214, 299; publication of, 131, 135; site of, 92–94. *See also* Books; Meteorology; Moon; Navigation; Observers

Observatories: architecture and layout of, 4, 22, 45–46, 61, 99, 121, 130–31, 313, 332; computing rooms of, 10, 15, 50; culture of, 8–25, 228; directors of, 8, 11, 13, 16, 33, 41, 44–45, 49–50, 60, 78, 86, 90, 95, 106, 137, 148, 162–68, 204–6, 226, 232, 237, 243, 285–86, 289, 310, 328–29, 336; discipline and, 10, 16, 20, 48–49, 64–67, 124, 129, 176, 183–84, 192, 217, 269, 278, 309–10; as laboratories of visuality, 20; libraries and, 10, 46, 50–51, 103, 313, 330; naval, 16, 38, 41, 60, 148–49, 161–68, 225–26, 229–30, 232, 237; networks of, 34, 39, 51, 66–67, 76–79, 127–28, 139, 162–63, 255, 272; personnel of, 7, 25–26, 40, 48, 118–20, 129, 139, 165, 168, 176,

211, 215, 302; as places, 1, 4, 7–8, 22, 44–45, 119–20; routine in, 10, 165, 232, 237; as temples, 1, 46; Tychonic vs. Hevelian, 52; visitors to, 22, 33, 36, 45–48, 136, 164, 206, 285, 312, 329–35; women and, 232. *See also* Instruments; Popular observatories; Telescopes

Observatory sciences, 2–8, 11–20, 23–25, 33, 58, 69, 104–6, 122, 125–28, 138–39, 199–200, 217, 231, 255–57, 269, 278, 286, 307

Observatory techniques, 1–7, 11–21, 25–26, 36, 39, 42–43, 46, 50–51, 59, 118–28, 148–49, 167, 199–200, 204–5, 217–19, 227–28, 241, 255, 307, 318–21

Observers, 4, 40, 49, 92, 124, 128, 131, 138, 140, 175, 204–5, 207, 236, 257, 299; difference among, 10, 75, 271; expertise, 47, 129, 133, 213–14, 217; instructions to, 15, 106, 214, 296, 333; network of, 15–16, 18, 255, 272–73; nonhuman, 16, 66, 271; nonprofessional, 77, 107, 296, 321

Octant, 148, 150, 152, 163

On-board school, 157

Optics, 11–12, 139, 162, 228–34, 245–46, 288; instruments, 9–10, 20–22, 25, 49, 214, 320–22, 332, 337, 339; refractometry, 235–40. *See also* Circle; Instruments; Prism; Telescope

Ordnance Survey, 176, 183–84, 208

Oxford University, 131

Padua Observatory, 73, 75, 77

Palermo Observatory, 73–77

Panopticon, 20, 47, 126, 192

Parallax: lunar, 151, 156; solar, 229; stellar, 33, 39, 50, 126

Paramatta Observatory, 19, 36, 39, 118–24, 139

DAVID AUBIN is an associate professor at the
Université Pierre et Marie Curie–Paris 6 and
a member of the Institut de Mathématiques
de Jussieu.

CHARLOTTE BIGG is a research scientist at the
Centre National de la Recherche Scientifique/
Centre Alexandre Koyré, Paris.

H. OTTO SIBUM is the Hans Rausing Professor
of History of Science and director of the Office for
History of Science at Uppsala University, Sweden.

Library of Congress Cataloging-in-Publication Data
The heavens on earth : observatories and astronomy
in nineteenth-century science and culture / edited by
David Aubin, Charlotte Bigg, and H. Otto Sibum.
p. cm. — (Science and cultural theory)
Includes bibliographical references and index.
ISBN 978-0-8223-4628-9 (cloth : alk. paper)
ISBN 978-0-8223-4640-1 (pbk. : alk. paper)
1. Astronomy—History—19th century.
2. Nineteenth century. I. Aubin, David, 1967–
II. Bigg, Charlotte. III. Sibum, Heinz Otto, 1956–
IV. Series: Science and cultural theory.
QB32.H43 2010
520.9′034—dc22   2009041448